Land Bridges

Land Bridges

Ancient Environments, Plant Migrations, and New World Connections

ALAN GRAHAM

The University of Chicago Press
Chicago and London

The University of Chicago Press, Chicago 60637
The University of Chicago Press, Ltd., London
© 2018 by The University of Chicago
All rights reserved. No part of this book may be used or reproduced in any
manner whatsoever without written permission, except in the case of brief
quotations in critical articles and reviews. For more information, contact the
University of Chicago Press, 1427 E. 60th St., Chicago, IL 60637.
Published 2018
Printed in the United States of America

27 26 25 24 23 22 21 20 19 18   1 2 3 4 5

ISBN-13: 978-0-226-54415-1 (cloth)
ISBN-13: 978-0-226-54429-8 (paper)
ISBN-13: 978-0-226-54432-8 (e-book)
DOI: https://doi.org/10.7208/chicago/9780226544328.001.0001

Library of Congress Cataloging-in-Publication Data

Names: Graham, Alan, 1934– author.
Title: Land bridges : ancient environments, plant migrations, and New World
    connections / Alan Graham.
Description: Chicago : The University of Chicago Press, 2018. | Includes
    bibliographical references and index.
Identifiers: LCCN 2017032723 | ISBN 9780226544151 (cloth : alk. paper) |
    ISBN 9780226544298 (pbk. : alk. paper) | ISBN 9780226544328 (e-book)
Subjects: LCSH: Natural bridges—Western Hemisphere. | Landforms—
    Western Hemisphere. | Biogeography—Western Hemisphere. | Plant
    diversity—Western Hemisphere.
Classification: LCC GB401.5.G73 2018 | DDC 551.41—dc23
LC record available at https://lccn.loc.gov/2017032723

♾ This paper meets the requirements of ANSI/NISO Z39.48-1992
(Permanence of Paper).

To those who study the past and present biotas, climates, and geology of the New World and from those perspectives attempt to anticipate its future and plan its conservation

A well-known environmental campaigner in Honduras, Lesbia Yaneth Urquia, was murdered. There was widespread international outrage after her body was found abandoned on a rubbish dump. She was the second opponent of a giant dam project to be killed in four months.

—*Economist,* 16 July 2016

CONTENTS

Abbreviations, Time Scale, and Conversions / xi

Preface / xv
    Protocols and Organization / xv
    References / xxi

Introduction / 1
    References / 11
        Additional References / 15

PART I: BOREAL LAND BRIDGES

    Bering Land Bridge / 19
        Beringia / 19
        Background / 22
    References / 25
        Additional References / 27

ONE / West Beringia: Siberia and Kamchatka / 29
    Siberia / 29
        Geographic Setting and Climate / 29
        Geology / 37
        Modern Vegetation / 37
        Indigenous People / 41
    Kamchatka / 42
        Geographic Setting and Climate / 42
        Modern Vegetation / 48
        Indigenous People / 52
    References / 52
        Additional References / 55

TWO / East Beringia: Alaska, Northwestern North America, and the
Aleutian Connection / 57

Geographic Setting and Climate / 57
Geology / 61
Modern Vegetation / 65
Indigenous People / 68
Utilization of the Bering Land Bridge / 68
Peopling of America (from the West) / 76
References / 81
    Additional References / 85

THREE / North Atlantic Land Bridge: Northeastern North America,
Greenland, Iceland, Arctic Islands, Northwestern Europe / 89

Geographic Setting and Climate / 91
Geology / 103
Modern Vegetation / 107
Utilization of the North Atlantic Land Bridge / 110
    Modernization of the Flora / 116
    Biodiversity and Vegetation Density / 116
    Floristic Relationships between Eastern Asia and
        Eastern North America / 117
    Geofloras and the Madrean-Tethyan Hypothesis / 119
Indigenous People / 122
Peopling of America (from the East) / 122
References / 124
    Additional References / 129

PART II: EQUATORIAL LAND BRIDGES

FOUR / Antillean Land Bridge / 135

Stepping Stones or Lost Highway / 135
Geographic Setting and Climate / 139
Geology / 141
Modern Vegetation / 146
Indigenous People / 150
Utilization of the Antillean Land Bridge / 152
References / 160
    Additional References / 162

FIVE / Central American Land Bridge / 164

South and North of the CALB / 164
Geographic Setting and Climate / 178
Forging the Final Link: Geology / 179
Modern Vegetation / 184
Indigenous People / 189

Utilization of the CALB / 190
References / 197
    Additional References / 206

PART III: AUSTRAL LAND BRIDGE

SIX / Magellan Land Bridge: Cono del Sur and Antarctica / 213
Cono del Sur / 213
    Geographic Setting and Climate / 213
    Geology / 218
    Modern Vegetation / 219
    Indigenous People / 223
    Peopling of South America (from the North) / 223
Antarctica / 226
    Geographic Setting and Climate / 226
    Geology / 229
Land Bridges and Island Biogeography / 234
Utilization of the Magellan Land Bridge / 235
    Cono del Sur / 235
    Antarctica / 239
References / 246
    Additional References / 253

SEVEN / Case Studies / 259
Ferns and Allied Groups / 262
Gymnosperms / 263
Angiosperms / 264
    Monocotyledons / 264
    Dicotyledons / 268
References / 282
    Additional References / 289

EIGHT / Summary and Conclusions / 293
Events, Processes, and Responses / 294
    Bering Land Bridge / 294
    North Atlantic Land Bridge / 295
    Antillean Land Bridge / 296
    Central American Land Bridge / 297
    Magellan Land Bridge / 298
Conceptual Issues and Future Needs / 298
References / 302
    Additional References (Conservation) / 303
    Additional References (Selected Classical Literature) / 303

Acknowledgments / 305
Index / 307

ABBREVIATIONS, TIME SCALE, AND CONVERSIONS

I	Graham, A. 1999. Late Cretaceous and Cenozoic History of North American Vegetation. Oxford University Press, Oxford. Cited as I.
II	Graham, A. 2010. Late Cretaceous and Cenozoic History of Latin American Vegetation and Terrestrial Environments. Missouri Botanical Garden Press, St. Louis. Cited as II.
III	Graham, A. 2011. A Natural History of the New World. University of Chicago Press, Chicago. Cited as III.
ACC	Antarctic Circum-Current
ALB	Antilles Land Bridge
APG	Angiosperm Phylogeny Group
BCE	before the common era
BLB	Bering Land Bridge
BP	before present
ca.	approximately, about
CALB	Central American Land Bridge
CE	common era
CM	combustible metamorphic rocks
DSDP	Deep Sea Drilling Project
EECO	Early Eocene Climatic Optimum
EPICA	European Project for Ice Coring in Antarctica
GISP	Greeenland Ice Sheet Program
GPS	Global Positioning System
GRIP	European Greenland Ice Core Project
GSPC	Global Strategy for Plant Conservation
IODP	International Ocean Drilling Program
K/T	Cretaceous/Tertiary boundary (65 Ma)

ka	thousand years ago
kyr	thousand years
LGM	Last Glacial Maximum (21–18 kyr BP)
Ma	million years ago
MAP	mean annual precipitation
MAT	mean annual temperature
MLB	Magellan Land Bridge
MMCO	Middle Miocene Climatic Optimum
MPCO	Middle Pliocene Climatic Optimum
msl	mean sea level
NALB	North Atlantic Land Bridge
Neogene	Miocene and Pliocene
NOAA	National Oceanic and Atmospheric Administration
NRL	nearest living relative, modern analog method for reconstructing paleoclimates based on the ecology of similar living species
ODP	Ocean Drilling Program
Paleogene	Paleocene, Eocene, Oligocene
PECO	Paleocene-Eocene Climatic Optimum
PETM	Paleocene-Eocene Thermal Maximum
PGCO	Post-Glacial Climatic Optimum
ppmv	parts per million by volume
SCAR	Scientific Committee on Antarctic Research
s.d.	without date, date of publication not listed
Sv	Sverdrup, measure of volume of ocean current transport equal to 1 million cubic meters (0.001 km^3) per second or 264,000 US gallons per second
WAIS	West Antarctic Ice Sheet Program

Geologic Column
(Mesozoic and Cenozoic Eras)

	Quaternary Period	
	Holocene Epoch	11,500
	Pleistocene	2.6 MA
	Tertiary	
	Pliocene	
	Piacenzian Stage	3.6
	Zanclean	5.3
	Miocene	
	Messinian	7.2
	Tortonian	11.6
	Serravallian	13.6
Cenozoic Era	Langhian	15.9
	Burdigalian	20.4
	Aquitanian	23.0
	Oligocene	
	Chattian	28.4
	Rupelian	33.9
	Eocene	
	Priabonian	37.2
	Bartonian	40.4
	Lutetian	48.6
	Ypresian	55.8
	Paleocene	
	Thanetian	58.7
	Seldanian	61.7
	Danian	65.5
	Cretaceous	
	Senonian	
	Maastrichtian	70.6
	Campanian	83.5
	Santonian	85.8
	Coniacian	89.3
	Gallic	
	Turonian	93.5
Mesozoic Era	Cenomanian	99.6
	Albian	112.0
	Aptian	125.0
	Barremian	130.0
	Neocomian	
	Hauterivian	136.4
	Valanginian	140.2
	Berriasian	145.5
	Jurassic	
	Triassic	

Conversion table for United States customary and metric units of measurement

acre	0.4 hectare
hectare	2.7 acres
feet to meters	multiply by 0.3048
meters to feet	multiply by 3.2808
miles to kilometers	multiply by 1.6093
kilometers to miles	multiply by 0.6213
inches to millimeters	multiply by 25.4
millimeters to inches	multiply by 0.0393
mi^2 to km^2	multiply by 2.590
km^2 to mi^2	multiply by 0.3861
°F to °C	$5/9 \, (°F - 32°)$
°C to °F	$9/5 \, (°C + 32°)$

PREFACE

Land bridges are the causeways of diversity. When they form, organisms are introduced into a new patchwork of environments and associations, and when they founder, these organisms are separated into reproductively isolated populations. Land bridges are a factor in determining global climates by configuring the routes of moisture transport by wind and heat transport by ocean currents. As the equatorial connections developed in the Miocene and Pliocene ca. 10 to 3.5 Ma, the Gulf Stream was strengthened, bringing greater warmth to the North Atlantic, and when the austral connection separated, it facilitated formation of an Antarctic Circum-Current (ACC) in the Oligocene ca. 34–29 Ma that reduced heat transfer to the southern continent. Land bridges are a factor in the development of biogeographic patterns between geographically remote regions. The boreal connections across the North Atlantic and North Pacific oceans allowed exchange of organisms between North America, Europe, and Asia, and when these pathways were severed in the early Tertiary (North Atlantic Land Bridge) and fluctuated in the late Tertiary and Quaternary (Bering Land Bridge), biotic remnants were left disjunct in Eastern Asia and eastern North America. The aim of this book is to trace the formation and disruption of New World land bridges and describe the biotic, climatic, and biogeographic consequences.

Protocols and Organization

Nothing contributes so much . . . as a steady purpose—a point on which the soul may fix its intellectual eye.

—"Letter from R. Walton to Mrs. Saville, 11 December 17__," Mary Wollstonecraft Shelley, *Frankenstein* (1818)

The vegetation history of North America and Latin America is discussed in Graham (1999, 2010, and 2011). To reduce repetitive citation, these are referenced throughout this book as I, II, and III. Each contains illustrations of community types and plant fossils, and the North American book (I, chaps. 2–4) further includes discussions of cause and effect, context, and methods, principles, strengths, and limitations, which may be consulted as needed for background to the present work. After those summaries were completed, it was clear that another topic needed further consideration: namely, the land bridges that joined the New World to and separated it from adjacent regions during the Late Cretaceous and Cenozoic. The intent of the previous books was to outline the development of New World vegetation within the context of geologic and climatic change. The "steady purpose" here is a focus on land bridges.

Five connections are recognized (see Fig. 0.1): (1) the Bering Land Bridge (BLB) between Eurasia and northwestern North America; (2) the North Atlantic Land Bridge (NALB) across eastern North America, Greenland, Iceland, the Arctic Islands, and into Western Europe; (3) the Antillean Land Bridge (ALB, including the Bahamas) partially connecting eastern South America with the Yucatán Peninsula of Mexico (and the southeastern United States); (4) the Central American Land Bridge (CALB) between western South America and Central and North America; and (5) the Magellan Land Bridge (MLB) extending from southern South America through the Magellan Strait and Drake Passage to Antarctica and Australasia. Summaries are presented of the present geographic setting and climate, modern vegetation, indigenous people (with special attention to their impact on past and present vegetation), geologic history, and utilization, as reflected in appendices listing representative fossil plants for each region.

The appendices and tables that accompany this book can be found online: press.uchicago.edu/sites/graham/. The appendices are intended as a list of representative plants utilizing the land bridges at different points in time. Limitations are, first, that in most instances fossils (particularly fossil pollen) of some families, like the Cyperaceae, Poaceae (except *Zea*), Asteraceae, Amaranthaceae/Chenopodiaceae (except *Iresine*), and Ericaceae, cannot be identified to genus because of limited variation relative to the size of the group; and, second, that pollen of the important tropical family Lauraceae does not preserve. The appendices mostly do not include specimens identified only to family, those placed in *incertae sedis*, or nomenclatorial vagaries like *Polypodites*, *Celastrophyllum* (?), *Compositiphyllum*, *Dicotylophyllum* type, *Juglandiphyllites*, cf. *Laurophyllum* sp., and "*Lindera*." Appendices for the Boreal Land Bridges contain some works not summarized in I

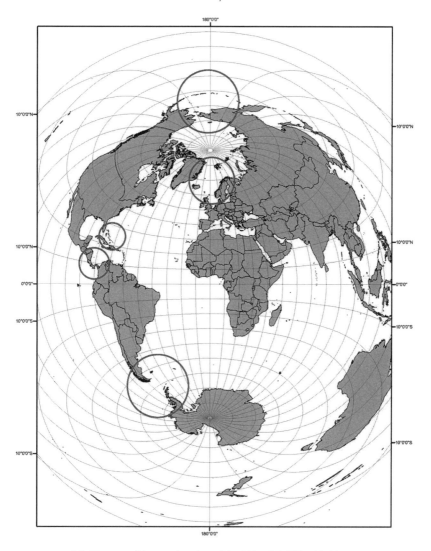

0.1. Diagram of the core location of the 5 New World land bridges; extended location as described in the text (chapters 1–6).

(1999), and an extensive list of references is provided in the legends. Appendices for the Equatorial and Austral Land Bridges are based on many studies cited in II (2010), so fewer and primarily recent publications are given in the legends. As far as possible, recent monographs, taxonomic revisions, and summaries of the geologic range of the fossils are used. Where such studies are not available, older records from Takhtajan et al. (1974 et seq., *Magnoliophyta Fossilia URSS*) and LaMotte (1952, *Catalogue of the Cenozoic Plants of North America through 1950*) are mentioned. An asterisk (*) indicates the plant has been reported as a fossil from North America (I, tables and index) or Latin America (II, appendices; *Catalog and Literature Guide for Cretaceous and Cenozoic Vascular Plants of the New World* online at http://www.mobot.org; Graham, 2012), or in other sources (as cited in II, prologue, p. xiii), including SCIENCE CHINA Earth Sciences (SCES; SCP @sci.scientific-direct.net) and the extensive databases now maintained by Carlos Jaramillo at the Smithsonian Tropical Research Institute, Panama.

The geologic age and the spelling of plant names are according to the original author unless otherwise stated. Family assignments and the names of extant plants are cited according to TROPICOS as used at the Missouri Botanical Garden based on the Angiosperm Phylogeny Group (APG) III (2009) system. The APG website (Stevens, 2001 et seq.) includes additional information and references on the historical biogeography of the various clades. The distinction between crown groups (derived) and stem groups (basal) is important phylogenetically and, as used here, carries a temporal connotation. Crown group implies reference to the recent taxa of a clade with particular interest in estimates of when and where they first appear. Stem group implies older/more ancient taxa. Some crown groups important specifically in the history of Amazonian vegetation are given in Hoorn et al. (2010, table S1).

Lists of widely used common names for extant plants of Europe are provided in Ozenda (1979); for those of Siberia, in Ukraintseva (1992, vii); of North America, in Oreme (2002, 276–77); and of Eurasia, in Shahgedanova (2008, 545–51). (These have been combined and augmented in online appendix 1 to include the equatorial and austral regions.)

Reference lists are given in each chapter with the primary purpose of updating those in my earlier works (I, II, III). About 1600 publications are listed, with about 750 from the post-2000 literature. In the references all authors are cited up to 5, after which *et al.* is used.

A choice of words: The term *ecosystem* is used in preference to *biome* because it better conveys the intended interaction between organisms, climate, and landscape. Geologists might object that *biome* implies that undue emphasis is placed on the living component, just as the use of *geome* might suggest to biologists that the primary concern is with geology. If either is the case, then use of *biome* or *geome* is appropriate; if not, *ecosystem* is considered the more inclusisve term.

The geologic epoch selected for beginning the survey is the Late Cretaceous, ca. 100 Ma. That is the time used in the other summaries, and it is when the angiosperms became dominant in most of the world's ecosystems (Friis et al., 2011; Rothwell and Stockey, 2016; Soltis et al., 2008). A time scale, based on the International Stratigraphic Chart of the International Commission on Stratigraphy, is given in the front matter. In the literature mention is occasionally made of magnetic anomalies and numbered chrons (Berggren et al., 1995). These are local variations in the Earth's magnetic field expressed as differences in the chemistry or magnetic polarities of the rocks and are used to correlate geologic events with vegetation history (see, for example, the section on Kamchatka in chapter 1, below). Paleogene refers to the Paleocene through the Oligocene, Neogene to the Miocene and Pliocene, and Cainozoic is the equivalent of Cenozoic. Plate tectonics and the drift of continents played a major role in the formation and dismantling of land bridges. In addition to other websites mentioned in the text, these movements can be followed on the Howard Hughes Medical Institute/Biointeractive website: go to http://www.hhmi.org/biointeractive/earthviewer and click on the subject of interest (the opening one allows viewing continental positions at different points in time; see also Explore Extinctions, Earth and Environments, etc.). The global paleotemperature and sea-level curves presented in chapters 2–6 are discussed in I (chap. 3, "Context") and are based on Zachos et al. (2001; see also Rohling et al., 2014; and Snyder, 2016).

The fullest account of the biology of the northern New World land bridges covers the Cretaceous and Paleogene of Beringia, through the works of Alexei Herman and Robert Spicer (see references to online appendix 2). Their treatment for Siberia, Kamchatka, and Alaska is augmented here by works on the Neogene and on the Aleutian Islands, Yukon Territory, British Columbia, and the Northwest Territories of Canada. Elsewhere, when material relevant to times and places is meager for the core areas (for example, the northeastern US portion of the NALB; see Fig. 0.1), the scope is also extended (for example, to include the Early to Middle Cretaceous of the mid-Atlantic states of North America). For the CALB, coverage encompasses northern South America to the southernmost states of Mexico, and for the MLB, it is from the Cono del Sur of Argentina, southern Brazil, Chile, Paraguay, and Uruguay southward to Antarctica.

Extant plants and communities of plants are the analogs used for interpreting fossil assemblages. Land bridges have been augmented throughout the Anthropogene by the early voyages of exploration, old and modern shipping lanes that follow the Earth's principal wind and ocean currents, widely used aviation corridors, and recent human-made barriers and connections (removal of vegetation, construction of canals and bridges; see II, fig. 3.94 and cover illustration; see also the discussion of the ALB in chapter 4, below). Thus, current conditions and past and present conservation and restoration efforts that modify the range and composition of vegetation, sometimes undetectably, in the vicinity of the land bridges are relevant (Martínez-Ramos et al., 2016; Smith et al., 2016; Spera et al., 2016; Graham, 2015; Harris et al., 2009; also chapter 4, below; for reviews see Duane, 2016; and Gaworecki, 2016a,b). Studies on the early potential human modification of local vegetation in the Old World, peopling of the New World, and the beginning of trade and sedentary agriculture include Clottes (2016) and Meltzer (2015). Others and those involving reclamation activities are listed at the end of each chapter under the heading "Additional References."

The Vikings early crossed the North Atlantic Ocean, Vitus Bering navigated the shoals and cloud-shrouded islands of the North Pacific, Vasco Nuñez de Balboa discovered for Europeans the vast and previously unknown Pacific Ocean, Ferdinand Magellan traversed the strait between the southern islands of South America (the surviving crew were the first to circumnavigate the Earth), and Ernest Shackleton's quest was to be the first to walk across the Antarctic continent. Their motivations were mixed, their success in achieving the goals of the expeditions varied, and the ultimate results were often tragic, but there is no doubt their discoveries clarified the extent, geography, habitability, conditions, and preparations needed to survive exploration of the New World land bridges and vicinity. These efforts are briefly recounted in chapters 1–6.

Measurements are in the metric system and in United States Customary Units where that seems useful. Electronic conversion tables are available at several websites, such as www.onlineconversion.com, including those for changing longitude and latitude from degrees and minutes to decimals, consistent with current usage. Conversions for commonly used measurements are given in the front matter. A list of abbreviations is also provided in the front matter, and specific abbreviations are explained the first time they are used in the text. Photographs are by Shirley A. Graham and the author unless otherwise credited. Sources and copyright holders granting permissions to use other materials are listed in the captions. In a few instances it proved impossible to reach a publisher, author, or estate after best-intent efforts. The source of such material is indicated in the captions.

A matter of style: Generalizing syntheses typically follow one of three approaches: an edited volume with topics addressed separately by specialists, a multiauthor treatment, or a single-author work. All have advantages and disadvantages. The previous texts (I, II, III) have been used in courses and seminars at several universities, and for that reason I have chosen to continue the single-author approach, which more readily allows for continuity between topics and facilitates an integrated presentation. In addition, it provides a consistent picture of New World vegetation and environmental history that can be augmented by discussions and other readings to provide alternative viewpoints. As an aid for students and the general reader, a number of websites, review articles, and programs from the public media are included in the Additional References sections at the end of each chapter. In commenting on a previous book (III, *A Natural History of the New World*), one reviewer said: "I found the numerous historical asides and personal anecdotes distracting (although to be fair I should acknowledge here that my graduate students had the reverse reaction and really enjoyed these parts)" (Fine, 2011, 51), and another (Waller, 2011, 357) noted: "The many asides and anecdotes might be annoying if they were any less amusing." The point is taken, and their numbers have been reduced in the present text. Nonetheless, it is my style, and the general reader and students are among the intended audience (see Graham, 2014, for a discussion of motivations and background influencing how topics are presented). Enlightening notes on use of the passive voice and other trends in scientific writing have recently appeared in *The Economist* (2016) and *Nature* (2016), also Hyland and Jiang (2017).

References

Angiosperm Phylogeny Group. 2009. An update of the Angiosperm Phylogeny Group classification for the orders and families of flowering plants: APG III. Botanical Journal of the Linnean Society 161: 105–21.

Berggren, W. A., D. V. Kent, M.-P. Aubry, and J. Hardenbol (eds.). 1995. Geochronology, Time Scales and Global Stratigraphic Correlation. SEPM (Society for Sedimentary Geology) Special Publication 54. Tulsa.

Clottes, J. 2016. What Is Paleolithic Art? Cave Paintings and the Dawn of Human Creativity. University of Chicago Press, Chicago.

Duane, D. 2016. The unnatural kingdom: If technology helps us save the wilderness, will the wilderness still be wild? New York Times, 13 March.

The Economist. 2016. Passive panic, in partial defense of an unloved grammatical tool. Johnson column. 2 July.

Fine, P. V. A. 2011. Book review: A 100 million year love affair with American plants: A Natural History of the New World, by Alan Graham. Frontiers of Biogeography 3: 50–51.

Friis, E. M., P. R. Crane, and K. R. Pedersen. 2011. Early Flowers and Angiosperm Evolution. Cambridge University Press, Cambridge.

Gaworecki, M. 2016a. Brazil's cerrado region: A new tropical deforestation hotspot. https://news.mongabay.com/2016/04/brazils-cerrado-region-a-new-tropical -deforestation-hotspot/.

———. 2016b. Human disturbances outside rainforests can jeopardize tropical biodiver-

sity, study confirms. https://news.mongabay.com/2016/04/even-human-disturbances
-outside-rainforests-can-jeopardize-tropical-biodiversity-study-confirms/. [Study by
Martínez-Ramos et al., 2016.]

Graham, A. 1999. Late Cretaceous and Cenozoic History of North American Vegetation.
Oxford University Press, Oxford. [Cited throughout this book as I.]

———. 2010. Late Cretaceous and Cenozoic History of Latin American Vegetation and
Terrestrial Environments. Missouri Botanical Garden Press, St. Louis. [Cited throughout this book as II.]

———. 2011. A Natural History of the New World. University of Chicago Press, Chicago.
[Cited throughout this book as III.]

———. 2012. Catalog and literature guide for Cretaceous and Cenozoic vascular plants of
the New World. Annals of the Missouri Botanical Garden 98: 539–41.

———. 2014. Academic Tapestries: Fashioning Teachers and Researchers Out of Events
and Experiences. Missouri Botanical Garden Press, St. Louis. [For a related discussion
of background and status in the presentation and acceptance of opinions, in this case
the early presence of humans in the New World, see Meltzer, 2015, 14.]

———. 2015. Past ecosystem dynamics in fashioning views on conserving extant New
World vegetation. Annals of the Missouri Botanical Garden 100: 150–58.

Harris, G., S. Thirgood, J. G. C. Hopcraft, J. P. G. M. Cromsigt, and J. Berger. 2009. Global
decline in aggregated migrations of large terrestrial mammals. Endangered Species
Research 7: 55–76.

Hoorn, C., et al. (+ 17 authors). 2010. Amazonia through time: Andean uplift, climate
change, landscape evolution, and biodiversity. Science 330: 927–31 (plus supporting
online material).

Hyland, K., and F. Jiang. 2017. Is academic writing becoming more informal? English for
Specific Purposes 45: 40–51.

LaMotte, R. S. 1952. Catalogue of the Cenozoic Plants of North America through 1950.
Geological Society of America Memoir 51. Geological Society of America, Boulder.

Martínez-Ramos, M., I. A. Ortiz-Rodríguez, D. Piñero, R. Dirzo, and J. Sarukhán. 2016.
Anthropogenic disturbances jeopardize biodiversity conservation within tropical rainforest reserves. Proceedings of the National Academy of Sciences USA 113:
5323–28.

Meltzer, D. J. 2015. The Great Paleolithic War. How Science Forged an Understanding of
America's Ice Age Past. University of Chicago Press, Chicago.

Nature. 2016. Editorial: Scientific language is becoming more informal. 539: 140. doi:
10.1038/539140a.

Orme, A. R. (ed.). 2002. The Physical Geography of North America. Oxford University
Press, Oxford.

Ozenda, P. 1979. Vegetation Map of the Council of Europe Member States. European
Committee for the Conservation of Nature and Natural Resources, Strasbourg. [Scale
1:3,000,000.]

Rohling, E. J., et al. (+ 6 authors). 2014. Sea-level and deep-sea-temperature variability over the past 5.3 million years. Nature 508: 477–82 (plus supplementary information).

Rothwell, G. W., and R. A. Stockey. 2016. Phylogenetic diversification of Early Cretaceous
seed plants: The compound seed cone of Doylea tetrahedrasperma. American Journal
of Botany 103: 923–37.

Shahgedanova, M. (ed.). 2008. The Physical Geography of Northern Eurasia. Oxford University Press, Oxford.

Smith, A. B., Q. G. Long, and M. A. Albrecht. 2016. Shifting targets: Spatial priorities for ex situ plant conservation depend on interactions between current threats, climate change, and uncertainty. Biodiversity and Conservation 25, 5: 905–22. doi: 10.1007/s10531-016-1097-7.

Snyder, C. W. 2016. Evolution of global temperature over the past two million years. Nature 538: 226–28.

Soltis, D. E., C. D. Bell, S. Kim, and P. S. Soltis. 2008. Origin and early evolution of angiosperms. Annals of the New York Academy of Sciences 1133: 3–25.

Spera, S. A., G. L. Galford, M. T. Coe, M. N. Macedo, and J. F. Mustard. 2016. Land-use change affects water recycling in Brazil's last agricultural frontier. Global Change Biology 22: 3405–13.

Stevens, P. F. 2001 et seq. Angiosperm Phylogeny website. http://www/mobot.org/MOBOT/research/APWeb.

Takhtajan, A., et al. (eds.). 1974 et seq. Magnoliophyta Fossilia URSS. Vol. 1. Magnoliaceae-Eucommiaceae. Nauka, Leningrad.

Ukraintseva, V. V. (ed. L. D. Agenbroad, J. I. Mead, and R. H. Hevly). 1992. Vegetation Cover and Environment of the "Mammoth Epoch" in Siberia. The Mammoth Site of Hot Springs, South Dakota.

Waller, D. M. 2011. Review: A Natural History of the New World: The Ecology and Evolution of Plants in the Americas, by Alan Graham, University of Chicago Press, Chicago. Quarterly Review of Biology 86: 357.

Zachos, J., M. Pagani, L. Sloan, E. Thomas, and K. Billups. 2001. Trends, rhythms and aberrations in global climate 65 Ma to present. Science 292: 686–93.

Introduction

In Barclay's 1509 *A Ship of Fools* is the adage, "there is tume and place for euery thynge." The beginning of the present book was a symposium "Floristics and Paleofloristics of Asia and Eastern North America," organized for the XI International Botanical Congress in Seattle (1969), and another for the American Institute of Biological Sciences in Bloomington, Indiana (1970), "Vegetation and Vegetational History of Northern Latin America" (Graham, 1972, 1973a). Before that, a project had been undertaken by the author to study Late Cretaceous and Tertiary vegetation in the northern neotropics under the title, *Studies in Neotropical Paleobotany* (for references, see Stevens et al., 2014, xxv–xxvi). Such studies had waned since the pioneering efforts of E. W. Berry, Arthur Hollick, and others in the early 1900s (literature in LaMotte, 1952; Graham, 1973b, 1979, 1982, 1986; http://www/mobot.org/mobot/research/CatalogFossil/catalog.shtml; Burnham and Graham, 1999, app. 1; Graham, 2012; and Jaramillo et al., 2014, table 4). Numerous misconceptions prevailed, especially about the stability of tropical environments and biotas through long intervals of geologic time, and it was clear that impressions (as of the 1970s) about the history of plant communities had to be reconsidered as a prelude to new initiatives. Researchers were invited to present summaries on the modern vegetation of the New World (Howard, 1973, Antilles; Rzedowski, 1973, Mexican dry regions; Gómez-Pompa, 1973, Veracruz, Mexico; Breedlove, 1973, Chiapas, Mexico; Porter, 1973, Panama) and adjacent territories (Yurtsev, 1972, Siberia). Integrated with reviews of the extant vegetation were papers on its history (Dilcher, 1973, southeastern US; Leopold and MacGinitie, 1972, Rocky Mountains; Wolfe, 1972, Alaska; Tanai, 1972, Japan; and Bartlett and Barghoorn, 1973, Panama).

To further the paleobotanical efforts and to secure support for future investigations, some predictability was needed for acquiring the data, given the time and expense of field studies in the tropics. Initial collections were made from localities with lenses of highly organic lignite, which often contains abundant, well-preserved plant microfossils and for which the geology and location are frequently well known because of its importance as a low-cost source of fuel. Some of the sites contained fruits, seeds, and leaves described earlier in the century (e.g., the Oligocene San Sebastian flora of Puerto Rico, Hollick, 1928), and the localities are still accessible. This was mentioned as an incentive for further study of the macrofossils. Recently, Paola Hernandez Gonzalez, an undergraduate at the University of Puerto Rico, Mayaguez, working with Hernán Santos, has undertaken a reexamination of the San Sebastian flora (pers. comm., 2015). Fossil floras have long been known from the boreal region, and new investigations extending from Mexico and the Antilles to Argentina and Antarctica by numerous authors are now reducing the earlier void in equatorial and austral areas.

In this discussion three basic assumptions are made. One is that patterns of distribution and diversity evident among modern plants and animals are residuals of history. Another is that efforts to understand these patterns without due attention to the fossil record will yield results that are equivocal. This is recognized in a body of recent literature advocating incorporation of past events and processes to better understand the nature and functioning of extant biotas. Drummond et al. (2006, 456) refer to the inevitable geohistorical circumstances prevailing at any particular point in time and space; Fukami (2015, 1) notes a historical contingency in the structure and function of communities; Uribe-Convers and Tank (2015, 1854) mention that upticks in diversification are attributable in part to colonization of new geographic environments via dispersification (dispersal + diversification; Moore and Donoghue, 2007). Land bridges enable this colonization to happen more readily. Gillson (2015) advocates using a long-term perspective in conservation efforts to develop multifunctional landscapes that can maintain biodiversity over time. The perspective emphasized in that book is the Holocene (11,500 yr ago) to the present. Here and elsewhere (Graham, 2015) the perspective is extended back to the Late Cretaceous and Tertiary. Siegel et al. (2015) recognize the fossil record as a valuable adjunct to archaeological studies concerned with the migration of early humans and their impact on the environment:

> Paleoenvironmental reconstruction is proving to be a more reliable method of identifying small-scale colonization events than archaeological data alone. (275)

The third assumption is that land bridges are important in facilitating the radiation, diversification, community assembly, and human influence on New World biotas. This history has not been fully explored, and it warrants further consideration.

Among organisms utilizing land bridges, plants have received comparatively little attention. Their role is essential for three principal reasons:

> First, plants are the conspicuous, conveniently stationary component of the Earth's ecosystems and are most often used to name or characterize the systems (boreal forest, mangroves, savanna). Second, plant fossils are abundant and have been used extensively to reveal the history of lineages and communities—their origin, evolution, migration, and the time relationships between them—and to reconstruct the Earth's paleoenvironments. Last, much of the literature and many natural history presentations intended for general audiences understandably feature charismatic, exotic, dangerous, attacking animals harassed by film crews and wannabe celebrity naturalists. This makes for good theater and it admirably focuses attention on selected treasures of the biological kingdom, but on another level, it gives only a partial picture of the Earth's ecosystems, and it is not a broad enough basis for a meaningful consideration of their structure, function, importance, conservation, and sustained management. For those seeking a fuller understanding of the Earth's web of life and a rational agenda for ensuring its continuation, this should be emphasized. (I, x)

NatureServe (2016) has announced a new, ecology-based classification for the extant US vegetation. The system used here was devised to accommodate both extant communities and their past analogs (I, 3) and it can be modified to incorporate regional variations along the New World land bridges:

Desert
Shrubland/chaparral-woodland-savanna (including thorn forest, thorn
    scrub, cerrado, steppe on cold slopes)
Grassland (pampas)
Mangrove
Beach/strand/dune
Freshwater herbaceous bog/marsh/swamp
Aquatic
Lowland tropical rain forest
Lower to upper montane broad-leaved forest (deciduous forest)

Coniferous forest
  Boreal
  Western montane
  Appalachian
Tundra
  Lowland
  Alpine (páramo)

To fully assess the role of time, place, and environment in the construction and use of New World connections, the fossil record and morphological and molecular profiles are central. However, it would be presumptuous to assume that any single approach will provide a full and accurate account of their use. This includes study of the incomplete fossil record; time-unconstrained or poorly constrained estimates of events derived from morphologically based taxonomies; statistical modeling necessitated by the lack of adequate field data and the rapid disappearance and extensive modification of the present biota; and relationships and explanations for distributions based on molecular evidence from extant organisms (Pyron, 2015). The fossil record alone is too fragmentary, and scenarios based solely on extant taxa through either a morphological, molecular, or statistical approach are insufficiently grounded in historical fact to be precise, dependable, and verifiable (Bell, 2015).

"In the case of plants, an adequate fossil record does not exist" (Boulter et al., 1972, quoted in Sanderson, 2015); "molecular clocks are like Santa Claus: everyone wants to believe in them, but no one really does" (Shaffer, pers. comm., in Sanderson, 2015; see Magallón, 2004, and Doyle et al., 2004, for the alternative viewpoint); and "statistical significance is a measure of probabilities; practical significance is a measure of effect" (Moore and McCade, 2003, 463). In other words, undue preoccupation with probabilities and the model can strip experience of its meaning and diminish the importance of reality. The approach must be multidisciplinary, and this is becoming standard in studies dealing with vegetation and lineage histories. Some of the more theoretical aspects of this history are explored in Beller et al. (2017), Burrows et al. (2014), Lamsdell et al. (2017), Marshall (2017), and Meyers (2017).

In addition to climate and physiography, there is an almost ethereal overlay of additional factors suspected to be important but insufficiently known and difficult to quantify. These include the distribution potential of each taxon at the time and under the conditions prevailing at that time (see, e.g., Clobert et al., 2012), the underestimated role of ocean currents

(De Queiroz, 2005), and the possibility that an organism might outdistance or outlast its pollinator or seed-dispersal agent and encounter local predators or pathogens (Fuller et al., 2012) that halt, impede, or temporarily alter the pace and direction of movement (Clobert et al., 2001; Dingle, 1996; Forget et al., 2005). The current loss of pollinators (Vogel, 2017) provides a modern analogy for the potential impact of this factor in the past (Burkle et al., 2013; Garibaldi et al., 2013; Tylianakis, 2013; Grimaldi, 1999). Another factor influencing past distributions is the diversity of soil types encountered along the way (Retallack, 2001; Zamotaev, 2008; see Fig. 1.6). For much of the area of the northern land bridges, glaciation has deposited a veneer of relatively uniform soils called gelisols (or gleysols, = clay), defined in the USDA classification system as shallow, highly organic, dark soils, sterile at depth, occurring in cold climates primarily of Siberia, Alaska, and northern Canada, and with permafrost to within 2 m of the surface (for a more detailed classification of North American soils, see Steila, 1993). Permafrost (shown for North America in Fig. 2.2; Nelson and Hinkel, 2002) is frozen throughout the year to a depth of up to several hundred meters, and polygonal or patterned ground is common. By contrast, regions beyond the far-north and south offer a diverse array of podzolic (= sandy) soils reflecting the composition of the underlying bedrock and decomposition products of the overlying vegetation. Certainly one barrier to the northern migration of tropical elements through the CALB was the edaphically dry (even though climatically moist) karst substrate of the Petén region of northeastern Nicaragua, Guatemala, and the Yucatán Peninsula of Mexico. Variability in soil types is a facet of the environment that plants encounter in the dispersal of their propagules regardless of the physiography or climate, or whether the regions were connected or separated by the coming and going of land bridges.

It is taken for granted that attempts to explain patterns of diversity and distribution in plants must consider the widely varied modes of dispersal, such as by migratory birds and terrestrial frugivores, but these include both present (Snow, 1985) and past vectors (Webb, 1986; Steadman, 1985); and for *Apeiba*, *Guazuma*, *Hymenaea*, and others, possibly the extinct Pleistocene megafauna (Janzen and Martin, 1982). The development of metabolic, defense, and anticompetition compounds (e.g., *Digitalis*) can allow invasion and persistence in new territories independent of or supplemental to other changes. In addition to dispersers and pollinators (modern and extinct), pathogens, and evolving chemical characteristics, coal fires (prominent, for example, across vast expanses of the Kuznetsk Basin in western Siberia) are another of the many factors influencing the past and present migration of plants and animals in special situations. The coals are Late Paleozoic in age and over time periodically have been ignited by natural forest or steppe fires caused by lightning and more recently and more frequently by human activity. The oldest presently dated fires are $1.7 +/-0.3$ Ma (early Pleistocene), but undoubtedly there were others much earlier resulting from rarer events like volcanism and meteorites as evident by the late Eocene Popigai Crater in Siberia (Bottomley et al., 1997) and the recent near-hit over Tunguska, Siberia, on 30 June 1908 (Chyba et al., 1993; Svetson, 1996; see Fig. 1.5). Of particular interest is that recent fires, and presumably the older ones, cluster at intervals. "As coal fires usually take place under warm dry conditions their dependence on climate opens up the possibility of using combustion metamorphic (CM) rocks as new climate indicators" (Sokol et al., 2014, 1043; see also Bond, 2015).

Trying to understand the multiple causes for the movement of organisms across land bridges also requires attention to the nebulous (and immensely difficult to quantify) effect of different sizes of the resource and target areas encountered during migration and changes in these areas through time. The changes may be independent or may happen concurrently with the coming and going of land bridges and alterations in landscape and climate, and they occur while evolutionary processes are modifying the organism's dispersal capacities and ecological parameters. It is intuitively logical that the relative size of the land at either end of the bridge is important, because size relates to the carrying capacity or biological density of the landscape and, therefore, to the diversity of organisms. Density and diversity translate into competition, and one of the consequences of competition, besides speciation and extinction, is migration (range expansion). Most New World land bridges are narrow compared to the regions at either end, and they act as filters. However, beyond the bridges the regions vary, and have varied in the past, in size, shape, and topography ,diminishing or intensifying the filtering process. The Bering, North Atlantic, and Antillean bridges connect lands of continental scale at both ends, while the Panama and Magellan bridges join lands that taper to a point on one side (southern Central America and southern South America) and open onto large continental expanses on the other (northern South America and Antarctica; Fig. 0.1). Topographic diversity is slight to moderate across most of western Beringia (Siberia) and extensive in eastern Beringia (Alaska and vicinity). The flat plains of Siberia open to the southeast onto the mountainous Kamchatka Peninsula, which is bounded by the Bering Sea and the Sea of Okhotsk (see Figs. P1.1, P1.2). In contrast,

Alaska and vicinity is mountainous and opens to the southeast onto the relatively flat and extensive Great Plains. Thus, after crossing the bridge, the migrants faced very different physiographic environments. The importance of similarities and differences in the size of the arrival and departure lands is raised as a point of awareness, but to my knowledge there are no models factoring in the effect.

Collectively, all these events and processes acting on organisms and communities of organisms determine their movement, diversification, and geographic affinities at any one moment in time. Unraveling this history is an immensely difficult task and should impart a reluctance to overinterpret, simplify, or underestimate any one source of information, including the fragmentary fossil record and temporally poorly constrained morphological taxonomies, molecular-based phylogenies, and modeling estimates.

Yet, there is often a tendency to treat casually or even bypass deeper-time history and the lack of fossils, even though modern organisms and communities of organisms have experienced alterations in climate and distribution throughout the Mesozoic (e.g., Ma et al., 2017) and Cenozoic (Mudelsee et al., 2014), 18–20 glacial-interglacial intervals in the past 2.5 Ma of the Pleistocene, and significant changes during the Holocene (e.g., the Medieval Warm Period ca. 800–1300 CE, the Little Ice Age ca. 1400–1850 CE, current oscillations of several thousands of years [Heinrich events] to a few thousand or even a few hundred years [Dansgaard-Oeschger or D-O events], centennial changes in CO_2 [Marcott et al., 2014], as well as sub-decadal El Niños/Las Niñas [Ritchie and MacDonald, 1986; I, 279]). The more recent and subtle of these cycles may have left an environmental imprint on distributions that has not been adequately explored. Regions within the New World have undergone separation (BLB, NALB), partial connection (ALB), full connection (CALB), and some limited drift into new climatic zones (MLB). Scenarios about origin, isolation, and interaction of biotas most likely to stand the test of time are those based on (1) modern floristic inventories, taxonomies, phylogenies, and models providing accurate biogeographic affinities and patterns of distribution; (2) paleobotanical information about the history of individual lineages, communities, and environments; and, eventually, (3) the seamless integration of past and present into a conceptual view of biotas as a continuum. Those without a sound historical component may be constructed provisionally out of necessity, but they lack the important element of time. The extant biota alone is too short a lens for viewing so complex a problem as dispersal and diversification.

Land bridges are important but they are not the only means of migration, and various plants and animals had other options. These include distributions that are residuals of ancient continental positions, crossings over the land by birds and/or wind, drift through or around the bridges by ocean currents, migration on land by means other than human transport (extant or extinct), more recent distributions by humans and the animals they followed, and combinations of all these. The tabulations are minimal inventories because if climates, landscape, soils, pollinators, dispersal vectors, distribution patterns, and paleontology reveal conditions favorable to a particular community (e.g., boreal forest) during a given interval of time (e.g., after the MMCO), then it is reasonable to assume that associates with a comparable ecology may have used the bridge even though direct evidence in the form of fossils is lacking. If conditions were notably different (e.g., tropical), then other means and routes must be assumed. This is important for asssessing views like the Madrean-Tethyan hypothesis proposed to explain similarities in dry vegetation between Mediterranean Europe and western North America via former continuity when suitable conditions existed neither across the NALB nor across the BLB in the Late Cretaceous and Cenozoic.

By the Quaternary broad regional patterns of distribution had mostly been established. In these more recent times the human impact on vegetation was profound, and its importance for interpreting paleocommunities has been noted. In turn, knowledge about plant resources, environments, timing of connections, preferred directions of movement, and vegetation types existing along the land bridges is important for understanding human migrations and the development of their cultures. New evidence suggests that the birthplace, dispersal pathways, and early differentiation of Indo-European languages may be associated with changes in the Eurasian steppe environment (Callaway, 2015; see also post to the bioRxiv.org preprint server, 10 February 2015; and W. Haak et al. [http://doi.org/z9d, 2015]). In other words, the spread and complexity of verbal and written communication was facilitated by the extent to which populations were willing and able to migrate unimpeded by physical, climatic, and biological barriers. Apparently, the extent of these barriers need not have been great to be effective. The Ural Mountains average only 1,900 m in height, and there are numerous valleys traversing the range. Yet they played a major role in the political and cultural development of a predominantly European-oriented and industrialized western Russia, and an Asian-oriented, relatively neglected, and politically independent eastern Russia. The effect of narrow ocean barriers is equally evident in the riveting account by Frost (2003) of the first contact in 1742 between eastern Europeans on the Vitus Bering expedition with Native Americans of the Pacific Northwest. The language of the Native Americans was mostly unintelligible even to speakers of several dialects from nearby Kamchatka. At the narrowest point the distance between the Seward Peninsula of Alaska and the Chuckchi Peninsula of Russia across the Being Strait is 85 km (65 mi), and between Attu Island of the US Aleutian Islands and the Commander Islands of Russia just off the Kamchatka Peninsula it is about 320 km (200 mi); the seafaring Americans were skilled in the use of boats; and there are intervening smaller islands with evidence of long use as temporary encampments. Yet, the ocean, the weather, and the unknown were sufficient to keep the populations separated and linguistically distinct for centuries. Breaking such barriers has always been a huge cultural step:

> To cross a bridge, a river or a border is to leave behind the familiar, personal and comfortable and enter the unknown, a different and strange world where, faced with another reality, we may well find ourselves bereft of home and identity. (Jean-Pierre Vernant, quoted in Badescu, 2007)

Crossing bridges is a long, dynamic, and immensely complicated process.

Land bridges are two-way streets, and study of extant biotas and the fossil record offers the most productive approach for conjuring explanations likely to stand the test of time. A requirement is that once preliminary accounts have been proposed, the scenarios must be assessed within the broadest possible context of independent data, and those data are readily available. Bottom-line summaries are often difficult, diffuse, and elusive, however, because supporting evidence comes from such diverse sources. Guo el al. (2012) suggest on the basis of nuclear gene and mitochondrial gene fragments from natricine snakes that there was an out-of-Asia crossing of Beringia at 27 Ma (mid- to late Oligocene; see review by Sanmartín et al., 2001); Li et al. (2015) use extant treefrogs (*Hyla*) as evidence for crossings of the Bering Land Bridge; Klompmaker et al. (2014) have examined the biogeographic implications for the hosts in patterns of parasite distribution, and Souza et al. (2015) have done the same for the parasites; Marwick (1998) uses fossil crocodilians for reconstructing the past climates; Zimmerman et al. (2015) use diatoms for inferring rates and direction of paleowind direction important for tracking dispersals; Fitzgerald et al. (1993) study paleocurrents to estimate the time of uplift of Denali (formerly Mount McKinley); Elias (2010) discusses the Quaternary entomology of Siberia and eastern Beringia; Martínez and Kutschker (2011) cite distinctive gravels in Patagonia as evidence of high-energy flooding (local habitat instability and the creation of open areas); and Solonevich and Vikhireva-Vasikova (1977) use coprolites and gastrointestinal contents to study the dietary habits of ancient humans from Paisley Caves in Oregon and mammoths in Siberia. As mentioned earlier, others have emphasized the importance of paleopedology and paleofires. Woodring (1966) noted long ago that a land bridge is a sea barrier, so patterns of similarity and dissimilarity across the marine realm should be reciprocal to any proposed terrestrial record or a plausible explanation provided for the lack of correlation. The use of multidisciplinary approaches and assessment of models through the broadest possible context are necessary to avoid inadvertent overemphasis of a particular discipline or a favored approach-du-jour, and this is the gradually emerging as the hallmark of modern investigations.

The other fragmented record: Land bridges have played an important role in the evolution and modification of ecosystems. The following chapters present views about their use, ranging from carefully considered, evidence-based opinions to less precise impressions, some insights, and a few wild guesses. In this regard, the fragmentary nature of the fossil record is widely acknowledged, and admonitions that it should be used with care, augmented with results from other methodologies, and interpreted conservativly within context are frequently made and well taken. Less well considered is the equal to greater fragmentary nature of the extant reservoir used to calculate evolutionary relationships, the age of clades, divergence times, and rates of radiation based on molecular evidence. Every time a species goes extinct—just like every time a plant fails to enter the fossil record, or a locality is destroyed or remains undiscovered—it fragments and reduces the database used for comparisions, calibration, and reconstruction. It is like removing all the specimens from the compartment of a herbarium case and attempting to revise a species or determine phylogenies based on the remaining portion. The partial representation of organisms in both the past and present biota is a limiting factor and a potential source of error for both approaches.

Assessing the use of land bridges is a difficult task, but important for understanding modern ecosystems and how they got that way, anticipating where they might be headed, and determining the causes. Difficult, but as noted by Barclay in 1509, with the proper approach and improved resources, maybe "there is a tume and place for every thynge."

References

Badescu, S. 2007. From One Shore to Another: Reflections on the Symbolism of the Bridge. Cambridge Scholars Publishing, Newcastle, UK.

Bartlett, A. S., and E. S. Barghoorn. 1973. Phytogeographic history of the Isthmus of Panama during the past 12,000 years (a history of vegetation, climate, and sea-level change). In A. Graham (ed.), Vegetation and Vegetational History of Northern Latin America. Elsevier, Amsterdam. Pp. 203–99.

Bell, C. D. 2015. Between a rock and a hard place: applications of the "molecular clock" in systematic biology. Systematic Botany 40: 6–13.

Beller, E., et al. (+ 7 authors). 2017. Toward principles of historical ecology. American Journal of Botany 104: 645–48.

Bond, W. J. 2015. Fires in the Cenozoic: a late flowering of flammable ecosystems. Frontiers in Plant Science, published online 5 January 2015. doi: 10.3389/fpls.2014.00749.

Bottomley, R., R. Grieve, D. York, and V. Masaitis. 1997. The age of the Popigai impact event and its relation to events at the Eocene-Oligocene boundary. Nature 388: 365–68.

Boulter, D., J. A. M. Ramshaw, E. W. Thompson, M. Richardson, and R. H. Brown. 1972. A phylogeny of higher plants based on amino-acid sequences of cytochrome-c and its biological implications. Proceedings of the Royal Society of London B (Biological Sciences) 181: 441–55.

Breedlove, D. E. 1973. The phytogeography and vegetation of Chiapas (Mexico). 1973. In A. Graham (ed.), Vegetation and Vegetational History of Northern Latin America. Elsevier, Amsterdam. Pp. 149–66.

Burkle, L. A., J. C. Martin, and T. M. Knight. 2013. Plant-pollinator interactions over 120 years: loss of species, co-occurrence, and function. Science 339: 1611–15.

Burnham, R. J., and A. Graham. 1999. The history of neotropical vegetation: new developments and status. Annals of the Missouri Botanical Garden 86: 546–89.

Burrows, M. T., et al. (+ 20 authors). 2014. Geographical limits to species-range shifts are suggested by climate velocity. Nature 507: 492–95.

Callaway, E. 2015. Language origin debate rekindled. Nature 518: 284–85.

Chyba, C. F., P. J. Thomas, and K. J. Zahnle. 1993. The 1908 Tunguska explosion: atmospheric disruption of a stony asteroid. Nature 361: 40–44.

Clobert, J., M. Baguette, T. G. Benton, and J. M. Bullock (eds.). 2012. Dispersal Ecology and Evolution. Oxford University Press, Oxford.

Clobert, J., E. Danchin, A. A. Dhondt, and J. D. Nichols (eds.). 2001. Dispersal. Oxford University Press, Oxford.

De Queiroz, A. 2005. The resurrection of oceanic dispersal in historical biogeography. Trends in Ecology and Evolution 20: 68–73.

Dilcher, D. L. 1973. A paleoclimatic interpretation of the Eocene floras of southeastern North America. In A. Graham (ed.), Vegetation and Vegetational History of Northern Latin America. Elsevier, Amsterdam. Pp. 39–60.

Dingle, H. 1996. Migration, the Biology of Life on the Move. Oxford University Press, Oxford.

Doyle, J. A., H. Sauquet, T. Scharaschkin, and A. Le Thomas. 2004. Phylogeny, molecular and fossil dating, and biogeographic history of Annonaceae and Myristicaceae (Magnoliales). International Journal of Plant Sciences 165 (supplement): S55–S67.

Drummond, A. J., S. Y. W. Ho, M. J. Phillips, and A. Rambaut. 2006. Relaxed phylogenetics and dating with confidence. PLoS One 4: e88. doi 10.1371/journal.pbio.0040088.

Elias, S. 2010. Advances in Quaternary Entomology. Elsevier, Amsterdam. [See especially chaps. 9 and 10, Siberia and eastern Beringia.]

Fitzgerald, P. D., E. Stump, and T. F. Redfield. 1993. Late Cenozoic uplift of Danali and its relation to relative plate motion and fault morphology. Science 259: 497–99.

Forget, P.-M., J. E. Lambert, P. E. Hulme, and S. B. Vander Wall (eds.). 2005. Seed Fate, Predation, Dispersal and Seedling Establishment. CABI Publishing, Oxfordshire, UK.

Frost, O. 2003. Bering: The Russian Discovery of America. Yale Uiversity Press, New Haven.

Fukami, T. 2015. Historical contingency in community assembly: integrating niches, species pools, and priority effects. Annual Review of Ecology, Evolution, and Systematics 46: 1–23.

Fuller, T., et al. (+ 7 authors). 2012. The ecology of emerging infectious diseases in migratory birds: an assessment of the role of climate change and priorities for future research. EcoHealth 9: 80–88.

Garibaldi, L. A., et al. (+ 49 authors). 2013. Wild pollinators enchance fruit set of crops regardless of honey bee abundance. Science 339: 1608–11.

Gillson, L. 2015. Biodiversity Conservation and Environmental Change, Using Palaeoecology to Manage Dynamic Landscapes in the Anthropocene. Oxford University Press, Oxford.

Gómez-Pompa, A. 1973. Ecology of the vegetation of Veracruz. In A. Graham (ed.), Vegetation and Vegetational History of Northern Latin America. Elsevier, Amsterdam. Pp. 73–148.

Graham, A. 1972. Floristics and Paleofloristics of Asia and Eastern North America. Elsevier, Amsterdam.

———. 1973a. Vegetation and Vegetational Hstory of Northern Latin America. Elsevier, Amsterdam.

———. 1973b. Literature on vegetational history in Latin America. In A. Graham (ed.), Vegetation and Vegetational History of Northern Latin America. Elsevier, Amsterdam. Pp. 315–60.

———. 1979. Literature on vegetational history in Latin America. Supplement I. Review of Palaeobotany and Palynology 27: 29–52.

———. 1982. Literature on vegetational history in Latin America. Supplement II. Review of Palaeobotany and Palynology 37: 185–223.

———. 1986. Literature on vegetational history in Latin America. Supplement III. Review of Palaeobotany and Palynology 48: 199–239.

———. 2012. Catalog and literature guide for Cretaceous and Cenozoic vascular plants of the New World. Annals of the Missouri Botanical Garden 98: 539–41.

———. 2015. Past ecosystem dynamics in fashioning views on conserving extant New World vegetation. Annals of the Missouri Botanical Garden 100: 150–58.

Grimaldi, D. 1999. The co-radiations of pollinating insects and angiosperms in the Cretaceous. Annals of the Missouri Botanical Garden 86: 373–406.

Guo, P., et al. (+ 7 authors). 2012. Out of Asia: natricine snakes support the Cenozoic Beringian dispersal hypothesis. Molecular Phylogenetics and Evolution 63: 825–33.

Hollick, A. 1928. Paleobotany of Porto Rico. Scientific Survey of Porto Rico and the Virgin Islands 7: 177–393. New York Academy of Sciences, New York.

Howard, R. A. 1973. The vegetation of the Antilles. In A. Graham (ed.), Vegetation and Vegetational History of Northern Latin America. Elsevier, Amsterdam. Pp. 1–38.

Janzen, D. H., and P. S. Martin. 1982. Neotropical anachronisms: the fruits the gomphotheres ate. Science 215: 19–27.

Jaramillo, C., et al. (+12 authors). 2014. Palynological record of the last 20 million years in Panama. In W. D. Stevens, O. M. Montiel, and P. H. Raven (eds.). Paleobotany and Biogeography, a Festschrift for Alan Graham in His 80th Year. Missouri Botanical Garden Press, St. Louis. Pp. 134–251.

Klompmaker, A. A., P. Artal, B. W. van Bakel, R. H. Fraaije, and J. W. Jagt. 2014. Parasites in the fossil record: a Cretaceous fauna with isopod-infested decapod crustaceans, infestation patterns through time, and a new ichnotaxon. PLoS One 9 (3): e92551.

LaMotte, R. S. 1952. Catalogue of the Cenozoic Plants of North America through 1950. Geological Society of America Memoir 51. Geological Society of America, Boulder.

Lamsdell, J. C., C. R. Congreve, M. J. Hopkins, A. Z. Krug, and M. E. Patzkowsky. 2017. Phylogenetic paleoecology: tree-thinking and ecology in deep time. Trends in Ecology and Evolution 32: 452–63.

Leopold, E. B., and H. D. MacGinitie. 1972. Development and affinities of Tertiary floras in the Rocky Mountains. In A. Graham (ed.), Floristics and Paleofloristics of Asia and Eastern North America. Elsevier, Amsterdam. Pp. 147–200.

Li, J.-T., J.-S. Wang, H.-H. Nian, S. N. Litvinchuk, J. Wang, Y. Li, D.-Q. Rao, and S. Klaus. 2015. Amphibians crossing the Bering Land Bridge: Evidence from Holarctic treefrogs (*Hyla*, Hylidae, Anura). Molecular Phylogenetics and Evolution 87: 80–90.

Ma, C., S. R. Meyers, and B. B. Sageman. 2017. Theory of chaotic orbital variations confirmed by Cretaceous geological evidence. Nature 542: 468–70.

Magallón, S. A. 2004. Dating lineages: molecular and paleontological approaches to the temporal branches of clades. International Journal of Plant Sciences 165 (supplement): S7–S21.

Marcott, S. A., et al. (+ 14 authors). 2014. Centennial-scale changes in the global carbon cycle during the last deglaciation. Nature 514: 616–19 (plus additional supplementary information).

Marshall, C. R. 2017. Five palaeobiological laws needed to understand the evolution of the living biota. Nature Ecology & Evolution 1. doi: 10.1038/s41559-017-0165.

Martínez, O. A., and A. Kutschker. 2011. The "Rodados Patagónicos" (Patagonian shingle formation) of eastern Patagonia: environmental conditions of gravel sedimentation. Biological Journal of the Linnean Society 103: 336–45.

Marwick, P. J. 1998. Fossil crocodilians as indicators of Late Cretaceous and Cenozoic climates: implications for using palaeontological data in reconstructing palaeoclimate. Palaeogeography, Palaeoclimatology, Palaeoecology 137: 205–71.

Meyers, S. 2017. Cracking the palaeoclimate code. Nature 1. doi: 10.1038/nature 22501.

Moore, B. R., and M. J. Donoghue. 2007. Correlates of diversification in the plant clade Dipsacales: geographic movement and evolutionary innovations. American Naturalist 170 (supplement): S28–S55.

Moore, D. S., and G. P. McCabe. 2003. Introduction to the Practice of Statistics. W. H. Freeman, New York.

Mudelsee, M., T. Bickert, C. H. Lear, and G. Lohmann. 2014. Cenozoic climate changes: a review based on time series analysis of marine benthic $\delta^{18}O$ records. Reviews of Geophysics 52: 333–74.

NatureServe. 2016. NatureServe and partners unveil adaptable, ecology-based U.S. national vegetation classification. http://www/natureserve.org/conservation-tools/ projects/us-national-vegetation-classification.

Nelson, F. E., and K. M. Hinkel. 2002. The far North: a geographic perspective on permafrost environments. In A. R. Orme (ed.), The Physical Geography of North America. Oxford University Press, Oxford. Pp. 249–69.

Porter, D. M. 1973. The vegetation of Panama: a review. In A. Graham (ed.), Vegetation and Vegetational History of Northern Latin America. Elsevier, Amsterdam. Pp. 167–202.

Pyron, R. A. 2015. Post-molecular systematics and the future of phylogenetics. Trends in Ecology and Evolution 30: 384–89.

Retallack, G. J. 2001. Soils of the Past: An Introduction to Paleopedology. Blackwell Science, Oxford.

Ritchie, J. C., and G. M. MacDonald. 1986. The patterns of post-glacial spread of white spruce. Journal of Biogeography 13: 527–40.

Rzedowski, J. 1973. Geographical relationships of the flora of Mexican dry regions. In A. Graham (ed.), Vegetation and Vegetational History of Northern Latin America. Elsevier, Amsterdam. Pp. 61–72.

Sanderson, M. J. 2015. Commentary: Back to the past: a new take on the timing of flowering plant diversification. New Phytologist 207: 257–59.

Sanmartín, I., H. Enghoff, and F. Ronquist. 2001. Patterns of animal dispersal, vicariance and diversification in the Holarctic. Biological Journal of the Linnean Society 73: 345–90.

Siegel, P. E., et al. (+ 7 authors). 2015. Paleoenvironmental evidence for first human colonization of the eastern Caribbean. Quaternary Science Reviews 129: 275–95.

Snow, D. W. 1985. Seed dispersal. In B. Campbell and E. Lack (eds.), Dictionary of Birds. Poyser, Calton, Staffordshire, UK.

Sokol, E. V., S. A. Novikova, D. V. Alekseev, and A. V. Travin. 2014. Natural coal fires in the Kuznetsk Basin: geologic causes, climate, and age. Russian Geology and Geophysics 55: 1043–64.

Solonevich, N. G., and V. V. Vikhireva-Vasilkova. 1977. Preliminary results of a study of plant remains from the gastrointestinal tract of the Shandrin mammoth (Yakutia). In O. A. Skarlato (ed.), Anthropogene Fauna and Flora of the Siberian Northeast. Nauka, Leningrad. Pp. 208–21. [See also pp. 277–80.]

Souza, M. R., S. Ul-Hasan, and D. B. Keeney. 2015. The impact of host dispersal on parasite biogeography. Frontiers of Biogeography 7: 135–37.

Steadman, D. W. 1985. Fossil birds. In B. Campbell and E. Lack (eds.), Dictionary of Birds. Poyser, Calton, Staffordshire, UK.

Steila, D. 1993. Soils. In Flora North America Editorial Committee (eds.), Flora of North America. Vol. 1, chap. 2, pp. 47–54.

Stevens, W. D., O. M. Montiel, and P. H. Raven (eds.). 2014. Paleobotany and Biogeography, a Festschrift for Alan Graham in His 80th Year. Missouri Botanical Garden Press, St. Louis.

Svetson, V. V. 1996. Total ablation of the debris from the 1908 Tunguska explosion. Nature 383: 697–99.

Tanai, T. 1972. Tertiary history of vegetation in Japan. In A. Graham (ed.), Floristics and Paleofloristics of Asia and Eastern North America. Elsevier, Amsterdam. Pp. 235–56.

Tylianakis, J. M. 2013. The global plight of pollinators. Science 339: 1532–33.

Uribe-Convers, S., and D. C. Tank. 2015. Shifts in diversification rates linked to biogeographic movement into new areas: an example of a recent radiation in the Andes. American Journal of Botany 102: 1854–69.

Vogel, G. 2017. Where have all the insects gone? Science 356: 576–79.

Webb, S. L. 1986. Potential role of passenger pigeons and other vertebrates in the rapid Holocene migrations of nut trees. Quaternary Research 26: 367–75.

Wolfe, J. A. 1972. An interpretation of Alaskan Tertiary Floras. In A. Graham (ed.), Floristics and Paleofloristics of Asia and Eastern North America. Elsevier, Amsterdam. Pp. 201–34.

Woodring, W. P. 1966. The Panama land bridge as a sea barrier. Proceedings of the American Philosophical Society 110: 425–33.

Yurtsev, B. A. 1972. Phytogeography of northeastern Asia and the problem of Transberingian floristic interrelations. In A. Graham (ed.), Floristics and Paleofloristics of Asia and Eastern North America. Elsevier, Amsterdam. Pp. 19–54.

Zamotaev, I. 2008. Soils. In M. Shahgedanova (ed.), The Physical Geography of Northern Eurasia. Oxford, UK. Pp. 103–21.

Zimmerman, C., G. Jouve, R. Pienitz, P. Francus, and N. I. Maidana. 2015. Late Glacial and Early Holocene cyclic changes in paleowind conditions and lake levels inferred from diatom assemblage shifts in Laguna Potrok Aike sediments (southern Patagonia, Argentina). Palaeogeography, Palaeoclimatology, Palaeoecology 427: 20–31.

Additional References

Angel, H. 2016. Pollination Power. University of Chicago Press, Chicago. [Images of the marmalade honeyfly (*Episyphrus balteatus*) on Sargent's lily (*Lilium sargentiae*) and the male cape sugarbird (*Promerops cafer*) on the pincushion protea *(Leucospermum cordifolium)* vividly capture the process.]

Carlucci, M. B., et al. (+ 22 authors). 2016. Phylogenetic composition and structure of tree communities shed light on historical processes influencing tropical rainforest diversity. Ecography. doi: 10.1111/ecog.02104.

Cernansky, R. 2016. Secrets of life in the soil. Nature 537: 298–300.

———. 2017. Biodiversity moves beyond counting species. Nature 546: 22–24.

Chen, W.-Y., T. Suzuki, and M. Lackner (eds.). 2017. Handbook of climate change mitigation and adaptation. Springer, Berlin.

Cheng, H., et al. (+ 13 authors). 2016. The Asian monsoon over the past 640,000 years and ice age terminations. Nature 534: 640–46 (plus additional supplementary information).

Costello, M. J., R. M. May, and N. E. Stork. 2013. Can we name Earth's species before they go extinct? Science 339: 413–16.

Cousens, R., and C. Dytham. 2008. Dispersal in Plants, a Population Perspective. Oxford University Press, Oxford.

Crepet, W. L., and K. C. Nixon. 1998. Fossil Clusiaceae from the Late Cretaceous (Turonian) of New Jersey and implications regarding the history of bee pollination. American Journal of Botany 85: 1122–33.

The Economist. 2016. Global warming: in the red. 28 May.

Feeley, K. J., M. R. Silman, and A. Duque. 2015. Where are the tropical plants? A call for better inclusion of tropical plants in studies investigating and predicting the effects of climate change. Frontiers of Biogeography 7: 174–77 (correspondence).

Fernández-Armesto, F. 2006. Pathfinders, A Global History of Exploration. W. W. Norton & Company, New York.

François, L., et al. (+ 6 authors). 2013. Testing palaeoclimate and palaeovegetation model reconstructions with palaeovegetation data: an application to the middle Miocene. Geophysical Research Abstracts 15: EGU2013-9114.

Fry, C. 2016. Seeds, a Natural History. University of Chicago Press, Chicago.

Goodwin, Z. A., D. J. Harris, D. Filer, J. R. I. Wood, and R. W. Scotland. 2015. Widespread mistaken identity in tropical plant collections. Current Biology 25: 1066–67.

Graham, A. 1993. History of the vegetation: Cretaceous (Maastrichtian)-Tertiary. In: Flora of North America Editorial Committee, Flora of North America North of Mexico. Oxford University Press, Oxford. Pp. 57–70.

Hackett, J. 2016. Scores of museum specimens carry a name that isn't theirs. http://www/scientificamerican.com/article/scores-of-museum-specimens-carry-a-name-that . . . [Review of original article by Goodwin et al., 2015.]

Katoh, S., et al. (+ 18 authors). 2016. New geological and palaeontological age constraint for the gorilla-human lineage split. Nature 530: 215–18 (plus additional supplementary information).

Kemp, T. S. 2016. The Origin of Higher Taxa, Palaeobiological, Developmental, and Ecological Perspectives. University of Chicago Press, Chicago.

Lewis, L. R., et al. (+ 9 authors). 2014. First evidence of bryophyte diaspores in the plumage of transequatorial migrant birds. PeerJ 2: e424. doi: 10.7717/peerj.424.

Magurran, A. E. 2016. How ecosystems change, conservation planning must accommodate changes in ecosystem composition to protect biodiversity. Science 351: 448–49.

Maldonado, C., et al. (+ 8 authors). 2015. Estimating species diversity and distribution in the era of Big Data: to what extent can we trust public databases? Global Ecology and Biogeography 24: 973–84. ["Taxonomic errors and geographical uncertainty of species occurrence records . . . put into question to what extent such data can be used to unveil correct patterns of biodiversity and distribution," p. 973.]

NOVA. 2015. Making of North America. PBS, 4, 11, 18, November.

Oliveira, F. M., et al. (+ 10 authors). 2014. Evidence of strong storm events possibly related to the Little Ice Age in sediments on the southern coast of Brazil. Palaeogeography, Palaeoclimatology, Palaeoecology 415: 233–39.

Pellens, R., and P. Grandcolas (eds.). 2016. Biodiversity Conservation and Phylogenetic Systematics: Preserving our Evolutionary Heritage in an Extinction Crisis. Springer, Berlin.

Potts, S. G., et al. (+ 9 authors). 2017. Safeguarding pollinators and their values to human well-being. Nature 540: 220–29.

Qian, H., and R. E. Ricklefs. 2004. Taxon richness and climate in angiosperms: is there a globally consistent relationship that precludes region effects? American Naturalist 163: 773–79.

———. 2008. Global concordance in diversity patterns of vascular plants and terrestrial vertebrates. Ecology Letters 11: 547–53.

Raven, P. H. 1972. Plant species disjunctions: a summary. Annals of the Missouri Botanical Garden 59: 234–46.

Raven, P. H., and D. I. Axelrod. 1974. Angiopserm biogeography and past continental movements. Annals of the Missouri Botanical Garden 61: 539–673.

Renner, S. S., and T. J. Givnish (organizers). 2004. Tropical Intercontinental Disjunctions. International Journal of Plant Sciences 165 (supplement): S51–S158.

Ricklefs, R. E. 2004. A comprehensive framework for global patterns in biodiversity. Ecology Letters 7: 1–15.

———. 2005. Historical and ecological dimensions of global patterns in plant diversity. Biologiske Skrifter 55: 583–603.

———. 2012. Naturalists, natural history, and the nature of biological diversity. American Naturalist 179: 423–35.

Ricklefs, R. E., and D. G. Jenkins. 2011. Biogeography and ecology: towads the integration of two disciplines. Philosophical Transactions of the Royal Society of London B (Biological Sciences) 366: 2438–48.

Scheffers, B. R., et al. (+ 16 authors). 2016. The broad footprint of climate change from genes to biomes to people. Science 354: aaf7671-1-11.

Seddon, A. W. R., M. Macias-Fauria, P. R. Long, D. Benz, and K. J. Willis. 2016. Sensitivity of global terrestrial ecosystems to climate variability. Nature 531: 229–32 (plus additional supplementary information).

Simpson, G. G. 1940. Mammals and land bridges. Journal of the Washington Academy of Sciences 30: 137–63. [Note further discussion of this classic paper in chapter 8.]

Skinner, B. J., and B. W. Murck. 2011. Blue Planet, an Introduction to Earth System Science. 3d ed. Wiley, New York.

Snyder, C. W. 2016. Evolution of global temperature over the past two million years. Nature. doi: 10.1038/nature19798.

Thorne, R. 2016. Tropical plant disjunctions: a personal reflection. International Journal of Plant Sciences 165 (Supplement): S137–38.

Van den Elzen, C. L., E. A. LaRue, and N. C. Emery. 2016. Oh, the places you'll go! Understanding the evolutionary interplay between dispersal and habitat adaptation as a driver of plant distribution. American Journal of Botany 103: 2015–18.

Weeks, B. C., S. Claramunt, and J. Cracraft. 2016. Integrating systematics and biogeography to disentangle the roles of history and ecology in biotic assembly. Journal of Biogeography. doi: 10.1111/jbg.12747.

Wen, J., R. H. Ree, S. M. Ickert-Bond, Z. Nie, and V. Funk. 2013. Biogeography: where do we go from here? Taxon 62: 912–27.

XKYZ website, retrieved 14 September 2016. Earth temperature timeline. http://xkcd.com/1732/?ftcamp=crm/email//nbe/FirstFTAsia/product.

Xing, Y., et al. (+ 12 authors). 2016. Testing the biases in the rich Cenozoic angiosperm macrofossil record. International Journal of Plant Sciences 177: 371–88.

PART ONE

Boreal Land Bridges

Bering Land Bridge

Beringia

In a quest for the fabled Northwest Passage between Asia and Europe in 1816,

> Kotzebue and his Russian crew initially enjoyed . . . a clear passage through the Bering Strait north of Alaska. On August 16th a sailor from the masthead saw only open sea to the east. Kotzebue clearly thought himself on the brink of a major geographical achievement, one that would place him in the ranks of Cortez and Cook: he "cherished the hope of discovering a passage into the Frozen Ocean, more particularly as the strait appeared to run without impediment to the horizon" [Kotzebue, 1821]. He was foiled, in the end, only by the shallowness of the local waters ahead that made further northeastward navigation impossible. (Wood, 2014, 128)

The shallowness of the waters through the Bering Strait has long been a factor in determining continuity between Asia and North America, resulting from a combination of changes in sea level (isostacy) and elevation of the land (orogeny). As defined here, Beringia extends from approximately 75°N to 58°N latitude and from 120°W to 100°E longitude (Figs. P1.1, P1.2). To the west are Siberia and the Kamchatka Peninsula of Russia, and to the east are Alaska and the adjacent Yukon Territory, Northwest Territories, and the province of British Columbia, Canada. South of the BLB proper is the 1200-mile-long chain of 69 mostly volcanic peaks of the Aleutian Islands, separating the Pacific Ocean to the south from the Bering Sea and the Arctic Ocean to the north. The westernmost United States as

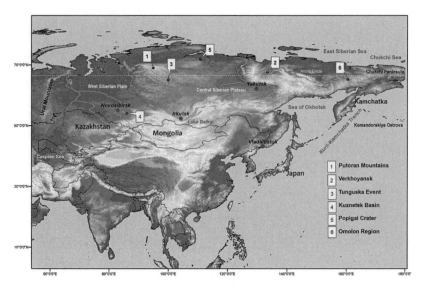

P1.1. Index map of place names and physiographic features for
Siberia, Kamchatka, and vicinity, western Beringia.

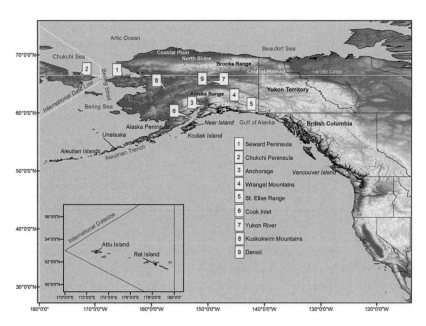

P1.2. Index map of place names and physiographic features
for Alaska and vicinity, eastern Beringia.

defined by longitude is at Amatignak Island, or, as miles from the mainland, it is the uninhabited island of Attu in the Near Island Group of the Aleutian chain (173°E, 1100 mi/1770 km). Farther to the west are the Komandorsklye Ostrova (Commander) Islands belonging to Russia. The International Date Line runs between Attu and the Commander Islands.

Vast areas of Siberia and Alaska are sparsely populated, and locations are commonly cited by longitude and latitude using GPS instead of by distance from remote population centers. Yakutsk is said to be at 62/129 (62°02′N 129°44′E) rather than 1600 km northeast of Irkutsk. At sea in the early days of exploration, determining latitude was relatively easy because it could be read off reference charts by noting the altitude of the sun at noon. Longitude was more difficult to estimate, and estimation of longitude became possible only after instruments were developed that could keep accurate time for years in the corrosive saline environment on often heavily rolling ships. Having recorded the time at the starting point, and knowing the time at the ship's current position, the hours under sail could be calculated and from the ship's average speed a guess could be made of its position. Another problem is that the lines of longitude are 60 mi apart at the equator and converge toward the poles, which created a further difficulty with the "dead reckoning" method. In 1714 the British government established the Longitude Act, offering a prize of 20,000 pounds on the grounds that "nothing is so much wanted and desired at sea, as the discovery of the longitude, for the safety and quickness of voyages, the preservation of ships, and the lives of men" (Sobel and Andrewes, 1995). The prize was awarded to Yorkshire carpenter John Harrison for his marine chronometer, but that was not until 1773 and long after Vitus Bering first sailed the treacherous waters of Beringia in 1728 guided by stars, rumor, and notoriously faulty maps. Field studies in Beringia are still difficult because of extreme cold, intense winter storms, and prolonged darkness. Today, when visiting a site with the enticing name of Bering Land Bridge National Preserve on the Seward Peninsula, visitors are advised that it is the nation's most remote and least visited national park, providing "solitude not often available in other parts of the country." Park headquarters are in Nome, which itself is unconnected by paved roads to any major city, like the state capital Juneau. Access to the preserve is by chartered floatplane or by conventional aircraft utilizing a dirt landing strip. There are no accommodations within the preserve, and travel is on foot or by snowmobile. Visitors must bring their own food, tents, fuel, emergency supplies, and communication devices. They are cautioned to be prepared for sudden and extreme changes in weather, to provide an itinerary of their whereabouts, and to be especially clear about arrangements for departure. In spite of these difficulties, those who study the geology and natural history of Beringia mostly agree that the scenic beauty of the surroundings and the often novel and important research results make confronting the isolation and physical challenges worth the effort.

Background

Longstanding interest in the biology and geology of Beringia and the surrounding land is evident by the number of books, international symposia, and edited volumes devoted to the region over the years. Among the earliest are *Outline of the History of Arctic and Boreal Biota during the Quaternary Period* by Eric Hultén (1937; Fig. P1.3) and *The Bering Land Bridge*

P1.3. Eric Hultén, prominent Swedish botanist, biogeographer, and early student of the Arctic vegetation. http://people.wku .edu/charles.smith/chronob/HULT1894.htm.

by David M. Hopkins (1967), based on a meeting of the VII Congress of the International Association for Quaternary Research in Boulder, Colorado, in 1965. These were followed by *Arctic Geology* (Second International Symposium on Arctic Geology, San Francisco, 1971; Pitcher, 1973), *Paleoecology of Beringia* (Hopkins et al., 1982), and *Beringia in the Cenozoic Era* (All-Union Symposium, "The Bering Land Bridge and Its Role in the History of Holarctic Floras and Faunas in the Late Cenozoic"; Kontrimavichus, 1984). There are summaries of the fossil plants (Boulter and Fisher, 1994), archaeology (West, 1981), mammoths (Haynes, 1991), and the "Mammoth Epoch" in Siberia (Ukraintseva, 1993). Interest is due to the long fascination with this remote land that formerly connected North America and Asia; early recognition of the biological similarities between eastern North America and central Asia, particularly the Hubei (formerly Hupeh) and Sichuan (Szechwan) provinces of China (e.g., Halenius, 1750; Gray, 1840, 1846; see Graham, 1966, 1971a); and more recently the discovery of vast oil and natural gas reserves on the North Slope of Alaska and in Siberia.

Interest in the biogeography and vegetation history lulled after Asa Gray's work in the mid-1800s and was revived when, in July 1971, the Geological Institute of the Academy of Sciences of the USSR hosted the Third International Palynological Conference in the academic city of Akademgorodok just south of Novosibirsk. The conference included presentations on the stratigraphy and correlation of economically important strata and, hence, was of interest to economic geologists and to those in the petroleum industry, as well as the academic world. However, it was at the height of the Cold War, epitomized by the frequently cited and variously translated phrase, "We will bury you," uttered by Soviet Premier Nikita Khrushchev at the Polish Embassy in Moscow in 1956. This pronouncement was followed by the launch of the world's first satellite (Sputnik) by the Soviet Union on 4 October 1957; Khrushchev's alleged shoe-banging incident at the United Nations in 1960 (the widely circulated photograph was faked); and the Cuban missile crisis in October 1962. The tensions between East and West had an effect on the selection of attendees even at non–military-related scientific conferences. When invitations were sent out in 1969 for the Russian meeting, I was one of the few North Americans issued a visa and allowed to attend. Since I was a relative newcomer to the field, this was surprising considering the number of well-known botanists and geologists anxious to interact with Russian colleagues long isolated behind the Iron Curtain. It was clear that careful attention had been given to the selection of foreign delegates. Members from the Western bloc nations were mostly limited to two per country so they could not dominate the conference or sway its adopted policies, recommendations, and resolutions.

My attendance was likely facilitated by the fact that as part of the XI International Botanical Congress in Seattle in 1969, I had organized the symposium "Floristics and Paleofloristics of Asia and Eastern North America" in collaboration with the Japan–United States Cooperative Science Program (Graham, 1971b). I invited the Academy of Sciences of the USSR to provide a speaker on the vegetation of the USSR, and they sent B. A. Yurtsev of the Komarov Botanical Institute in Leningrad. His paper was one of the few up-to-date, English-language summaries available at the time on the vegetation of Russia (Yurtsev, 1972). Once the conference in Akademgorodok was under way, isolated in the Hotel Gold Valley (Fig. P1.4), overt supervision of Westerners eased, but they were kept apart from their Soviet counterparts who at precisely 11:00 p.m. were shuttled back to their separate dormitories. This was symbolic of the difficulties in exchange of scientific information between East and West in the 1970s. Translations were provided for the papers but rather than translating simultaneously, the translator came to the podium, stood beside the speaker, and after each sentence repeated the sentence in the alternate language. Every 20-minute paper took 40 minutes, and the day sessions scheduled to end at 6:00 p.m. continued past midnight.

P1.4. Gold Valley Hotel, Akademgorodok, July 1971.

Monitoring the interaction between foreigners and the townspeople was less severe in the evening at the Hotel Gold Valley. One night a huge, bearded Russian workman came into the bar, took a bottle of vodka and two glasses from the shelf, and sat down. He poured two drinks, shouted, "Prost!" and repeated the ritual several times, then got up and left. The bartender said it was the locals' way of welcoming strangers.

At dusk, as the sun set over the River Ob, there was a magnificent view of the densely forested landscape from the upper floor of the hotel. The taiga or boreal forest extended from horizon to horizon with only a single, narrow road running in a nearly straight line for 2000 miles between Novosibirsk and Moscow. It was easy to envision plants and animals of the past moving virtually unimpeded as climates changed. Then, in the Quaternary, they began moving preferentially eastward, especially the huge herds of bison and woolly mammoth, in response to cold conditions that developed slightly earlier and more intensely on the vast interior continental Siberian plain than on the more restricted and topographically diverse landscape of Alaska. Finally, sometime after about 18,500 years ago, the megafauna was followed by human hunters and gatherers. An important additional legacy of these movements were the plant propagules the human migrants inadvertently or deliberately introduced into the new lands over the millennia (van Kleunen et al., 2015; Rejmánek, 2015) and those attached to or within the fauna they were following. To judge from recent invasives like *Frangula alnus* (De Kort et al., 2016), *Lythrum salicaria*, and *Pueraria lobata*, the effect was probably local rapid simplification and decrease in biodiversity at places in the new land. Such introductions are not easily recognizable where time has been insufficient for morphological differences to develop between the original and introduced populations (see chapter 4). However, if microsatellite analyses could detect divergences as recent as 18,000 years ago, this might help in corroborating the time humans and some plants crossed into the New World. A similar approach has been used to detect recent expansion of *Rhizophora* in Florida and the Caribbean islands (Kennedy, 2014), and it would be an interesting possibility to pursue for plants of the land bridges.

References

Boulter, M. C., and H. C. Fisher (eds.). 1994. Cenozoic Plants and Climates of the Arctic. Springer, Berlin.

De Kort, H., J. Mergeay, H. Jacquemyn, and O. Honnay. 2016. Transatlantic invasion routes and adaptive potential in North American populations of the invasive glossy buckthorn, *Frangula alnus*. Annals of Botany 118: 1089–99.

Graham, A. 1966. Plantae Rariores Camschatcensis: a translation of the dissertation of Jonas P. Halenius, 1750. Brittonia 18: 131–39.

———. 1971a. Outline of the origin and historical recognition of floristic affinities between Asia and eastern North America. In A. Graham (ed.), Floristics and Paleofloristics of Asia and Eastern North America. Elsevier, Amsterdam. [See appendix for reprints of the 1840 and 1946 papers of Asa Gray (below) dealing with the floristic similarities between eastern Asia and eastern North America.]

——— (ed.). 1971b. Floristics and Paleofloristics of Asia and Eastern North America. Elsevier, Amsterdam.

Gray, A. 1840. Dr. Siebold, Flora Japonica (review). American Journal of Science and Arts 39: 175–76.

———. 1846. Analogy between the flora of Japan and that of the United States. American Journal of Science and Arts 2 (2): 135–36.

Halenius, J. P. 1750. Plantae Rariores Camschatcenses. Thesis, University of Uppsala, Uppsala.

Haynes, G. 1991. Mammoths, Mastodons, and Elephants: Biology, Behavior and the Fossil Record. Cambridge University Press, Cambridge.

Hopkins, D. M. (ed.). 1967. The Bering Land Bridge. 1967. Stanford University Press, Stanford.

Hopkins, D. M., J. V. Matthews, Jr., C. E. Schweger, and S. B. Young (eds.). 1982. Paleoecology of Beringia. Academic Press, New York.

Hultén, E. 1937. Outline of the History of Arctic and Boreal Biota during the Quaternary Period. Bokförlags Aktiebolaget Thule, Stockholm.

Kennedy, J. P. 2014. Postglacial expansion of Rhizophora mangle L. in the Caribbean Sea and Florida. MS thesis, Florida Atlantic University, Boca Raton, FL. ProQuest, Ann Arbor, MI.

Kontrimavichus, V. L. (R. Chatravarty, translator; V. S. Kothekar, general editor.) 1984. Beringia in the Cenozoic. Oxonian Press, New Delhi.

Kotzebue, O. von. 1821. A Voyage of discovery into the South Sea and Beering's [sic] Straits, for the purpose of Exploring a North-East Passage, Undertaken in the Years 1815–18, at the expense of His Highness . . . Count Romanzoff, in the ship Rurick, under the command of the lieutenant in the Russian imperial navy, Otto von Kotzebue. Longman, Hurst, Rees, Orme & Brown, London. [Translated from the German by H. E. Lloyd; digitized by Google from a copy in the New York Public Library and uploaded to the Internet Archive by user tpb.]

Pitcher, M. G. (ed.). 1973. Arctic Geology. American Association of Petroleum Geologists, Tulsa.

Rejmánek, M. 2015. Global trends in plant naturalization. Nature 525: 39–40.

Sobel, D., and W. J. H. Andrewes. 1995. The Illustrated Longitude: The True Story of a Lone Genius Who Solved the Greatest Scientific Problem of His Time. Walker & Company, New York.

Ukraintseva, V. V. (Edited by L. D. Agenbroad, J. I. Mead, and R. H. Hevly.) 1993. Vegetation Cover and Environment of the "Mammoth Epoch" in Siberia. Mammoth Site of Hot Springs, SD.

Van Kleunen, M., et al. (+ 40 authors). 2015. Global exchange and accumulation of nonnative plants. Nature 525: 100–103.

West, F. H. 1981. The Archaeology of Beringia. Columbia University Press, New York.

Wood, G. D. 2014. Tambora, the Eruption That Changed the World. Princeton University Press, Princeton.

Yurtsev, B. A. 1972. Phytogeography of northeastern Asia and the problem of Transberingian floristic interrelations. In A. Graham (ed.), Floristics and Paleofloristics of Asia and Eastern North America. Elsevier, Amsterdam. Pp. 19–54.

Additional References

Alder, K. 2002. The Measure of All things: The Seven-year Odyssey and Hidden Error That Transformed the World. Free Press, New York.

Hough, R. 1994. Captain James Cook, a Biogeography. W. W. Norton & Company, New York.

Linden, E. 2006. The Winds of Change: Climate, Weather, and the Destruction of Civilizations. Simon & Schuster, New York.

Preston, D., and M. Preston. 2004. A Pirate of Exquisite Mind: Explorer, Naturalist, and Buccaneer: The Life of William Dampier. Walker & Company, New York.

Thomas, N. 2003. Cook: The Extraordinary Voyages of Captain James Cook. Walker & Company, New York.

ONE

West Beringia: Siberia and Kamchatka

Siberia

Geographic Setting and Climate

The vast area of Siberia has a population of about 40 million people (27% of Russia's total population) sparsely distributed across 13 million km² (77% of the country). It extends from the Ural Mountains east to the Pacific Ocean and from the Arctic Ocean south to Kazakhstan and the Mongolia and China borders (see Fig. P1.1). The Ural Mountains run north and south for a distance of ca. 2500 km, and the highest peak is Mount Narodnaia in the north, at 1900 m. The major cities of Moscow, Leningrad, and Kiev are all west of the Urals; hence, the tendency of Peter the Great (1672–1725, tsar from 1680 until his death, and born in Moscow) to essentially ignore Siberia and fill important political and administrative positions with Europeans. One of these was Vitus Jonassen Bering (Fig. 1.1). He was Danish, eventually elected to the Russian nobility, and appointed commander of both Kamchatka expeditions (1727–30 and 1741–42). He was further charged with the secret tasks of assessing the economic potential of Siberia and Kamchatka, exploring the Icy or White Sea (Arctic Ocean), determining the point at which Russia was closest to America, and, if possible, establishing contact with Native Americans living along the coast of the poorly known Big Land (Alaska; see Fig. P1.2):

"At one o'clock in the afternoon on July 16, 1741 Captain-Commander Vitus Bering and his seventy-seven men aboard the *St. Peter* saw North America unveiled to the north all across the horizon as a curtain of clouds and drizzle lifted. First they saw a band of thick, dark forest on a narrow plain along the littoral. And finally, in very clear weather and with the sun shining, they saw

1.1. Bust of Vitus Jonassen Bering in the Museum of Regional Studies, Petropavlosk, Russia.

a huge mountain range of perpetual snow topped by a single volcanic peak rising almost from sea level to a point higher than any mountain they had seen in Siberia." It was Mount St. Elias (5485 m) in the St. Elias Mountain Range on Cape St. Elias in the Gulf of Alaska. Four days later "at six o'clock in the morning of July 20 [they] dropped anchor over a muddy blue bottom. On this day would occur the first documented European landing on the northwest coast of North America." (Frost, 2003, 144,149)

Considering Bering's prominent social standing and his contributions to the science and politics of the day, it is surprising how quickly he disappeared from later historical accounts. Even the alleged portrait of him circulated until recently was actually a portrait of his great-uncle Vitus Pedersen Bering (Fig. 1.2). It shows a portly person obviously ill-suited to winter treks of 6000 miles on foot across Siberia to Kamchatka, the rigors of multiyear sea voyages in the early 1700s frequently facing starvation and mutinous crews, and lacking the physical qualities to inspire confidence in the fateful decisions required of a commander. His actual likeness (Fig. 1.1) conveys these qualities and shows a determined countenance reminiscent of Oswald Heer (see Fig. 3.9) and Ernest Shackleton (see Fig. 6.5). The second expedition began in May 1741. How little the geography was understood is evident in Bering's decision to sail south seeking a part of North America closer to Kamchatka. He reached Kayak Island in the Bay of Alaska at 59°N, when the nearest connection was to the north at 65°N. It was mid-July, too late to safely begin the trip back, and Bering was in poor health from the rigors of previous expeditions, but there was no choice. The ship foundered, so they built another and headed back to sea. By this time Bering was too ill to command. He died on 8 December 1742 and was laid in a shallow grave in the "unbearable Arctic winter, surrounded by crates full of his court clothes and wigs" (Fernández-Armesto, 2007). Bering's burial site was unknown for two and half centuries until 1991, when it was discovered through the joint efforts of a Danish-Russian expedition and a memorial erected on what is now Bering Island in the

1.2. Widely circulated picture supposedly of Vitus Jonassen Bering and now known to be a portrait of his great-uncle, Vitus Pedersen Bering. From Wikipedia.

Commander Island group. The motivations of early explorers like Bering, and Errnest Shackleton on Antarctica (see chapter 6), were often mixed, but there is consensus about his scientific contributions:

> Bering's story excels fiction for human interest and farce for human foibles. He had overmighty ambitions. Dreams of grandeur made him desert his Danish homeland for Russsian service and risk his life for the tsarina's reward in the uncharted corner of the world that now bears his name. Bering's was a breakthrough era in scientific knowledge of the boreal world, not only because of his own discoveries, but also because of the survey he commanded of the Arctic coast from the White Sea to Kamchatka, and the expedition of Pierre-Louis de Maupertuis to Finland to determine the length of a degree on the surface of the globe near the Arctic Circle. The effects of his work only became enshrined in uncontested maps nearly half a century after his death. (Fernández-Armesto, 2007)

With the death of Peter the Great and loss of his dominant influence, there was turmoil during the subsequent reign of his wife Catherine I, and especially among his niece Anna Ivanovana, infant son Ivan IV, and daughter Elizabeth. One result of this political and social instability was growing resentment and suspicion of foreigners. A monopolistic concession was granted in 1799 to the nationally controlled Russian-American Company to explore Siberia and Kamchatka with less outside scrutiny. This began a legacy of isolation and secrecy about the land that has lingered even into modern times.

In the northern Ural Mountains the MAT at low elevations is $-5\,°C$, and in the south it is $0.5\,°C$. Although the elevation is moderate, it is sufficient to cast a rainshadow to the east, which is augmented by the Siberian High Pressure System coming from the north and characterized by low winter precipitation. The combined effect on the east-facing northern slopes is a cold desert steppe (Fig. 1.3) and south toward Kazakstan a dry deciduous forest. Rainfall is 600 mm on the west side of the northern Urals and 400 mm to the east.

Immediately beyond the Urals to the east is the low-lying West Siberian Plain (see Fig. P1.1) consisting mostly of eroded alluvial deposits of Cenozoic age. In a region extending 6000 by 700 km the average elevation is less than 250 m and over half the land is swamp and flood plain. Novosibirsk is the largest city, with a population of about 1.5 million people. Farther east is the Central Siberian Plateau consisting of a series of slightly higher uplands tilted moderately northward. The average elevation is about 500 m,

1.3. Stony desert steppe typical of the southern Urals. From A. Chibilyov, 2008, in M. Shahgedanova (ed.), The Physical Geography of Northern Eurasia, Oxford University Press, Oxford. Used with permission.

reaching a maximum of 1700 m in the Putoran Mountains. In these mountains the vegetation is alpine tundra at the top and a *Picea-Abies* forest on the slopes and in the valleys. The uplands of the plateau continue eastward into the Chukchi Peninsula (Chukotka).

The modern physiography of Siberia is due to a stable underlying cratonic platform, uplift of the Ural Mountains in the late Paleozoic providing a source of eroded sediments transported eastward, and extensive glaciations beginning in the late Pliocene and extending south to the Ob River near Novosibirsk. The ice sheets carved basins into the lowlands and formed cirques at the tops of mountains during times of glacial advance. Upon retreat meltwater filled the cirques to form tarns or glacial lakes in the uplands, and blocks of detached ice melted to fill basins in the lowlands. Lake Baikal (Fig. 1.4) is a glacially modified structural depression with a depth of 1713 m; it is the deepest lake in the world, holding 84% of Russia's and 20% of the world's supply of fresh water. Glaciers beveled the highlands and further filled in the lowlands to create the relatively flat topography with an extensive system of meandering rivers like the Angara flowing out of Lake Baikal, and the Ob, Yenisei, and Lena rivers all flowing northward and northwestward into the Arctic and Pacific oceans. The

1.4. Lake Baikal, Siberia, Russia.

southern parts of the rivers thaw before the northern outlets, causing water to back up in the spring and early summer, contributing to the extensive swamplands of the West Siberian Plain.

Rainfall across Siberia is generally low. It is only 15 mm along the Arctic Coast, and precipitation is in the form of snow for about 250 days, which also provides a highly reflective heat surface contributing to the cold climate. Snow cover decreases to 120 days in the south. In the mid-continent region rainfall is further reduced as winds lose moisture blowing eastward across the land. The continental climate is characterized by the usual winter/summer seasonal variation in temperature, while the interior location and far-northern position of Siberia means these variations are extreme. The average annual temperature is just below 0°C, the January average on the West Siberian Plain is −27°C, and in eastern Siberia at Yakutsk it is −43°C. The tiny town of Verkhoyansk (population 1300) holds the world record for seasonal temperature differences, with a remarkable range of −68°C to 37°C. The coldest settlement is Oymyakon in the Republic of Sakha where on 6 February 1933 the temperature reached −89.9°C. The harshness of the environment was vividly demonstrated in 2012 when Verkhoyansk was attacked by a pack of 400 wolves facing starvation from decline in their principal food of blue hare (*Telegraph*, 24 March 2015). Citizens had to patrol the town on snowmobiles until government troops could arrive. As summarized by Shahgedanova (2008, 70), "three main factors are responsible for the formation of the Eurasian climates: the northern position of the continent, its remarkable size, and the arrangement of the mountain systems." Central Siberia was the site of a meteorite event at Tunguska on 30 June 1908 at 7:15 a.m. when a near-miss felled millions of trees across the region (Fig. 1.5). There is no crater, but the scar is still visible from the ground and on satellite photographs.

There are several consequences of the relatively low, flat topography of Siberia relevant to the vegetation and environmental history of the region. One is that a sea-level rise of only 50 m would inundate much of the land from the Arctic Ocean to Kazakstan in the south and from the Ural Mountains in the east to the Caspian Sea in the west. Sea levels were higher by 150–200 m in the Cretaceous (Haq et al., 1987; Miller, 2009, 884) and were lower by an estimated 120 m during the 18–20 glacial maxima. Another consequence of the comparatively flat interior landscape is that the extreme climate, including day length with prolonged periods of darkness, and the increasingly low incidence of light toward the north, has resulted in comparatively few plants capable of coping with the environment. Thus, the current assemblage and distribution of the lowland coniferous vegetation in Siberia is of relatively recent origin, probably after the Middle Miocene Climatic Optimum (MMCO) and the Middle Pliocene Climatic

1.5. Impact of 1908 Tunguska explosion on vegetation near Podkamennaya Tunguska, Siberia, Russia. From Wikipedia, photograph by Leonid Kulik.

Optimum (MPCO) in late Pliocene to Pleistocene times. That is to say, the current version is 11,000–12,000 yr old (late glacial–Holocene transition); throughout, it has been comparatively simple in composition (nine principal tree genera; see section on modern vegetation, below). The cold climate and low-lying topography have further allowed the accumulation of diverse sediments useful for tracing the biotic and environmental history. Fossils preserved in peats, lignites, volcanic shales, lacustrine, and near-shore clay, silt, and mudstones provide an extensive record of the vegetation. This is especially true for the Quaternary with extensive bog and lake deposits, but there are also lignites and coals of late Mesozoic and Tertiary age containing fossils and inorganic isotopic compounds such as O^{16}/O^{18} and C^{12}/C^{14} important for estimating paleotemperatures and making age determinations independent of the fossils (I, chap. 3).

The soils of Arctic and sub-Arctic Russia are discussed by Zamotaev (2008; Fig. 1.6). They are mostly clayey gleysols with an active layer to a depth of only 50–150 cm. The effective rainfall is low (15 mm), and much of the region is a polar desert covering ca. 1.8 million km^2 (8.1%) of Russia. To the south the principal soil types are various sandy podzols with a deeper organic layer covering ca. 8.3 million km^2 (38%) of the country, and the mean annual precipitation (MAP) is generally above 700 mm.

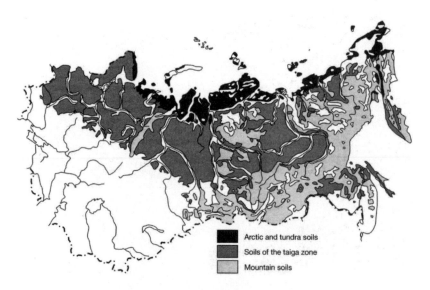

Arctic and tundra soils
Soils of the taiga zone
Mountain soils

1.6. Soil types in cold regions of Russia. From I. Zamotaev, 2008, in M. Shahgedanova (ed.), The Physical Geography of Northern Eurasia, Oxford University Press, Oxford. Used with permission.

Geology

The average elevation of the Ural Mountains is less than 2000 m because much of the uplift was in the Pennsylvanian Period and they have undergone 300 Ma of erosion, much like the Appalachian Mountains of eastern North America. The Ural Mountains consist of a series of parallel ridges and valleys reflecting east-west compression forces during a tectonic interval known as the Uralian Orogeny (Pennsylvanian and Early Permian), when the eastern edge of the Eurasian Plate began colliding with an ancient Asian block called Kazakhstania. The cover of erosional and marine depositional sediments leveling the topography has accumulated to a depth of ca. 1.5 km. By the early Tertiary low-lying terrestrial conditions were present, with lignite deposition extending into the Pliocene (Koronovsky, 2008). The rift that formed Lake Baikal formed ca. 25 Ma.

Modern Vegetation

The extant vegetation of the New World can be arranged into plant formations in a system developed especially for discussing vegetation history (see the preface). Those pertaining to the boreal land bridges are defined and briefly characterized below. The figures refer to illustrations in the previously mentioned publications (I, II, III):

Tundra (I, fig 1.2): A treeless vegetation dominated by lichens, bryophytes, herbs (grasses, sedges), small shrubs, and dwarf forms of trees, and characterized by cold climates in polar regions (Arctic tundra) and high altitudes farther south (alpine tundra; II, figs. 3.135–3.137). In the most extreme environments where much of the ground is barren, it is referred to as a desert tundra. A few individual or small stands of taller trees may occur locally in protected areas. It is a dry environment with effective precipitation often less than 25 mm, and permafrost is common.

Taiga or boreal coniferous forest (I, fig. 1.3): A community of the high-northern latitudes consisting primarily of evergreen gymnosperms with enclaves of bog vegetation, deciduous trees in drier habitats, and *Populus* and *Salix* along streamsides and in other moist habitats. There is frequently a diffuse ecotone with tundra to the north and deciduous forest to the south. The coniferous forest may occur as a montane community at progressively higher altitudes toward the south (I, figs. 1.5, 1.6; II, fig. 3.43). The nine principal genera are *Abies, Larix, Picea, Pinus, Tsuga, Alnus, Betula, Populus,* and *Salix.*

Deciduous forest (I, figs. 1.9, 1.10; II, fig. 3.36; Graham, 2011, fig. 9): A community of predominantly deciduous angiosperms at mid-latitudes in regions of moderate cold-warm seasonality and at higher elevations toward equatorial regions with wet-dry seasonality. Compared to the taiga, the deciduous forest is more diverse, with 60 or more dominants (e.g., *Acer, Carpinus, Carya, Cornus, Corylus, Fagus, Juglans, Liquidambar, Liriodendron, Nyssa, Ostrya, Quercus, Ulmus*). It is the modern expression of Chaney's (1959) Arcto-Tertiary Geoflora (see chapter 3).

Grassland (prairie, steppe; I, figs. 1.1, 1.11): Often an edaphic, climatic, or slope/exposure-controlled community on calcareous soils in an environment too dry for trees and too moist for desert.

Freshwater herbaceous bog/marsh/swamp including polder (Dutch), riparian (*Alnus, Salix*), and valley vegetation (e.g., Po Valley, *Quercus robur, Populus, Salix*; II, fig. 3.24).

Aquatic: (II, fig. 3.25; III, fig. 2.25a).

The vegetation of Russia is catalogued in the 30-volume *Flora USSR* (Komarov, 1934 et seq.) as described in *Vascular Plants of Russia and Adjacent States (the Former USSR)* by Czerepanov (1995; see also Kirpicznikov, 1969). The number of vascular plants is estimated at 21,770 species. There are also the 14-volume *Flora of Siberia* (Krasnoborov, 2000 et seq.) listing 4200 species; the *Flora of Kamschatka and the Adjacent Islands* (Hultén, 1927, 1168 species); other works by Hultén (1958, 1962, 1986); and analyses by Yurtsev (1982) of the relic xerophytic vegetation in Beringia (see also Yurtsev, 1984a,b).

The ridge and valley province and the extensive north-south range of the Ural Mountains provide a diversity of habitats for relic taxa and introductions, and a resevoir of different communities available for expansion during intervals of climatic change. To the north and east is the treeless vegetation of the polar desert and tundra, consisting of about 1700 species adapted to this cold climate with only about 25 mm of annual precipitation. Tundra has virtually 100% ground cover, while the polar desert generally has less than 50% (Bliss, 1981; Barbour and Christensen, 1993). The tundra environment is often made harsher by high winds that can blow 50 mph or more for days. Although biodiversity is low, many of the plants that have adapted are circumboreal in distribution or widely represented by closely related species pairs. The High Arctic zone fide Yurtsev (1982; a polar desert) is the most extreme and depauperate, often with patches of open ground and extensive populations of *Eriophorum variegatum*. To the south mosses, lichens, grasses (Poaceae), reeds (Juncaceae), and sedges

(Cyperaceae) are prominent, along with *Andromedia, Anemone patens, Arcto-staphylos, Cassiope tetragona, Calluna, Empetrum nigrum, Erica, Kalmia, Ledum palustre, L. decumber, Pyrola grandiflora, Vaccinium,* and dwarf trees of *Betula exilis* (Fig. 1.7), *Salix arctica* (Fig. 1.8), and *Populus balsamifera* (Mead et al., 1993).

Southward beyond the permafrost the vegetation grades into the taiga or boreal forest. It varies from a dark and moist phase with dominants of *Picea* and *Abies* to a lighter and drier phase with *Pinus* and *Cladonia* lichens on the forest floor (Fig. 1.9). Bordering the tundra, the taiga often occurs

1.7. *Betula exilis,* Kamchatka, Russia.

1.8. *Salix arctica,*
Kamchatka, Russia.

1.9. Dry phase of taiga with *Pinus, Picea, Cladonia*
lichen substrate, near Lake Baikal, Siberia, Russia.

in patches then consolidates southward into a continuous cover of needle-
leaved forest, increasingly mixed farther south with deciduous trees. In the
extensive wetlands of the West Siberian Plain taiga forest mostly grows on
the better drained uplands and along sandy river margins. The trees are
Picea obovata, Abies sibirica, Pinus sibirica, P. sylvestris, and *Larix sibirica,* with
Cornus alba, Ledum palustre, Sambucus racemosa, Sorbus sibirica, Vaccinium,
and rare *Tilia sibirica* in the undergrowth. The East Siberian taiga covers
nearly one-fourth of Russia, extending between the watersheds of the Lena
and Yenisey rivers, and includes about 2300 species of vascular plants. The
principal trees are *Picea sibirica* mixed with *Tsuga, Abies sibirica, Larix dahurica,* and *Pinus sibirica;* broadleaf trees of *Alnus fruticosa* and *A. hirsuta,
Betula costata, B. ermanii, B. excilis, B. nana,* and *Ulmus sibirica;* with *Populus
suaveolans* and *Salix excilis, S. lantana, S. fuscescens, S. myrtilloides,* and *S. repens* growing in marshes and along streams. Shrubs of *Artemisia oppulenta,
Rhododendron aureum, Myrica tomentosa, Vaccinium oxycoccus,* and *V. myrtillus* grow in the understory. These few trees collectively make up most of
the taiga forest in this climatically challenging, edaphically rather uniform,
and topographically limited landscape. *Sorbus, Crataegus,* and others are
added on warmer, well-drained sites, and these increase farther south in
the broadleaved deciduous forest. Also bordering the taiga in areas with
greater topographic relief is a grassy steppe on cold, dry, open slopes. The

East Siberian taiga supports an extensive endemic biota with many rare and endangered plant species. Among those listed by the World Wildlife Fund (https://www/worldwildlife.org/ecoregions/pa0601) are *Adenophora jacutica, Artemisia czekanowskiana, Senecio lenensis, Caltha serotina, Cotoneaster lucidus, Potentilla jacutica, Draba sambykii, Megadenia bardunovii, Gastrolychis angustifolia* ssp. *tenella* (= *Silene*), *Juncus longirostris, Oxytropis calva, O. leucantha, Papaver variegatum, Polygonum amgense, Salix saposhnikovii, Thymus evenkiensis, Viola alexandroviana,* and the orchids *Calypso bulbosa* and *Orchis militaris.*

These communities constitute the vegetation types of Siberia that expanded and contracted during the many intervals of Late Cretaceous and Cenozoic climate change.

Indigenous People

The exensive plains and forests of Siberia have long provided habitats for nomadic tribes of hunters and gatherers, while lakesides and river margins supported seasonal to more permanent settlements. Humans were present 45,000 years ago, as evidenced by the oldest known human genome, which was sequenced from a male thighbone found eroding from alluvial deposits along the River Irtysh in the village of Ust'-Ishim of western Siberia (Fu et al., 2014; Pääbo, 2014). The fingerbone of a child from Denisova Cave in the Altai Mountains southeast of Novosibirsk is estimated from its DNA to be 41,000 years old and the result of interbreeding between Neanderthals and modern humans (Krause et al., 2010). Also noteworthy is the regeneration of *Silene stenophylla* from fruit tissue reported to be 31,800 years old preserved in burrows buried in the Siberian permafrost (Yashina et al., 2012). The absence of significant physiographic barriers across wide areas allowed for the migration of plants and animals, including people, as changing climates and shifting resources dictated. There is archaeological evidence that ancestors of the modern Khanty and Mansi ethno-territorial groups have inhabited the western taiga since at least 2000 BCE (Bronze Age; Balalaeva et al., s.d.), and their descendants, the Chukchi Eskimos, are among the prominent native groups now adapted to the conditions of far-eastern Siberia. Extending the interaction between climate and human history into modern times, Russia's defeat of both the French under Napoleon in 1812 and the Germans under Hitler in 1941 was due, in part, to the retreating Russian armies luring these unprepared foreign troops into the unsurvivable winters of Siberia.

Kamchatka

Geographic Setting and Climate

The Kamchatka Peninsula projects southward from the far East Siberian Plain for 1250 km between the Sea of Okhotsk on the west and the Bering Sea on the east. It is farther from North America than Chukotka is from the Seward Peninsula but virtually connected to the New World by the Aleutian Islands running from the Commander Islands of Russia to the Alaska Peninsula (Figs. P1.1, P1.2). Approximately 325,000 people live in Kamchatka, mostly concentrated in two large cities, Petropavlovsk-Kamchatsky and Yelizovo. Petropavlovsk-Kamchatsky was founded in 1740 by Vitus Bering (Frost, 2003; Lauridsen, 1889 [see BiblioLife English translation]). Petropavlovsk has a population of 180,000 people, and Yelizovo has another 40,000, so the intervening region is sparsely populated. The winters are long, cold, dark, and stormy. In coastal areas MAP reaches 2700 mm (110 in). In describing the region, Frost (2003, xvii) notes "how little the land has changed over the centuries in all these remote places, still largely uninhabited as part of the establishment of modern ecological preserves . . . or as a result of inclement weather, declining resources, or desolate landscapes." During a visit to a geothermal plant near the Valley of Geysers in July 2015, I could clearly see the harsh toll these conditions take on habitations, vehicles, and infrastructure. New buildings emerge from construction sites seemingly already with a tired and worn look. Along the highway, poles 12 m tall mark the edge of the road, and in winter snow accumulates to near the very top.

The complex geology of Kamchatka is due in part to far-distant tectonic events. These will be discussed later with regard to the NALB, but suffice it to say that some regional compression resulted from the overall westward movement of North America away from the Mid-Atlantic Ridge. Closer to home, the first stage in the opening of the Canadian Basin was extension and rifting in the Kimmeridgian (Late Jurassic, ca. 155 Ma; Schreider et al., 2013). Spreading developed at 151 Ma, with the principal opening of the basin between 151 and 145 Ma (M22–10Ar). In the immediate vicinity of Beringia, contact and subduction of the northwestern edge of the Pacific Plate beneath the Eurasian Plate was the prominent tectonic force. For example, in Arctic Russia a contact known as the South Anyual Suture is the result of a collision between the Alaska and Chukotka subplates with the Eurasian Plate in the Early to Middle Cretaceous. It was an important time and an immensely complicated process. The

1.10. Volcanic mountain landscape of Kamchatka, Russia.

1.11. Fumaroles in the Valley of the Geysers, Kamchatka, Russia.

uncertainty in age among the many sedimentary basins still prevents establishing a unified and detailed chronostratigraphic or lithostratigraphic sequence for much of the region (Miller et al., 2002).

Kamchatka is an earthquake-prone and highly active volcanic territory (Kozhurin et al., 2006; Figs. 1.10, 1.11). At the Institute of Volcanology in Petropavlovsk as of 2015 more than 260 volcanoes are recognized for Kamchatka, of which approximately 29 are currently active. This is about the same as for Alaska. Klyuchevskaya Sopka Volcano at 4750 m is the tallest in the Northern Hemisphere as measured from sea level. Throughout the Late Cretaceous and Tertiary ash falls and lava flows blocked streams, creating numerous lakes suitable for the preservation of fossils (Figs. 1.12, 1.13). As with Arctic Russia, the intense volcanism is due to the subduc-

1.12. Plant macrofossil from the Eocene of Kamchatka:
Sequoia affinis. From L. Budantsev, 1997, Komarov
Botanical Institute, Leningrad. Used with permission.

tion of the Pacific Plate (Okhotsk Subplate) beneath the Eurasian Plate
along the 10,500 m deep Kuril-Kamchatka Trench (see Fig. P1.1), which
joins the Japan Trench as part of the Pacific Ring of Fire. Volcanogenic sediments of Mesozoic age have accumulated in the trench to a thickness of
over 10 km, and those of Cenozoic age are over 14 km thick. Subduction
is occurring today at the rate of ca. 6–7 cm/yr (Koronovsky, 2008), rapid
when compared, for example, to the Nazca Plate, which is dipping under
northwestern South America at 3.1–3.7 cm/yr to form the Northern Andes.
Pulses of rapid subduction have left three volcanic zones that are progressively younger toward the Pacific. These are the West Kamchatka (Eocene),
Mid-Kamchatka (late Oligocene-[middle Miocene]-Quaternary), and Kuril-
Kamchatka zone (Recent and ongoing; Avdeiko et al., 2007). The latest
intervals of rapid subduction in the far-northern part of the peninsula,
and, therefore, the time of origin for much of its modern topography, are
7 Ma, 5 Ma, and 2 Ma (Lander and Shapiro, 2007). Details of the slab-type
subduction at the Kuril-Kamchatka Trench are described by Davaille and

Lees (2004) and Koulakov et al. (2011) and are further discussed at http://
volcanohotspot.wordpress.com/2015/03/18/tectonics-of-the-kamchatka
-peninsula. Recent earthquakes of 9.3 and 8.2 have been recorded. In addition to subduction, the Eurasian, Pacific, and North American plates form
a triple junction in the Beringian region, further contributing to the crustal
deformation and elevation of the high mountain peaks.

The oldest portion of the Pacific Plate to be subducted into the Asian
trenches is Early Cretaceous (ca. 140 Ma), so uplift of the land surrounding
the North Pacific Basin began at about this time. At 100 Ma beds of chalk
were being deposited across much of eastern Russia, indicating the presence
of relatively deep marine waters. The Arctic Ocean extended across most
of central Russia with the exception of the Ural highlands, and through
much of Beringia except for scattered and temporary highlands, and it was

1.13. Plant macrofossil from the Eocene of Kamchatka:
Ulmus latiserrata. From L. Budantsev, 1997, Komarov
Botanical Institute, Leningrad. Used with permission.

continuous with the Pacific Ocean. There was the Canadian Basin across Canada and the central United States between the emerging foothills of the Rocky Mountains and the ancient Appalachian Mountains. Another basin crossed southward into Central Europe connecting with the Tethys Sea. This was the sea that divided the landmass of Pangaea into northern Laurasia and southern Gondwana. The Mediterranean Sea is a remnant. Thus, the Cretaceous landscape imposed different migrational constraints than those of the Cenozoic when these seas had mostly retreated. Terrestrial organisms could move around the southern end of the central North American, European, and Russian epicontinental seas as they regressed. During this interval, however, the Arctic/Bering Sea remained mostly a barrier between eastern Russia and northwestern North America. Regional uplift had yet to occur. This was accomplished south of the converging Chuchi Alaska/Seward peninsulas in what would become the Aleutian Islands as the leading edge of the Pacific Plate became molten at depth in the Kuril-Kamchatka and Aleutian trenches. Domes formed, fissures developed, and lava moved to the surface along the cracks. This began in the Cretaceous and intensified in the Paleocene and early Eocene, so the beginning of the subaerial Aleutian-Bering sector of the land bridge as a direct migratory route dates from approximately this time. Before that, widely distributed organisms like *Dennstaedtia, Equisetum, Osmunda, Ginkgo,* and others (see online appendices 2 and 3) were already established across the Northern Hemisphere, and their present occurrences are in part the residuals of more ancient continental configuration and fragmentation.

In addition to convergence, subduction, and volcanism that created pathways for migration, other tectonic factors actually transported organisms and their fossils to new sites by (1) displacement of continental blocks (cratons); and (2) accretion of smaller exotic terranes onto distant continental margins. Northward displacement of western Beringia since the Cretaceous was not extensive—perhaps 1–2° of latitude. Yet there are numerous terranes representing island-arcs, marine fore-arc and continental back-arc structures, and small ocean basins accreted onto the margin of Beringia (Zinkevich and Tsukanov, 1993, 1994). Since the Cretaceous these have come mostly from the east and southeast rather than over great distances from the south.

Together with the shaping of the physical landscape, climate has played a major role in determining the movement of organisms across Beringia (see paleotemperature curves in the section on utilization of the BLB in chapter 2; see also I, 89; and Graham, 2011, 348, with reference to vegetation history [based on Miller et al., 1987; Zachos et al., 2001; and others]).

The asteroid impact at 65 Ma caused a drop in temperatures from mid-Cretaceous highs. Then, global climates warmed and rainfall increased during the Paleocene-Eocene Thermal Maximum (PETM, ca. 56 Ma) and the Early Eocene Climatic Optimum (EECO, ca. 56–50 Ma, the warmest in the Cenozoic Era), and tropical biotas became widely distributed until about the middle Eocene. There was a temperature decline in the middle Eocene that stabilized through the Oligocene until the Middle Miocene Climate Optimum (MMCO, 15 Ma), allowing subtropical and warm temperate biotas to consolidate and expand while tropical biotas retreated toward equatorial regions. From the late Miocene through the early Pliocene temperate to cool-temperate biotas were prominent, then, with the exception of the Middle Pliocene Climatic Optimum (MPCO, 3.5 Ma), cold-temperate assemblages were favored in the high northern and southern latitudes and elsewhere at high altitudes. Consistent with these trends is evidence from paleosols in Arctic Canada near the east coast of Axel Heiberg Island. Schnitzer et al. (1990a,b) found a gleysolic paleosol at ca. 45 Ma, which they interpret as indicating a climate that was warm and moist. From the Yukon Territory Tarnocai and Schweger (1991) describe Late Tertiary paleosols suggesting cooler climates but still warmer than the present, followed by early Pleistocene cryogenic soils of cold, glacial climates. These broad global trends based on edaphic evidence provide a valuable context independent of paleontology for reading the record throughout Beringia during the Late Cretaceous and Cenozoic. The message from these readings is that land continuity, separation, and physiography were important in facilitating the spread and diversification of plants within western Beringia and between it and immediately adjacent regions, but the separation was not great and the available time was vast. Within this context, the overriding factor seems to have been climate—tropical to warm-temperate in the Paleogene; a respite for adaptation, assemblage, and organization in the Oligocene; and cool to cold-temperate in the Neogene. This is consistent with environmental recontstructions for the boreal region generally, as presented in chapters 2 and 3. The record was quite different from those of the equatorial land bridges to be discussed in chapters 4 and 5, more similar but still distinct from the history of the austral land bridge considered in chapter 6.

In addition to soils and fossil plants, there are records of extinct animals long associated with the BLB (see, e.g., Markova et al., 2008). Most famous is the woolly mammoth (*Mammuthus primigenius*) that roamed the interface between the tundra and taiga-steppe in Siberia during the Pleistocene, shifting its range as climates changed and resources diminished. The plants consumed by these mammoths have been identified from their stomach contents and are listed in a table included in the online materials for this book (see also van Geel, 2008, 2011a,b). More recently a nearly complete cadaver of *Bison priscus* (steppe bison) dated at 9300 BP or early Holocene has been reported from caves in Sakha (Boeskorov et al., 2014). As previously noted, this was the site of the coldest modern temperature ever recorded ($-89.9\,^\circ$C). (See also the mention of *Bison* remains found in the Yukon, 13,300 yr old, in chapter 3.) Other remains at Sakha included *Equus* (wild horse) and *Mammuthus*, as western Beringia continues to yield new data on the history of the fauna and environments of the Quaternary.

Modern Vegetation

The vegetation of Kamchatka is similar, especially at the generic level, to that of the mountainous regions of Siberia. The high diversity and endemism is due to the partitioning and isolating effect of the extensive topographic relief (see Figs. 1.10, 1.11). There are about 1170 species of plants. *Betula ermanii* and *Alnus fruticosa* are common in the cold, wet northern tundra zone, while in the more continental climates of the Central Depression, *Larix cajanderi*, *Picea jezoensis*, and thickets of *Pinus pumila* are found at mid-altitudes together with more widespread azonal plants of *Betula exilis*, *Populus suaveolens*, *Salix arctica*, *Alnus hirsuta*, and two color forms of *Rhododendron*. River valleys have forests of *Populus suaveolens*, *Salix udensis*, *S. rorida*, and *Chosenia arbutifolia* (Krestov and Omelko, 2008). The remote

1.14. *Juncus* in mud pot, Valley of the Geysers, Kamchatka, Russia.

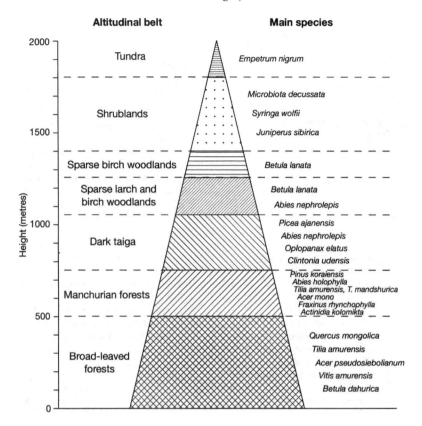

Altitudinal belt **Main species**

Tundra — Empetrum nigrum

Shrublands — Microbiota decussata / Syringa wolfii / Juniperus sibirica

Sparse birch woodlands — Betula lanata

Sparse larch and birch woodlands — Betula lanata / Abies nephrolepis

Dark taiga — Picea ajanensis / Abies nephrolepis / Oplopanax elatus / Clintonia udensis

Manchurian forests — Pinus koraiensis / Abies holophylla / Tilia amurensis, T. mandshurica / Acer mono / Fraxinus rhynchophylla / Actinidia kolomikta

Broad-leaved forests — Quercus mongolica / Tilia amurensis / Acer pseudosiebolianum / Vitis amurensis / Betula dahurica

1.15. Vertical zonation of vegetation, Kamchatka and vicinity (maritime provinces), Russia. From A. Ivanov, 2008, in M. Shahgedanova (ed.), The Physical Geography of Northern Eurasia, Oxford University Press, Oxford. Used with permission.

Kronotsky Nature Reserve located on the Pacific Coast of central Kamchatka has several active volcanoes rising to 3500 m elevation. There are mud pots with some remarkably persistent plants (Fig. 1.14), as well as geyser fields, and the climate is harsh (Quammen, 2009). A total of 767 species of vascular plants are listed and representatives are arranged vertically, as shown in Fig. 1.15 (see also Berg, 1959). Those cited as endangered or vulnerable in the Red Data Book of Kamchatka Krai and the Russian Federation (http://kronoki.org/territories/reserve/flora) are listed in a table included in the online materials for this book.

In 2015, on a three-day trip from Kamchatka to Khabarovsk on the Amur River, and to Irkutsk and Lake Baikal in Siberia, on the Trans-Siberian Express with Victor Kuzevanov, director of the Irkutsk Botanic Garden, we had the opportunity to compare the landscape with that observed around Novosibirsk nearly half a century earlier. Notable was the intensified lumbering of birch (Fig. 1.16). There has also been an effort to develop tourism at particularly scenic locations. Around the several sites designated collectively as "The Volcanoes of Kamchatka," the vegetation is better preserved because these were made a UNESCO World Heritage Site in 1996. Study of the vegetation throughout Kamchatka and Siberia is difficult, often requiring specialized and expensive transportation (Figs. 1.17, 1.18).

1.16. Birch logs harvested from the northern Russian forests and a major export of the country.

1.17. Blue Horse ex-military vehicle reassigned for field studies in Kamchatka, Russia.

1.18. Helicopter used for transport into remote areas of Kamchatka, Russia.

1.19. Escort for field work
in the Valley of the Geysers,
Kamchatka, Russia.

Associated with the tundra and taiga are animals that form a striking part of the ecosystem. Among these is the huge (540 kg/1200 pound) Kamchatka brown bear (*Ursus arctus beringianus*), related to *U. arctus horribilis* across the strait in Alaska, that warrants caution when traveling through isolated areas (Fig. 1.19). Others are the very large *Alces alces burulini* (Chukotka moose), *Canis lupus albus* (tundra wolf), *Gulo gulo* (wolverine), *Lemmus* (lemming), *Lepus timidus* (mountain hare), *Lutra lutra* (Eurasian otter), *Lynx lynx wrangeli* (East Siberian lynx), *Marmota* (marmot), *Martes zibellina* (sable), *Mustela ermine kaneii* and *M. nivalis pygmaea* (East Siberian stoat or weasel, Siberian least weasel), *Ovis nivicola* (Kamchatka snow or bighorn sheep), *Rangifer tarandus* (reindeer), *Spermophilus parryii* (Arctic ground squirrel), and *Vulpes lagopus* and *V. vulpes beringiana* (arctic, Anadyr fox). In ancient times as this fauna prospered, the people prospered, and when they moved, the people followed.

Indigenous People

Early residents of Kamchatka were the Itelmens and later the Koryak, but their numbers were reduced to around 1900 individuals by a smallpox epidemic in 1768 and the rest have been assimilated into the Russian culture. At the small settlement of Sosnovka (362 km south of Kirov; 56°15′N, 51°18′E) descendants of the Itelmens are attempting to maintain the language and preserve remnants of their heritage through lectures and village restoration.

References

Avdeiko, G. P., D. P. Savelyev, A. A. Palueva, and S. V. Popruzhenko. 2007. Evolution of the Kurile-Kamchatkan Volcanic Arcs and dynamics of the Kamchatka-Aleutian Junction. In J. C. Eichelberger, E. Gordeev, P. Izbekov, M. Kasahara, and J. Lees (eds.), Volcanism and Subduction: The Kamchatka Region. Geophysical Monograph Series 172. American Geophysical Union, Washington, DC. Pp. 37–55.

Balalaeva, O., T. Isayeva, and O. Starodubova. s.d. Hunters of West Siberian Taiga. Climate Change Adaptations: Traditional Knowledge of Indigenous Peoples Inhabiting the Arctic and Far North Hunters of West Siberian Taiga. UNESCO Institute for Information Technologies in Education, Paris.

Barbour, M. G., and N. L. Christensen. 1993. Vegetation. In Flora of North America Editorial Committee (eds.), Flora of North America. Oxford University Press, Oxford. Pp. 97–131 (chap. 5).

Berg, L. S. 1959. Die geographischen Zonen der Sowjetunion. 2 vols. (Vol. 1, 437 pp. + 105 photographs; vol. 2, 604 pp. + 13 photographs.) Teubner Verlag, Leipzig.

Bliss, L. C. 1981. North American and Scandanavian tundras and polar deserts. In L. C. Bliss, O. W. Heal, and J. J. Moore (eds.), Tundra Ecosystems: A Comparative Analysis. International Biological Program Synthesis Series Number 25. Springer, Berlin.

Boeskorov, G. G., O. R. Potapova, E. N. Mashchenko, A. V. Protopopov, T. V. Kuznetsova, L. Agenbroad, and A. N. Tikhonov. 2014. Preliminary analyses of the frozen mummies of mammoth (*Mammuthus primigenius*), bison (*Bison pricus*) and horse (*Equus* sp.) from the Yana-Indigirka lowland, Yakutia, Russia. Integrative Zoology 9: 471–80.

Budantsev, L. Y. 1997. Late Eocene Flora of Western Kamchatka. Russian Academy of Sciences, Proceedings of the Komarov Botanical Institute, St. Petersburg.

Chaney, R. W. 1959. Miocene floras of the Columbia Plateau: part 1: composition and interpretation. Carnegie Institution of Washington Contributions to Paleontology 617: 1–134.

Chibilyov, A. 2008. Steppe and forest-steppe. In M. Shahgedanova (ed.), The Physical Geography of Northeastern Eurasia. Oxford University Press, Oxford. Pp. 248–66.

Czerepanov, S. K. 1995. Vascular Plants of Russia and Adjacent States (the Former USSR). Cambridge University Press, Cambridge.

Davaille, A., and J. M. Lees. 2004. Thermal modeling of subducted plates: tear and hotspot at the Kamchatka corner. Earth and Planetary Science Letters 226: 293–304.

Fernández-Armesto, F. 2007. Pathfinders: A Global History of Exploration. W. W. Norton & Company, New York.

Frost, O. W. 2003. Bering: The Russian Discovery of America. Yale University Press, New Haven.

Fu, Q., et al. (+ 27 authors). 2014. Genome sequence of a 45,000-year-old modern human from western Siberia. Nature 514: 445–49.

Graham, A. 2011. The age and diversification of terrestrial New World ecosystems through Cretaceous and Cenozoic time. American Journal of Botany 98: 336–51.

Haq, B. U., J. Hardenbol, and P. R. Vail. 1987. Chronology of fluctuating sea levels since the Triassic (250 million years ago to present). Science 235: 1156–67.

Hultén, E. 1927. Flora of Kamchatka and the Adjacent Islands. Almqvist & Wilsell, Stockholm.

———. 1958. The Amphi-Atlantic Plants and Their Phytogeographical Connections. Almqvist & Wiksell, Stockholm.

———. 1962. The Circumpolar Plants. Almqvist & Wiksell, Stockholm.

———. 1986. Atlas of North European Vascular Plants North of the Tropic of Cancer. Koeltz, Königstein.

Ivanov, A. 2008. The Far East. In M. Shahgedanova (ed.), The Physical Geography of Northern Eurasia. Oxford University Press, Oxford. Pp. 422–47.

Kirpicznikov, M. E. 1969. The flora of the U.S.S.R. Taxon 18: 685–708.

Komarov, V. L. (Initial editor-in-chief and subsequent volume editors.) 1934 et seq. Flora of the USSR. Botanicheskii Institut (Akademia Nauk SSSR), Moscow. [English translation: Israel Program for Scientific Translations, Springfield, VA; Smithsonian Institution Libraries, Washington, DC.]

Koronovsky, N. 2008. Tectonics and Geology. In Maria Shahgedanova (ed.), The Physical Geography of Northern Eurasia. Oxford University Press, Oxford. Pp. 1–35.

Koulakov, I. Yu., N. L. Dobretsov, N. A. Bushenkova, and A. V. Yakovlev. 2011. Slab shape in subduction zones beneath the Kurile-Kamchatka and Aleutian arcs based on regional tomography results. Russian Geology and Geophysics 52: 650–67.

Kozhurin, A. (+ 13 authors). 2006. Trenching studies of active faults in Kamchatka, eastern Russia: palaeoseismic, tectonic and hazard implications. Tectonophysics 417: 285–304.

Krasnoborov, I. M. (Series editor; P. M. Rao, translator.) 2000 et seq. Flora of Siberia. Science Publishers, Enfield, NH.

Krause, J., Q. Fu, J. M. Good, B. Viola, M. V. Shunkov, A. P. Derevianko, and S. Pääbo. 2010. The complete mitochondrial DNA genome of an unknown hominin from southern Siberia. Nature 464: 894–97.

Krestov, P. V., and A. M. Omelko. 2008. Vegetation and natural habitats of Kamchatka. Berichte der Reinhold-Tüxen-Gesellschaft (RTG) 20: 195–218.

Lander, A. V., and M. N. Shapiro. 2007. The origin of the modern Kamchatka subduction zone. In J. C. Eichelberger, E. Gordeev, P. Izbekov, M. Kasahara, and J. Lees (eds.), Volcanism and Subduction: The Kamchatka Region. Geophysical Monograph Series 172. American Geophysical Union, Washington, DC. Pp. 57–64.

Lauridsen, P. 1889. Vitis Bering: The Discoverer of the Bering Strait. S. C. Griggs, Chicago. [BiblioLife English translation.]

Markova, A. K., T. van Kolfschoten, S. Bohncke, P. A. Kosintsev, J. Mol, A. Yu. Puzachenko, A. N. Simakova, N. G. Smirnov, A. Verpoorte, and I. B. Golovachev. 2008. Evolution of European Ecosystems during Pleistocene-Holocene Transition (24–8 kyr BP). KMK Scientific Press, Moscow.

Mead, J. I., L. D. Agenbroad, and R. H. Hevly. 1993. Paleoenvironments, mammoths, and Siberia: an introduction. In L. D. Agenbroad, J. I. Mead, and R. H. Hevly (eds.), Vegetation Cover and Environment of the "Mammoth Epoch" in Siberia by V. Ukraintseva. Mammoth Site of Hot Springs, SD. Pp. i–xiv.

Miller, E., A. Grantz, and S. L. Klemperer. 2002. Preface. In E. Miller, A. Grantz, and S. L. Klemperer (eds.), Tectonic Evolution of the Bering Shelf–Chukchi Sea–Arctic Margin and Adjacent Landmasses. Geological Society of America Special Paper 360. Geological Society of America, Boulder. Pp. v–ix.

Miller, K. G. 2009. Sea level change, last 250 million years. In V. Gornitz (ed.), Encyclopedia of Paleoclimatology and Ancient Environments. Springer Netherlands. Pp. 879–87.

Miller, K. G., R. G. Fairbanks, and G. S. Mountain. 1987. Tertiary oxygen isotope synthesis, sea level history, and continental margin erosion. Paleoceanography 2: 1–19.

Pääbo, S. 2014. Neanderthal Man: In Search of Lost Genomes. Basic Books, Philadelphia.

Quammen, D. 2009. Fragile Russian wilderness: the Kronotsky Nature Reserve is best appreciated from afar. National Geographic 215: 62. http://bgm.nationalgeographic.com/2009/01/Russia-wilderness.quammen-text.

Schnitzer, M., C. Tarnocai, P. Schuppli, R. Hempfling, R. Müller, and H.-R. Schulten. 1990a. Paleoenvironmental organic indicators in Eocene paleosols from Arctic Canada. Fresenius Journal of Analytical Chemistry 337: 882–84.

Schnitzer, M., C. Tarnocai, P. Schuppli, and H.-R. Schulten. 1990b. Nature of the organic matter in Tertiary paleosols in the Canadian Arctic. Soil Science 149: 257–67.

Schreider, Al. A., L. I. Lobokovsky, and A. A. Schreider. 2013. Kinematic model of the opening of the Canadian Basin, Arctic Ocean. Oceanology 53: 481–90. [Original Russian text in Okeanologiya, 2013, 53: 539–49.]

Shahgedanova, M. (ed.). 2008. The Physical Geography of Northern Eurasia. Oxford University Press, Oxford.

Solonevich, N. G., and V. V. Vikhireva-Vasilkova. 1977. Preliminary results of a study of plant remains from the gastrointestinal tract of the Shandrin Mammoth (Yakutia). In O. A. Skarlato (ed.), Anthropogene Fauna and Flora of the Siberian Northeast. Nauka, Leningrad. Pp. 208–21. [In Russian.]

Tarnocai, C., and C. E. Schweger. 1991. Late Tertiary and early Pleistocene paleosols in northwestern Canada. Arctic 44: 1–11.

Ukraintseva, V. V. (Edited by L. D. Agenbroad, J. I. Mead, and R. H. Hevly.) 1992. Vegetation Cover and Environment of the "Mammoth Epoch" in Siberia. Mammoth Site of Hot Springs, SD.

van Geel, B., et al. (+ 15 authors). 2008. The ecological implications of a Yakutian mammoth's last meal. Quaternary Research 69: 361–76.

van Geel, B., et al. (+ 8 authors). 2011a. Mycological evidence of coprophagy from the feces of an Alaskan Late Glacial mammoth. Quaternary Science Rviews 30: 2289–303.

van Geel, B., et al. (+ 9 authors). 2011b. Palaeo-environmental and dietary analysis of intestinal contents of a mammoth calf (Yamal Peninsula, northwest Siberia). Quaternary Science Reviews 30: 3935–46.

Yashina, S., S. Gubin, S. Maksimovich, A. Yashina, E. Gakhova, and D. Gilichinsky. 2012. Regeneration of whole fertile plants from 30,000-y-old fruit tissue buried in Siberian permafrost. Proceedings of the National Academy of Sciences USA 109: 4008–13. [See also S. Pappas interview with Jane Shen-Miller, 20 February 2012; http://www/livescience.com/ search ancient-plants-resurrected-siberian-permafrost.html.]

Yurtsev, B. 1972. Phytogeography of Northeastern Asia and the problem of Transberingian floristic interrrelations. In A. Graham (ed.), Floristics and Paleofloristics of Asia and Eastern North America. Elsevier, Amsterdam. Pp. 19–54.

———. 1982. Relicts of the xerophyte vegetation of Beringia. In D. M. Hopkins, J. V. Matthews, Jr., C. E. Schweger, and S. B. Young (eds.), Paleoecology of Beringia. Academic Press, London. Pp. 157–78.

———. 1984a. Problems of the Late Cenozoic paleogeography of Beringia in light of phytogeographic evidence. In V. L. Kontrimavichus (ed.), Beringia in the Cenozoic Era. Oxonian Press, New Delhi. Pp. 129–53.

———. 1984b. Beringia and its biota in the Late Cenozoic: a synthesis. In: V. L. Kontrimavichus (ed.), Beringia in the Cenozoic Era. Oxonian Press, New Delhi. Pp. 261–78.

Zachos, J., M. Pagani, L. Sloan, E. Thomas, and K. Billups. 2001. Trends, rhythms, and aberrations in global climate 65 Ma to present. Science 292: 686–93.

Zamotaev, I. 2008. Soils. In M. Shahgedanova (ed.), The Physical Geography of Northern Eurasia. Oxford University Press, Oxford. Pp. 103–21.

Zinkevich, V. P., and N. V. Tsukanov. 1993. Accretionary tectonics of Kamchatka. International Geological Review 35: 953–73.

———. 1994. Tectonics and geodynamics of the southern part of Koryak Highlands and Kamchatka. In ICAM-94 (International Conference on Arctic Margins, Magadan, Russia) Proceedings: Regional Terranes. Pp. 169–75.

Additional References

Andreev, A. A., et al. (+ 7 authors). 2014. Late Pliocene and early Pleistocene vegetation history of northeastern Russian Arctic inferred from Lake El'gygytgyn pollen record. Climate of the Past 10: 1017–39.

Budantsev, L. Y. 1992. Early stages of formation and dispersal of the temperate flora in the boreal region. Botanical Review 58: 1–48.

Daly, R. J., D. W. Jolley, and R. A. Spicer. 2011. The role of angiosperms in Palaeocene Arctic ecosystems: a palynological study from the Alaskan North Slope. Palaeogeography, Palaeoclimatology, Palaeoecology 309: 374–82.

Fedorov, A. V., C. M. Brierley, K. T. Lawrence, Z. Liu, P. S. Denkens, and A. C. Ravelo. 2013. Patterns and mechanisms of early Pliocene warmth. Nature 496: 43–49 (plus additional supplementary information).

Foreman, B. Z., M. T. Clementz, and P. L. Heller. 2013. Evaluation of paleoclimatic conditions east and west of the southern Canadian Cordillera in the mid-late Paleocene using bulk organic δ13C records. Palaeogeography, Palaeoclimatology, Palaeoecology 376: 103–13.

Gorbach, N. V., and M. V. Portnyagin. 2011. Geology and petrology of the lava complex of young Shiveluch Volcano, Kamchatka. Petrology 19: 134–66.

Herman, A. B., et al. (+ 8 authors). 2017. Eocene–early Oligocene climate and vegetation change in southern China: evidence from the Maoming Basin. Palaeogeography, Palaeoclimatology, Palaeoecology 479: 126–37.

Liu, K. 1988. Quaternary history of temperate forests of China. Quaternary Science Reviews 7: 1–20.

Mallard, C., N. Coltice, M. Seton, R. D. Müller, and P. J. Tackley. 2016. Subduction controls the distribution and fragmentation of Earth's tectonic plates. Nature 535: 140–43.

Natal'in, B. A. 2004. Phanerozoic tectonic evolution of the Chukotka-Arctic Alaska Block: problems of the rotational model. Smithsonian/NASA Astrophysics Data System. http://adsabs.harvard.edu/abs/2004AGUFMGP43C. .04N; also available at https://pangea.stanford.edu.

Pagani, L., et al. (+ 29 listed authors et al.). 2016. Geonomic analyses on migration events during the peopling of Eurasia. Nature 538: 238–42.

PBS (Public Broadcasting System). 2015a. First Peoples: Asians. June.

———. 2015b. Globe Trekker, Tough Trains: Siberia. 26 May. [Provides a visually informative view of the winter landscape and weather at and just above the Arctic Circle.]

Popova, S., T. Utescher, D. V. Gromyko, A. A. Burch, and V. Mosbrugger. 2012. Palaeoclimate evolution in Siberia and the Russian Far East from Oligocene to Pliocene: Evidence from fruit and seed floras. Turkish Journal of Earth Sciences 21: 315–34.

Popova, S., T. Utescher, D. V. Gromyko, V. Mosbrugger, E. Herzog, and L. Francois. 2013. Vegetation change in Siberia and the northeast of Russia during the Cenozoic cooling: a study based on diversity of plant functional types. Palaios 28: 418–32.

Robyns, J. 2015. The rapid and startling decline of world's vast boreal forests. Yale Environment 360. http://e360.yale.edu/.

Stockey, R. A., and G. W. Rothwell. 1997. The aquatic angiosperm *Trapago angulata* from the Upper Cretaceous (Maastrichtian) St. Mary River Formation of southern Alberta. International Journal of Plant Sciences 158: 83–94.

Stockey, R. A., G. W. Rothwell, and K. R. Johnson. 2007. *Cobbania corrugata* Gen. et Comb. Nov. (Araceae): a floating aquatic monocot from the Upper Cretaceous of western North America. American Journal of Botany 94: 609–24.

Syabryaj, S., T. Utescher, S. Molchanoff, and A. A. Bruch. 2007. Vegetation and palaeoclimate in the Miocene of Ukraine. Palaeogeography, Palaeoclimatology, Palaeoecology 253: 153–68.

Thubron, C. 2001. In Siberia. HarperCollins, New York. [Describes a 15,000-mile trip from the Yenisei River to Lake Baikal and beyond.]

Vikulin, S. 2014. New insights about "Haselbach-Complex" species in Oligocene floras of Russian Plain (Pasekovo, Zmiev, Svetlogorski). Abstract of conference paper on European Palaeobotany, Padua, Italy.

Yakubov, V. 2007. Plants of Kamchatka: The Field Guide. B. B. Rkyóob, Teket, Russia.

TWO

East Beringia: Alaska, Northwestern North America, and the Aleutian Connection

It was the summer of 1977, and there were some wonderful things in *The New Yorker*: . . . John McPhee's series about Alaska, *Coming into the Country*. I was dumbstruck by the sweep of his subject matter—Alaska: "If it is not God's country, God should try and get it, a place so beautiful it beggars description."

—Mary Norris, *Between You and Me: Confessions of a Comma Queen*

Geographic Setting and Climate

The United States purchased the 1.7 million km² of Alaska and the Aleutian Islands from Russia in 1867 for $7.2 million. At the time it was regarded as Secretary of State William Seward's folly and is now considered one of the greatest land deals of all time. The principal part of this eastern component of Beringia (Alaska, Yukon Territory, British Columbia and vicinity; Figs. P1.2, 2.1) lies in ca. 50°–70°N latitude, essentially the same as the western component (Siberia, Russian Arctic Islands, Kamchatka, 50°–75°N). Under the rubric that size matters Alaska is +500 miles east to west (from the Yukon Territory/British Columbia border to the eastern edge of the Seward Peninsula, excluding the principal bays that penetrate inland) and +800 miles north to south (from Point Barrow to the northern edge of the Alaska Peninsula), which is just less than 400,000 mi². By contrast, Siberia extends east-west for ca. 3000 miles from the Ural Mountains to the Bering Strait and north-south ca. 1700 miles from the Arctic Ocean to the Mongolian border, which is ca 5,100,000 mi² (Fig. P1.1). Among the reasons this difference matters is that continentality is more extensive in Siberia, making for greater extremes in climate, and changes in these extremes over time are a principal factor in determining the movement of organisms (away from the extremes).

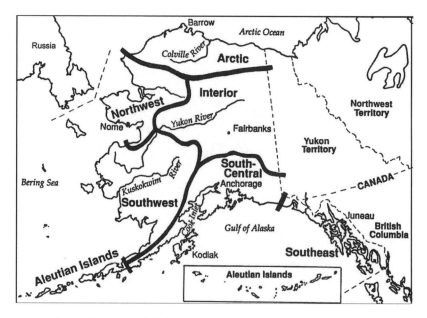

2.1. Geographic regions of Alaska. From L. Viereck et al., 1992, The Alaska Vegetation Classification. U.S. Department of Agriculture, Forest Service, Washington, DC.

In Alaska and the adjacent territories the annual range in temperature across the entire region is from ca. $-62\,°C$ to $+37\,°C$. Recall that at the settlement of Oymyakon in Siberia in February 1933, the low reached $-89.9\,°C$.

The main geographic provinces of Alaska are the Arctic and Pacific coastal plains bordering the Arctic Ocean/Beaufort Sea on the north and the Chukchi Sea/Bering Strait/Bering Sea on the west; the North Slope; the Brooks Range; the intermontane Central Plateau dissected by the Yukon River; Alaska Range; Alaska Peninsula; and the Aleutian Islands.

The water-logged alluvial Arctic Coastal Plain is the northwestern terminus of the vast interior central lowlands of North America that extend up along the north face of the Brooks Range as the tundra-covered North Slope of Alaska. It is characterized by numerous ephemeral lakes, bogs, and extensive permafrost (Fig. 2.2). The region is best known for the incompatible presence of extensive petroleum reserves at Prudhoe Bay, the Arctic National Wildlife Refuge, and the Trans-Alaska Pipeline. The northern coastal lowlands wrap around the western side of Alaska and extend southward to the Aleutian Islands. The Kuskokwim Mountains (1600 m at Dillingham High Point) are in southwestern Alaska, and the Kuskokwim River flows out onto the Pacific Coastal Plain, contributing to the water-logged conditions, numerous lakes, and spring flooding along the plain.

The Brooks Range is a structural continuation of the Rocky Mountains and Coast Ranges to the south. It was named for Alfred H. Brooks of the US Geological Survey, Division of Alaskan Mineral Resources, who gained continued support for scientific investigations by suggesting that the gold mines that gave rise to the gold rush of 1896–99 would remain productive for years to come (McPhee, 1976, 358). The highest elevations in the Brooks Range are Mount Isto at 2736 m and Mount Chamberlin at 2749 m. The Alaska Range is also part of the Rocky Mountains system and covers an extensive area of south-central Alaska. The highest elevation in the New World is at Denali (formerly Mount McKinley; 6168 m). Bordering the Alaska Range on the southeast are the Wrangell Mountains (Mount Blackburn, 4996 m; Mount Wrangell, 4717 m), named after Ferdinand Friedrich George Ludwig von Wrangel (1796–1870), manager of the Russian-American Company and of Russian Alaska from 1829 to 1834. To the south along the coast are the St. Elias Mountains (Mount St. Elias, 5959 m). This was the peak sighted by Vitus Bering in 1741 in his first view

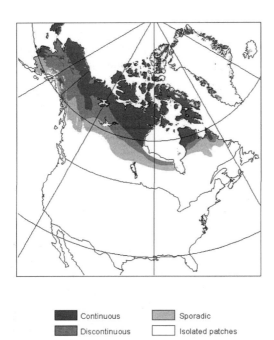

Continuous   Sporadic

Discontinuous   Isolated patches

2.2. Distribution of permafrost in North America.
From F. Nelson and K. Hinkle, 2002, in A. Orme (ed.),
The Physical Geography of North America, Oxford
University Press, Oxford. Used with permission.

of the North American continent. The south-facing slopes of these three mountains are harsh environments receiving the brunt of storms coming off the Gulf of Alaska.

Between the Brooks Range and the Alaska Range is the Central Plateau, with elevations between 915 and 1524 m. As an intermontane region it is drier than the surrounding area, with an annual rainfall of 608 mm (24 in), and being in the interior, it experiences extremes of annual temperature between 37.7°C (100°F) and −62°C (−80°F). There is scattered permafrost and numerous muskegs (bogs).

The Alaska Peninsula is ca. 800 km long, extending from 55°N to 60°N and bisected by the Aleutian Range. The highest peaks are Mount Redoubt at 3108 m and Pavlof Volcano on the southwestern edge of the peninsula at 2515 m. The region is very active volcanically, remote, and there are no roads. The south side closest to the subduction zone is especially rugged, with deep, clear, offshore waters, while to the north fill from erosion has created a flat terrane and shallow, muddy coastal waters. Temperatures in the winter are between −11°C and 1°C, and in the summer they range from 6°C to 15°C. MAP is from 610 mm to 1650 mm, and storms are frequent.

The Aleutian Islands (Fig. 2.3) were referred to by American soldiers during World War II as a cold, damp hell and by American schoolteachers as a flower garden with an advanced native culture (Eichelberger, 2007, 1). The islands essentially connect the Alaska Peninsula with Kamchatka through a distance of ca. 2200 km. Although most of the islands are separated by only a few miles, Eichelberger (2007, 1) provides an interesting account of events that have impeded field studies, including World War II and the Cold War when Alaska became an armed camp, atomic weapons were being tested in the Aleutians, and Kamchatka was off limits to most Russians. Even today field work is complicated. Chadwick (2008, xv) notes:

> Travelers can fly from Unalaska to Akutan, Nikolski, or Atka as long as they are willing to charter a Grumman Goose or a Metro and adjust to the less-than-daily flight schedule. However, to go from Unalaska to Sand Point, only 200 air miles away, travelers must fly nearly 800 miles to Anchorage, change planes, and then fly to Sand Point for a total of more than 2000 miles at a cost of over $1500.

Makushin Volcano on Unalaska is 1900 m high, and Mount Cleveland in the central Aleutians is 1730 m. The climate is controlled mostly by the Aleutian Low Pressure System, which generates an annual rainfall of ca. 2000 mm, and fog is nearly constant. In the marine environment of the

2.3. The Aleutian Islands. NASA image, Visible Earth, Jeff
Schmaltz, Rapid Response Team, NASA/GSFC.

Aleutians temperature extremes are muted; the range is from $-3\,°$C to $9\,°$C.
Precipitation ranges from 79 mm in February to 216 mm in October, for a
total of 1518 mm annually.

Geology

For purposes of comparison, western Beringia (Siberia and Kamchatka)
is underlain by a stable cratonic platform of ancient Paleozoic crystalline
rocks later folded by Mesozoic east-west compression (the Ural Mountains), and overlain to the east by thick Cenozoic aeolian (wind) and alluvial (river) erosion deposits (most of Siberia). Farther east there are late
Mesozoic and Cenozoic volcanics and accreted exotic terranes resulting
from subduction of the Pacific Plate along the Kuril-Kamchatka Trench
(Kamchatka).

By contrast, eastern Beringia (Alaska and adjacent territories) is located
along the edge of the North American craton/interior platform (Bally, 1989;
Bally and Palmer, 1989; Bally et al., 1989, fig. 1). It was formed by ancient
Paleozoic exotic (transported) terranes subsequently wielded together and

deformed in the Mesozoic by north-south compression forces paralleling the Pacific margin and creating an east-west trend in structures (e.g., the Brooks Range). Later there was accretion of exotic terranes resulting from the north-to-northwest subduction of the Pacific Plate along the Aleutian Trench. (For a brief characterization of terranes as they relate to vegetation history, see I, 64–65). Fitzgerald et al. (1993, 497) note that "the accretion of exotic terranes reshaped the margin of western North America" and "in southern Alaska subsequent thrusting and strike-slip faulting dissected these terranes" to form the local landscape (see also Brennan et al., 2011). There are important geological differences between western (Siberian) and eastern (Alaskan) Beringia as well as similarities. Wolf et al. (2002) discuss thickness and velocity estimates for the crust between the Seward and Chukchi peninsulas, and these are consistent with regional studies indicating structural continuity from Alaska to eastern Russia. Neither of the stable cratonic platforms has undergone extensive northward displacement since Late Cretaceous time. Therefore, the history of the vegetation as reconstructed from floras in the interior has been due mostly to topographic alterations and climatic trends acting on evolutionary variations in the ecological tolerances of the individual plant species.

Events far to the east, including movements in the North Atlantic along the Mid-Atlantic Ridge and opening of the Canadian Basin, had some effect on the physiography of Beringia, especially as reconstructed from a Rotational Opening Model (Mickey et al., 2002; Miller et al., 2002; but see Lane, 1997). For example, by these calculations the Omolon Massif of northeastern Russia has undergone a 30°–40° counterclockwise rotation relative to Siberia. That is, as the NALB was being wrenched apart in the Late Jurassic through the Early Cretaceous (ca. 140–100 Ma), and principally from about the Valanginian (131 Ma) to the Albian (100 Ma; Lawver et al., 1990), subduction was occurring along the North Pacific rim. Beringia was being forged in the vicinity of the Chukchi and Bering Seas by forcing of the Alaska/Chukotka Block against northeastern Siberia in the late Neocomian (100 Ma) and accretion of the Alaska Range between 100 Ma and 55 Ma. The Arctic Ocean was an isolated body of water until the opening of the Canadian Basin in the Jurassic/Cretaceous. Later in the Paleogene the BLB was structurally completed but remained only partially subaerial during most of the early Tertiary, and the Arctic and Pacific oceans were intermittently connected by shallow seas through net-negative subsidence of the land (e.g., ca. 7.5–5 Ma). Beringia was completed and extensive but periodically disrupted physically by ca. 3.5 Ma through the rise and fall of sea level resulting from glaciation.

The Brooks Range consists of early Paleozoic rocks folded by compressive forces in the early Late Cretaceous (Barremian/Aptian; ca. 125 Ma) with development of the Colville foredeep (or North Slope Basin) beginning ca. 130–126 Ma. The current topographic diversity and altitudinally zoned climates of the region date from this time, and tectonic activity continues to the present.

In the region of the Alaska Range there is little evidence of significant highlands between 16 Ma and 6 Ma. Apatite fission-track analyses from the western flank of Denali indicate it was formed by rapid uplift of more than a kilometer per million years, beginning 6 Ma (Fitzgerald et al., 1993). Studies of paleocurrents show that in the Oligocene and Miocene flow was predominantly southward over a subdued topography of ca. 0.2 km above mean sea level (MSL). In the Pliocene at 6 Ma the flow was reversed to the north, reflecting the rise of the Alaska Range to an average elevation of 2800 m. The Denali Fault runs along the southern edge of the mountains and delimits them from physiographic provinces to the south. Continuing slippage along the fault is responsible for the numerous and severe earthquakes in the region. Ridgway and Flesch (2007, 1055) characterize the southern margin of Alaska as "arguably the most tectonically complex part of the plate boundary that defines western North America . . . by an active subduction zone, two active volcanic arcs (the Aleutian and Wrangell arcs), some of the largest strike-slip fault systems on Earth (the Denali-Tintina Faults [northwestern British Columbia to central Alaska]), and an allochthonous crustal block, the Yakutat microplate, that is currently colliding in southeastern Alaska."

The Wrangell Mountains, along with parts of the St. Elias Mountains, consist of a series of lava flows that have been erupting periodically since the late Oligocene (ca. 26 Ma). They rest on a basement complex of volcanic arc terranes of Late Paleozoic age, ca. 300 Ma, and transported northward from tropical latitudes far to the south adjacent to the western coast of ancient North America. Docking time is estimated at ca. 100 Ma. The subduction that brought coastal terranes to southern Alaska is estimated at a rate of 15–25 cm/yr between 80 and 45 Ma (Oldow et al., 1989). Other terranes were added to form the complex known as the Wrangell Terranes, covering an expanse of 10,400 km^2. The latest was the Yakutat block, ca. 26 Ma, producing the current volcanic field (http://www.nature.nps .gov/geology/parks/wrst). The most recent eruptions of Mt. Wrangell were in 1784, 1884, and 1900. The St. Elias Mountains are mostly nonvolcanic and rose from compression forces beginning ca. 10 Ma.

The Aleutian Range of the central Alaska Peninsula is formed of folded

Mesozoic sedimentary rocks with granitic intrusions and late Tertiatry to Recent volcanics. These are stratovolcanoes like Mount Fuji in Japan, built up by layers of explosively erupted and ejected material, as opposed to gradual-sloping shield volcanoes like Mauna Loa resulting from successive flows of lava.

The proto-Aleutian arc was initiated in the Eocene between ca. 55 and 50 Ma (Oldow et al., 1989), with the arc itself (Figs. P1.2, 2.3) forming principally in the middle Eocene, ca. 46 Ma (Jicha et al., 2006). It continues to exhibit westward arc-continent collision and strong volcanic activity, with the western end currently colliding with the Kamchatka Peninsula (Geist and Scholl, 1994). Bogoslof Volcano and Fire Island in the Aleutians emerged in 1796 and 1883, respectively.

In summary, geological investigations in Beringia reveal a low-lying Late Cretaceous landscape *structurally* (in a geomorphological sense) continuous between northwestern North America and northeastern Asia. Emphasis is on *structurally* because that does not mean broad and continuous subaerial land extended all the way across the region of the Bering Land Bridge in Late Cretaceous and early Tertiary times. The landscape was in a state of rapid change in a geological sense (Shephard et al., 2013), but as a generalization, after the Cretaceous some of the region, including the Aleutians, from subduction after the early Eocene, was above sea level from compression and physical disruption. Greatest continuity of the bridge began episodically and primarily with the glacial/interglacial cycles of the late Pliocene and Quaternary. For example, melting of the Greenland Ice Sheet beginning at the Mio-Pliocene boundary raised global sea level by ca. 30%, although the Arctic Ocean remained mostly isolated and the Bering Strait was just beginning to open. Ice-free conditions prevailed in the early Pliocene until the sea and then ice expanded onto the land for the first time ca. 4 Ma (Knies et al., 2014; Gladenkov and Gladenkov, 2004).

This scenario provides a broad conceptual framework for envisioning movement of plants and animals across the Bering Land Bridge, but the data are usually insufficient to confirm the actual movement in detail for any one species. For example, it may be inferred from the fossil record that an organism was present first on the Asian side of the bridge then appeared later in North America. The most parsimonious explanation would be that it crossed from west to east. However, when specifics (i.e., fossils) are sought to confirm such movement, they are often lacking. If the example used is primarily an upland species of *Pinus*, *Quercus*, or *Ginkgo*, as opposed to a primarily lowland species of *Taxodium*, *Nyssa*, or *Salix*, the question is whether such uplands were commonplace or even in existence at the time crossing is proposed. In most instances the vagaries of Late Cretaceous and early Tertiary paleophysiographic reconstructions for the northern latitudes allow for alternate routes. The entity may have originated somewhere to the south in Kazakhstan or northern China and radiated east as far as the margin of Beringia in Siberia, Chukotka, or Kamchatka and west across the NALB to the margin of Beringia in northwestern North America, without ever crossing the BLB. Or it may have done both, and at different times. Whatever explanation is offered for the disjunct distribution of an organism, for example, in eastern Asia and eastern North America, it may not be possible to confirm it from the records presently available. This is the situation with most species, and explanations are complicated by the fact that much of the former land bridge is now underwater (see Fig. 2.10). The devil is in the details, and many of the details are frequently lacking. Given the choices, however, and following the tenets of Occam's Razor and the null hypothesis, it is likely that most organisms on both sides of the bridge, especially if earlier on one side than the other, even given the ambiguities of the fossil record, achieved this distribution by the simplest route. Unless there is compelling evidence to the contrary, alternative models invoking more complicated explanations are even more speculative. This emphasizes the value of multiple approaches, broad context, and "case studies" (see chapter 7) where a sound taxonomy and a good geologic record are available for establishing a baseline for other species where the data are less complete.

Modern Vegetation

The flora of Arctic North America is discussed in the *Flora of Alaska and Neighboring Territories* by Eric Hultén (1968; Fig. P1.3), *Anderson's Flora of Alaska and Adjacent Parts of Canada* in the various revisions by Welsh (1974), the ongoing Flora North America Editorial Committee's (1993) *Flora North America, North of Mexico* (http://www/floranorthamerica.org), the Centres of Plant Diversity project (e.g., vol. 3, *The Americas*; Davis et al., 1994–97), *Arctic Flora of Canada and Alaska* (Canadian Museum of Nature/Musée Canadien de la Nature; http://arcticplants.myspecies.info/taxonomy/term/2840), and the projected Global Strategy for Plant Conservation (GSPC) *World Flora Online* (http://www/plants2020.net/info _consortium). There is also a planned inventory of Alaska's national parklands (Parker, 2006). The number of species is difficult to estimate except that it is probably significantly more than Hultén's (1968) listing of 1559 for Alaska (1.7 million km^2) and adjacent territories (Yukon, 482,443 km^2; British Columbia, 944,735 km^2; total, 2,145,178 km^2). For comparison, the estimate for the flora of Siberia is 4200 species (13.1 million km^2) and for Kamchatka 1168 species (270,000 km^2) with some overlap of species (see table in online supplementary materials).

For general discussion of the modern vegetation of eastern Beringia, and Alaska in particular, the classification of Viereck et al. (1992) is used, along with other references listed at the end of this chapter. They recognize four vegetation types (Fig. 2.4). The lowland tundra (I, fig. 1.2) is well developed as a coastal plain community in northern Alaska. *Eriophorum angustifolium* and *E. vaginatum* together with *Carex aquafilis* are prominent. Both genera belong to the family Cyperaceae in which the microfossils especially are identifiable only to family. The same is true for most of the grasses (Poaceae), which are often dominant in tundra but define other communities, such as savanna and prairie, and families in which different genera are characteristic of distinct vegetation types (e.g., Ericaceae—*Vaccinium* [bogs], *Dryas* [shrub tundra]).

2.4. Vegetation types of Alaska. U.S. Department of Agriculture, Forest Service. Based on Viereck et al., 1992.

The variable moist tundra occupies low-altitude cold localities, upland and alpine tundra is found on drier sites, and each grades into other kinds of vegetation through a high shrub or brush tundra. In all instances the defining features are a treeless vegetation of cold climates. *Eriophorum vaginatum* is common along with the shrub tundra element *Dryas octopetala*.

The coastal forests are composed mostly of *Picea sitchensis* on wet sites associated with *Calamagrostis, Circaea, Cornus, Lysichiton, Oplopanax,* and *Rubus; Tsuga heterophylla* is widespread on drier ground with *Dryopteris* and *Vaccinium;* and *Thuja plicata* grows mostly south of latitude 57°N. Of these only *Thuja* is absent from west Beringia where it is widely cultivated. A fossil called *Thuites napanensis* is listed for the Eocene of western Beringia (see appendix 2 in online supplementary materials).

The interior/boreal forest or taiga (I, fig. 1.3) is the most extensive forest type in eastern Beringia, and in Alaska it covers ca. 880,595 km² (340,000 mi²). The dominants are *Abies balsamea, Larix laricina, Picea glauca,* and *Picea mariana,* intermingled with deciduous trees like *Alnus crispa* and *A. incana, Betula papyrifera,* and *Myrica gale.* Along the coast *Chamaecyparis nootkatensis* grows from the Cascades of Oregon and Washington to British Columbia and Prince Wiliam Sound in Alaska. *Populus balsamifera, P. tremuloides,* and *Salix* grow along streams moving onto the adjacent uplands in moist habitats, with grass-sedge-*Sphagnum* bogs in wet areas. Nondwarf *Pinus* and *Juniperus* occur along the southern margin. McPhee (1976) aptly describes the taiga at its northern limits as having a digital arrangement fingering into the tundra along protected valleys and streams. All these genera except *Chamaecyparis* are currently present in both east and west Beringia; *Chamaecyparis* sp. is reported in the Paleocene of western Beringia (see appendix 2 in online supplementary materials).

The Aleutian Islands are covered by a low shrub tundra of *Empetrum* together with the grasses *Calamagrostis nutkaënsis, Deschampsia beringensis,* and *Elymus arenarius.* On these islands and on much of the Alaska Peninsula the potential natural vegetation is alpine heath tundra with scattered dwarf *Betula* and *Salix.* A prominent nonvascular component of all tundra is the bryophytes (e.g., *Calliergon, Campyphyllum, Polytrichum, Sphagnum*) and lichens (ca. 1078 species in the Russian Arctic). The lichen *Cladonia rangiferina* is abundant, along with *Alectoria, Arctoparamella, Centrara, Haematoma,* and *Thamnolia.* The fossil record for tracing the history of bryophytes and lichens (and the algae and fungi) is not extensive (see Taylor et al., 2009, 2014; also Graham 1962, 1971, 1981; Kesling and Graham, 1962).

Indigenous People

In the old days,

> The ancient people lived in a less complex environment. When the morning light came, people got up. When darkness overcame them, they went to bed. (Dirks, 2008, 3)

Then things changed dramatically. The indigenous human population in Alaska dropped from 15,000 to 2500 within 50 years of the first contact with foreigners from the Vitus Bering expedition in 1741, then fell further when later explorers subjugated and relocated them to hunt and process furs of the vast number of seals, Arctic fox, and sea otter. They were further displaced for military reasons with the American presence beginning in 1867. There are numerous indigenous groups in Alaska recognized in part by language differences. The Iñupiat people include the Inuits of Alaska, the Eskimos (a circumboreal group), Tsimshian (coastal British Columbia and southern Alaska), and the Aleutes, a name often applied to all groups inhabiting the Alaska Peninsula and the Aleutian Islands. The Iñupiat of the Bering Sea region are divided into two subgroups of hunters and gatherers based on whether they live near the ocean and subsist primarily on fish or live further inland and use a wider variety of food. The distinction is important because two theories about the migration of the first humans into the New World from Eurasia hold that they were seafarers who either moved southward along the coast and/or went inland and followed ice-free corridors to the south (see section on the peopling of the Americas, below). They are contributing DNA to a study to help trace the origin and early migration of the first Americans. The Aleutes are important in this regard because they have a 9000-year history of residency in the region.

Utilization of the Bering Land Bridge

There are ca. 1410 taxa listed in appendix 2 (see the online supplementary materials), and those with the most credible identifications at least to the generic level are used to estimate movements across the BLB. The paleotemperature and sea-level environment is given in Fig. 2.5. The current paleobotanical database has all the foibles inherent to identifications in old and often unrevised collections made by multiple investigators over the centuries. One author's *Equisetites, Gleichenites, Ginkgoites,* and *Corylites* may be another's *Equisetum, Gleichenia, Ginkgo,* and *Corylus.* The names

Araliaephyllum, Celastrophyllum, Cercidiphyllum, Daphnophyllum, and *Lauro-phyllum* may be intended to imply biological relationship or used only to augment descriptions by providing examples of a comparable morphology. Some selectivity was exercised in compiling the online appendices in an effort to enhance their reliability; e.g., most material nebulously referred to as *Celastrophyllum*(?), cf. *Laurophyllum* sp., and *"Lindera"* was not included. The list was further edited when used to estimate similarities between west and east Beringia as other weaknesses or shortcomings became evident. These are pointed out as notations and footnotes in the online appendices and tables. This was the case, for example, with the Cretaceous occurrences of *Pinus* in appendix 2 (see footnote 6) in the online supplementary ma-

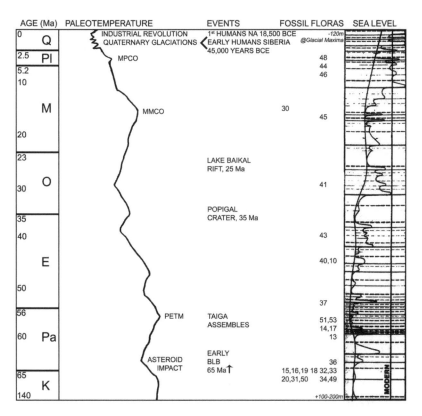

2.5. Paleotemperature and sea-level curves, events, and fossil floras for Beringia. Numbers refer to the principal fossil floras placed at approximate age as listed in online appendix 2 and references; left side, Siberia/Kamchatka; right side, northwestern North America. Q= Quaternary, Pl= Pliocene, M= Miocene, O= Oligocene, E= Eocene, Pa= Paleocene, K= Cretaceous, PETM= Paleocene-Eocene Thermal Maximum, MMCO= Middle Miocene Climatic Optimum, MPCO= Middle Pliocene Climatic Optimum.

terials. *Pinus* has now been reported from the Early Cretaceous of Nova Scotia 140–133 Ma (Falcon-Lang et al., 2016). Decisions are needed when the presence of a taxon in one of the two regions is based solely on an old and unrevised record (e.g., *Hausmania* in the Cretaceous of North America; *Aristolochia* in the Cretaceous of Russia). Each editorial device is an effort to produce a more reliable list from a very heterogeneous assortment. Even with these precautions, the online appendices are works in progress, and the data should be used with caution.

A need for improvement: In this modern era it is remarkable that the reliability of identifications in museum collections is so uneven, and that paleobotanical and related databases are so incomplete, especially given the important role of fossils in our efforts to estimate climate change and to calibrate phylogenies based on molecular evidence: "Scores of museum specimens carry a name that isn't theirs" (Hackett, 2016). With regard to estimating such important matters as biodiversity and distributions, the question is raised, "To what extent can we trust public databases?" (Maldonaldo et al., 2015). The answer is that "erroneous records affect special patterns . . . leading to an overestimation of spatial richness" and incorrectly locating hotspots at places and within political boundaries set or changed after the collections were made.

It is tantalizing to fantasize about a coordinated effort to verify the identification and stratigraphy of sets of fossil specimens in collections that have been examined by specialists in morphological taxonomy, molecular systematics, and paleobotany, leading to a selective online inventory of fewer but more reliable records.

Within this cautionary context, generalizations at several levels can be made from the listings. Lebedev (1992) and Herman (2013) mention the sequence of appearance of major plant groups in the Cretaceous and Paleogene. In the Early Cretaceous (Berriasian, Valanginian, Hauterivian; ca. 145–130 Ma) there is still a residue of ferns and gymnosperms from the late Paleozoic (Mississippian, Pennsylvanian (= Carboniferous), Permian) and early Mesozoic (Triassic and Jurassic) like *Aldania*, *Ctenis*, *Czekanowski*, *Nilssonia* (Fig. 2.6), and *Podozamites* (Fig. 2.7), and in the Jurassic *Ginkgo*-like fossils that are more modern looking by the Cretaceous (Fig. 2.8). As expected, in the earliest Cretaceous there are no unequivocal angiosperms (see Friis et al., 2011, for a global fossil record of flowering plants). Then, in the late Hauterivian and Barremian (134–125 Ma) the angiosperm *Menispermites* appears together with still diverse and abundant gymnosperms like *Cephalotaxopsis*, *Neozamites*, and *Plagiophyllum*. Near the Early/Late Cretaceous boundary (Albian, 112–100 Ma) the ancient cycadophytes

2.6. Plant fossil from the Upper Cretaceous of Alaska: *Nilsonnia serotina*. From Hollick, 1930, U.S. Geological Survey, Washington, DC.

2.7. Plant fossil from the Upper Cretaceous of Alaska: *Podozamites lanceolata*. From Hollick, 1930, U.S. Geological Survey, Washington, DC.

disappear and more modern conifers appear (e.g., *Metasequoia*) along with increasingly abundant and diverse angiosperms like the platanoids, *Quereuxia, Trochodendroides,* and *Macclintockia* (Bozukov, 2005; Fig. 2.9). The last-named is of unknown affinities ranging in age from the Albian to the middle Miocene and found at sites in the Ukraine and northern Russia (Novosibirsk or the New Siberian Islands, western Kamchatka, Yakutia), Alaska (Hollick, 1930), and Europe (see NALB). After that, modern conifers and angiosperms dominate in Beringia. In very cold and, therefore, very dry environments gymnosperms of the boreal forest (taiga) are well suited, having retained physiological and morphological legacies of an early diversification in the dry environments of the Permian (I, 8). Trees typically do not occur in the tundra or steppe where high wind, among other factors, works against the arborescent growth form. There, herbaceous and dwarf woody angiosperms, along with lichens and bryophytes, are best adapted. The general sequence of major vascular plant groups in the analyses of Lebedev (1992), Herman (2013), and in online appendix 2 for Beringia from cycadophytes and early conifers to modern conifers and angiosperms parallels the record described by Friis et al. (2011) for other parts of the world.

Some impressions also emerge about the movement of plants through the BLB in the Late Cretaceous and Cenozoic. In online table 2.2 taxa are listed from the 52 references cited in appendix 2 (see online supplemen-

2.8. Plant fossil from the Upper Cretaceous of Alaska: *Ginkgo pseudoadiantoides.* From Hollick, 1930, U.S. Geological Survey, Washington, DC.

2.9. Plant fossil from the Upper Cretaceous of Alaska: *Macclintockia alaskana*. From Hollick, 1930, U.S. Geological Survey, Washington, DC.

tary materials). The percentage representation of genera in the various plant groups is given below, and it increases for the angiosperms from the Cretaceous through the Tertiary as they diversified taxonomically and ecologically, dispersal mechanisms became more efficient, net land continuity increased (e.g., drainage of the Interior Basin of North America, in the Amazon Basin of northern South America), and climates prompted movement into new locales:

Cretaceous—141 total genera (east and west Beringia)
    Ferns and allied groups—21 (16%)
    Gymnosperms—51 (36%)
    Angiosperms—69 (48%)
Paleocene—67 genera
    Ferns and allied groups—8 or (12%)
    Gymnosperms—7 (10%)
    Angiosperms—52 (78%)
Eocene—131 genera
    Ferns and allied groups—14 (11%)
    Gymnosperms—16 (12%)
    Angiosperms—101 (77%)

Oligocene—56 genera
   Ferns and allied groups—2 (3%)
   Gymnosperms—7 (13%)
   Angiosperms—47 (84%)
Miocene—72 genera
   Ferns and allied groups—5 (7%)
   Gymnosperms—8 (11%)
   Angiosperms—59 (82%)
Pliocene—38 genera
   Ferns and allied groups—3 (8%)
   Gymnosperms—4 (11%)
   Angiosperms—31 (81%)

From these data, even in their present state, the rise to prominence of the angiosperms in the ecosystems of Beringia is evident: from ca. 48% in the Cretaceous (gymnosperms 36%) to 78% in the Paleocene, 77% in the Eocene, 84% in the Oligocene, 82% in the Miocene, and 81% in the Pliocene (gymnosperms to 11%).

Another use of the compilation is discerning whether the earliest occurrence of a particular genus in the region is from east or west Beringia, which may reflect its approximate place of origin and the direction(s) of subsequent migration. For example:

Earliest in west Beringia—*Schizaea* (presently tropical North America, southern continents), *Agathis* (Indochina, Malaysia, New Zealand), *Podocarpus* (widespread), *Nelumbo* (widespread), *Nuphar* (widespread), *Potamogeton* (widespread), *Craigia* (China), *Tetracentron* (Burma, China).

Earliest in east Beringia—*Dennstaedtia* (widespread), *Torreya* (eastern Asia, eastern North America).

In estimating the extent to which the BLB was continuous and fully operational, the percentage similarities of genera on each side from the Late Cretaceous through the Tertiary are of interest:

Cretaceous—141 of 214 genera or 66% present on both sides (i.e., 34% different) before the Beringia connection was fully established (they were already widespread in the boreal region).

Paleocene—67 of 84 genera or 73%

Eocene—131 of 175 or 75%

Oligocene—56 of 68 or 82%

Miocene—72 of 86 or 83%
Pliocene—38 of 49 or 80%

Even with the tentative nature of the database, it is clear that east and west Beringia have shared a high percentage of plants since the Late Cretaceous and that the percentage increased and was sustained through the Tertiary. The angiosperms dominate not only because they were the most abundant in most ecosystems but also because they had the most effective means for widespread dispersal. The data reflect an angiosperm-prominent vegetation through about the middle Miocene, after which modern conifers began to assemble and formed the boreal forest or taiga. Taking the record at face value, there is no *Picea* or *Tsuga*, and although some *Abies* and *Pinus* are reported in the Cretaceous (see appendix 2, table 2.3, in the online supplementary materials), they are mixed with *Agathis, Ginkgo, Glyptostrobus, Libocedrus, Taxodium,* and other nonboreal forest gymnosperms. In the Paleocene even *Abies* and *Pinus* are not reported. It is possible the boreal forest began assembling in Eocene highlands just to the south (I, 214), as suggested for the cool-temperate wasp fauna found in the Eocene of the Okanagan Highlands of British Columbia (Archibald and Rasnitsyn, 2015). From 66% (Cretaceous) to 83% (Miocene) of mainly tree and shrub genera were on or in the vicinity of the BLB through the Late Cretaceous and Tertiary. Almost all the plant taxa, or their close relations, and all the vegetation types mentioned in this discussion occur at present on both sides of the BLB. They either crossed on the bridge or had dispersal units enabling them to move over, around, or through it, or they used alternate routes. Alternate routes were available, and the history of the mammoths shows how important an accurate taxonomy is to determining these routes:

> The Columbian mammoth *Mammuthus columbi* was thought to have evolved in North America from a more primitive Eurasian immigrant. The earliest American mammoths (1.5 million years), however, resemble the advanced Eurasian *M. trogontherii* that crossed the Bering land bridge around that time, giving rise directly to *M. columbi*. Wooly mammoth *M. primigenius* later evolved in Beringia and spread into Europe and North America. The lineage illustrates the dynamic interplay of local adaptation, dispersal, and gene flow in the evolution of a widely distributed species complex. (Lister and Sher, 2015)

The conclusion is that many distributions for which the bridge is noted (e.g., between eastern Asia and eastern North America; presence of the

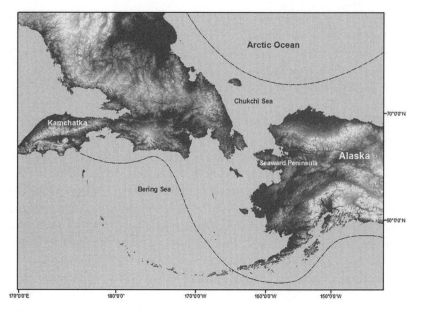

2.10. The Bering Land Bridge at its maximum extent during the
Cenozoic. National Park Service graphic, Bering Land Bridge National
Preserve. http://www/nps.gov/nr/feature/indian/s003/park.htm.

large Quaternary megafauna and the people that followed them), were attained by direct migration across Beringia, which at its maximum was over
1600 km (1000 mi) wide (Fig. 2.10), but that there were alternative means
and routes. In other words, given the relatively short distances involved
and the considerable time available, it was an important but not an all-
or-nothing pathway for migrations in the boreal realm. As discussed for
western Beringia, climate has been a highly prominent and perhaps overriding factor. Before assessing the strength of the connection further, it
will be necessary to consider the history of the North Atlantic Land Bridge
(NALB) as an alternate or joint pathway for migrations across the Northern
Hemisphere.

Peopling of America (from the West)

This is still a controversial and even contentious topic, particularly with
regard to timing of the initial crossing from Eurasia, whether the migrants mostly followed a coastal or an inland route, and identification of
the subsequent patterns of dispersal and differentiation into the current
indigenous people of east and west Beringia, and later elsewhere in the

New World. The earlier and now resolved issue of an American Paleolithic (Meltzer, 2006, 2009, 2015a,b,c; Raghavan et al., 2015) is characterized by Numbers in the back-jacket blurb of Meltzer's (2015a) book *The Great Paleolithic War* as having been "red in tooth and claw." Hinsley, in another cover blurb for the book, further concludes that in such situations, "status matters. Controversy in science is settled chiefly when those most competent to judge, and in a position to do so, decide it is time to settle it." It takes time to settle. There is a recent report of "unexpectedly early Americans" (130,000 yr ago) in California (Holden et al., 2017), which will undoubtedly be vigorously debated (Hovers, 2017; The Economist, 2017; Zimmer, 2017).

New information is rapidly forthcoming, and some opinions are gaining support over others. Records and the habits of the current indigenous people of Siberia and Kamchatka suggest their that ancient forebears were mostly content to stay close to home, migrate seasonally, and subsist by hunting and gathering the resources locally available. This is in contrast to the ancestors of the American Eskimos and Aleuts, who left to their descendants a legacy of boldly going to meet Vitus Bering at sea in boats, eagerly seeking trade, going on board the Russian ships, and inviting the newcomers into their camps. It is probable that the ancestral Asian Russians would have needed a nudge to start moving eastward into unfamiliar territory that in the late Quaternary was devoid of people. This was the time of the last glaciation, which reached a maximum ca. 22,000 years BP, began retreating at18,500 BP, and had regressed significantly by the Holocene at 11,500 BP. The traditional picture and a still plausible one is that the incentive to move came soon after 18,000 BP with the intense cold on the vast and relatively flat Siberian plain of continental interior west Beringia (documented as at least 13,000 BP [Heintzman et al., 2016] and now depicted on the general diagram in deMenocal and Stringer [2016] as 15,000 BP; see also Timmerman and Friedrich, 2016). Glaciers were more continuous than in the topographically diverse east Beringia where they developed somewhat later, as in Europe (Saarnisto and Lunkla, 2004). Sea levels were 100 m lower than at present (the current shallow depth is 30–50 m), and people had been living in or around west Beringia for some 45,000 years as indicated by the genomes sequenced from the human thighbone and fingerbone found in western Siberia, mentioned earlier. Throughout the Tertiary parts of the Arctic Ocean were moderately deep (at present a maximum of 5502 m and averaging 987 m) and there was open water (Spielhagen et al., 2004). Closer to the shoals of the Bering Strait, however, there was continuous or near-continuous low-elevation land up to 1000 km wide for much of the latest Tertiary (Fig. 2.10). During periods of glacial advance, migrating herds of animals moved eastward and then through ice-free corridors and along the western edge of the Cordilleran Ice Sheet of North America, depending on how completely the ice covered the shoreline. It is worth recalling that the Central Plateau of Alaska is comparatively dry (MAP 24 in), and moisture as well as cold is a preresquite for glacier formation. McPhee (1976, 194) describes the weather at a site on the south bank of the Yukon River near the US–Canada border (64°47′N, 141°125′W; 370 mi by road from Fairbanks): "During a night in January, 1975, the column went down to seventy-two degrees below zero. While this is the coldest part of Alaska, it is also the hottest and among the driest. Less than four feet of snow will fall in all of a winter, and about a foot of total precipitation (a figure comparable to New Mexico's) across the year." Assuming the net movement of animal resources was to the east, the people had little option but to follow.

The operative phrase is "net movement," for the dispersal of organisms was occurring in both directions throughout the Cenozoic. An example of an early east-to-west migration resulting in a considerable revamping of previous phylogenetic and biogeographic concepts is the discovery of Camelid fossils in the middle Pliocene of Ellesmere Island (Rybczynski et al., 2013), which was widely covered in the media (see, e.g., the TED [Technology Entertainment Design] program aired on PBS, 30 March 2016; PBS, 2015). Fossils were previously known from Plio-Pleistocene deposits in the Yukon (Harington, 2011). The Ellesmere Island specimen was originally thought to be a piece of fossil wood (plant fossils are abundant on Ellesmere Island; Matthews and Fyles, 2000), but collagen preserved in the Arctic cold revealed it was the tibia of a camel, and cosmic nuclide dating placed it in the mid-Pliocene warm interval at 3.4 Ma. Camels originated in North America during the Eocene ca. 45 Ma, moved westward to Asia, Europe, and North Africa via the Bering Land Bridge and southward across the Central American Land Bridge into South America at ca. 3 Ma (llamas, alpacas, vicuñas; Rybczynski et al., 2013), then became extinct in North America, leaving surviving representatives in Africa and South America. The Eocene was well after the breakup of Gondwana at the end of the Cretaceous, leaving the current geographic relationship challenging to explain without an adequate fossil record and leading to various and innovative theories. The conventional wisdom was that the hump and large, flat, hair-covered feet were adapations for water storage in a hot and arid environment and for walking across loose sandy substrate. It now seems that the hump originally may have been for fat storage in the long, cold, dark environments of the High Arctic with limited seasonal food reserves and the foot structure for walking across snow and ice.

The preferential movement in the Quaternary, however, was mostly west to east and then south. With regard to people (Meltzer, 2015a, especially 382-87 [chap. 10]), the Clovis were established in Arizona by 12,800–13,100 BP, as shown by the discovery of an arrow point adjacent to a bison tooth at Murray Springs (see III, fig. 2.1) and the slightly younger famous find of an arrowhead in the rib cage of a bison at Folsom, New Mexico. The Clovis were among the earliest humans in the New World, but they were not the first. In Washington state the skeleton of a mastodon with a spear embedded in the rib cage was dated at ca. 13,800 years old (Waters et al., 2011a); coprolites from Paisley Caves in southern Oregon identified as human were dated at 14,300–14,000 years old (Jenkins et al., 2012); tools 13,200–15,500 years old are known from Texas (Waters et al., 2011b); and a human skeleton (the Eve of Naharon) found in a cave in Yucatán is ca. 13,000–12,000 years old (Amador, 2011; Barclay, 2008; Largent, 2005). At Monte Verde in southern Chile a campsite 14,600 years old (Dillehay, 1989) implies the crossing of Beringia (Hoffecker et al., 1993, 51) long before the Clovis date of 12,800–13,100 years old. The latest contribution is the discovery of stone tools at Monte Verde possibly as old as 17,000 and 19,000 years (Dillehay, 2015; Gibbons, 2015). All this suggests the possibility, based on geological, DNA, and revised archaeological evidence, that humans crossed into North America via Beringia sometime shortly after 18,000 years ago. Whether they initially took a coastal or an interior route or both is under discussion, but recent findings (Heintzman et al., 2016; Pedersen et al., 2016; see review by Callaway, 2016) document the coastal route though not necessarily excluding, soon afterward or contemporaneously, an interior migration as well.

The topography of the new landscape undoubtedly played a role in the subsequent differentiation of the human communities. On the plains of mid-continent North America there are currently about 33 tribes of Native Americans, of which 16 were originally nomadic (e.g., Apache, Arapaho, Blackfoot, Cheyenne, Comanche, Crow) and 17 were semi-sedentary (e.g., Dakota, Iowa, Kaw, Mandan, Omaha, Osage, Pawnee). There were no insurmountable physical barriers separating these tribes, and interaction between them was frequent. The same was true on the plains and plateaus of Siberia where there are also about 30 tribes (www.survivalinternational .org.tribes/siberian).

After initial migration into the New World the first migrants found conditions quite different in the tectonically active far Pacific Northwest of

North America than on the flat plains of Siberia. Subduction, volcanism, and plate movements had long provided an extensive number of separate valleys and lowlands isolating the people into distinct communities. Here today the diversity is high compared to the Plains. There are ca. 17 principal coastal groups (e.g., Tsimshian, Chinook) and 60 or more on the inland plateaus (e.g., Cathlamet, Coeur d'Alene, Okanagan, Nez Perce)—that is, more than double the number of native indigenous groups on the American and Russian plains. Thus, climate and topography likely played a significant role both in the initial migration of people from Siberia across Beringia into North America and in their subsequent differentiation into subgroups. This is not to say the first arrivals were of a single genetic lineage. A recent discovery of two infants buried at the Upper Sun River site in central Alaska dated at 11,500 yrs BP showed they had different mitochondrial (i.e., maternal) DNA, which means that the mothers belonged to genetically distinct populations (Tackney et al., 2015). According to a view known as the Beringian Standstill Model, when the earliest Americans began migrating south, they had already lived in east Beringia long enough to have formed at least two subgroups by 11,500 BP.

The evidence at present suggests the first people crossed soon after 18,000 BP: "the Bering Strait some 17,000 years ago is the most widely accepted time and place for the original human migration into the Americas" (Callaway, 2015). They followed coastal and possibly inland routes, diversified in the ecologically heterogeneous environment of northwestern North America, and had spread southward as far as southern Arizona by 12,800–13,000 BP. As hunters and gatherers they altered the vegetation only slightly. That would come later, around 7000–6000 BP, when they reached the fertile valleys of central Mexico and reradiated northward as farmers and cultivators of the land (see chapter 5).

This marked the beginning of the early Anthropocene. Waters et al. (2016) list the following criteria as defining the late Anthropocene (the beginning of the Industrial Revolution at ca. 1800, and the mid-twentieth-century Great Acceleration of population growth and industrialization):

· rapid global dissemination of novel materials (technofossils) like elemental aluminum, concrete, plastics, black carbon, inorganic ash spheres, and spherical carbonaceous particles,
· sedimentary fluxes in erosion products from deforestation and road construction, sediment retention behind dams, and amplified delta subsidence,
· geochemical signatures like elevated levels of polyaromatic hydrocarbons,

polychlorinated biphenyls, pesticide residues, and heavy isotopes of lead from gasoline,

· doubling of soil nitrogen and phosphorus from fertilizer use reflected in lake sediments and Greenland ice cores, and

· detonation of the Trinity atomic device at Alamogordo, New Mexico, on 16 July 1945 . . . and thermonuclear weapons tests generating a clear global signal from 1952 to 1980, the so-called bomb-spike, of excess ^{14}C, ^{239}PU, and other artificial radionuclides that peaked in 1964. (Waters et al., 2016)

The human impact on the modern analog biota has been as severe as the Great Dying at the end of the Permian or the bolide event closing the Cretaceous Period, and it needs to be taken into account when using the present-day biotas as analogs for reconstructing paleovegetation and paleoenvironments.

References

Amador, F. E. 2011. Skull in underwater cave may be earliest trace of first Americans. http://blogs.nationalgeographic.com/blogs/news/chiefeditor/2011/skull-in-mexico-cave-may-be-oldest-american-found.html.

Archibald, S. B., and A. P. Rasnitsyn. 2015. New early Eocene Siricomorpha (Hymenoptera: Symphyta: Pamphiliidae, Scricidae, Cephidae) from the Okanagan Highlands, western North America. Canadian Entomology. doi: 10.4039/tce.2015.55. [Notes that the wasps are characteristic of upper microthermal MATs but fossils found in the warm early Eocene suggest the presence of high, cool elevations.]

Bally, A. W. 1989. Phaneorzoic basins of North America. In A. W. Bally and A. R. Palmer (eds.), The Geology of North America, Vol. A: The Geology of North America: An Overview. Geological Society of America, Boulder. Pp. 397–446.

Bally, A. W., and A. R. Palmer (eds.). 1989. The Geology of North America, Vol. A: The Geology of North America: An Overview. Geological Society of America, Boulder.

Bally, A. W., C. R. Scotese, and M. I. Ross. 1989. North America: plate-tectonic setting and tectonic elements. In A. W. Bally and A. R. Palmer (eds.), The Geology of North America, Vol. A: The Geology of North America: An Overview. Geological Society of America, Boulder. Pp. 1–16.

Barclay, E. 2008. Oldest skeleton in Americas found in underwater cave? http://news.nationalgeographic.com/news/2008/09/080903-oldest-skeletons.html.

Bozukov, V. 2005. Macclintockia basinervis (Rossm.) Knobl. in Cenozoic sediments in the Rhodopes Mt. region (S. Bulgaria). Acta Palaeontologica Romaniae 5: 11–15.

Brennan, P. R. K., H. Gilbert, and K. D. Ridgway. 2011. Crustal structure across the central Alaska Range: anatomy of a Mesozoic collisional zone. Geochemistry, Geophysics, Geosystems 12: Q04010. doi: 10.1029/2011GC003519.

Callaway, E. 2015. South America settled in one go. Nature 520: 598–99.

———. 2016. Plant and animal DNA suggests first Americans took the coastal route. Nature News, 536: 138. doi: 10.1038/536138a.

Chadwick, J. 2008. Introduction: Plymouth Rock of Russian America. In K. F. Wilson and J. Richardson (eds.), The Aleutian Islands of Alaska: Living on the Edge. University of Alaska Press, Fairbanks. Pp. xv–xvi.

Davis, S. D., V. H. Heywood, O. Herrera-MacBryde, J. Villa-Lobos, and A. Hamilton (eds.). 1994–97. Centres of Plant Diversity: A Guide and Strategy for Their Conservation. 3 vols. World Wide Fund for Nature (WWF) and the World Conservation Union, IUCN Publications Unit, Cambridge University, Cambridge.

deMenocal, P. B., and C. Stringer. 2016. Human migration: climate and the peopling of the world. Nature. doi: 10,1038/19471.

Dillehay, T. D. 1989. Monte Verde: A Late Pleistocene Settlement in Chile. Vol. 1. Smithsonian Institution Press, Washington, DC.

———. 2015. New archaeological evidence for an early human presence at Monte Verde, Chile. PLoS One 10 (11): e0145923. doi: 10.1371/journal.pone.0141923.

Dirks, M. L. 2008. In the beginning. In K. F. Wilson and J. Richardson (eds.), The Aleutian Islands of Alaska. University of Alaska Press, Fairbanks. Pp. 3–7.

The Economist. 2017. Pre-prehistoric man. 29 April 2017.

Eichelberger, J. C. 2007. Introduction: Subduction's sharpest arrow. In J. C. Eichelberger, E. Gordeev, P. Izbekov, M. Kasahara, and J. Lees (eds.), Volcanism and Subduction: The Kamchatka Region. Geophysical Monographs 172. American Geophysical Union, Washington, DC. Pp. 1–2.

Falcon-Lang, H. J., V. Mages, and M. Collinson. 2016. The oldest Pinus and its preservation by fire. Geology 44: 303–6.

Fitzgerald, P. G., E. Stump, and T. F. Redfield. 1993. Late Cenozoic uplift of Denali and its relation to relative plate motion and fault morphology. Science 259: 497–99.

Flora North America Editorial Committee. 1993 et seq. Flora North America, North of Mexico. Oxford University Press, Oxford.

Friis, E. M., P. R. Crane, and K. R. Pedersen. 2011. Early Flowers and Angiosperm Evolution. Cambridge University Press, Cambridge.

Gahlot, G. S., and A. Gahlot (eds.). 2006. Flora of the U.S.S.R. Index to Volumes I–XXX (compiled by B. P. Uniyal). M/s.Bishen Singh Mahendra Pai Singh, Dehra Run, and A. R. Gantner Verlag K.G., Ruggell.

Geist, E. L., and D. W. Scholl. 1994. Large-scale deformation related to the collision of the Aleutian Arc with Kamchatka. Tectonics 13: 538–60.

Gibbons, A. 2015. Oldest stone tools in the Americas claimed in Chile. doi: 10.1126/science.aad7452.

Gladenkov, A. Y., and Y. B. Gladenkov. 2004. Onset of connections between the Pacific and Arctic Oceans through the Bering Strait in the Neogene. Stratigraphy and Geological Correlation 12: 175–87.

Graham, A. 1962. The role of fungal spores in palynology. Journal of Paleontology 36: 60–68.

———. 1971. The role of myxomycete spores in palynology (with a brief note on the morphology of certain algal zygospores). Review of Palaeobotany and Palynology 11: 89–99.

———. 1981. Un alga fosil Micratiniaceae de la Zona del Canal de Panama. Biotica 6: 229–32.

Hackett, J. 2016. Scores of museum specimens carry a name that isn't theirs. Scientific American 314 (3) (1 March).

Harington, C. R. 2011. Pleistocene vertebrates of the Yukon Territory. Quaternary Science Reviews 30: 2341–54.

Heintzman, P. D., et al. (+ 16 authors). 2016. Bison phylogeography constrains dispersal and viability of the Ice Free Corridor in western Canada. Proceedings of the National Academy of Sciences USA 113: 8057–63.

Herman, A. B. 2013. Albian-Paleocene flora of the North Pacific: systematic composition, palaeofloristics and phytostratigraphy. Stratigraphy and Geological Correlation 21: 689–747.

Hoffecker, J. F., W. R. Powers, and T. Goebel. 1993. The colonization of Beringia and the peopling of the New World. Science 259: 46–53.

Holden, S. R., et al. (+ 10 authors). 2017. A 130,000-year-old archaeological site in southern California, USA. Nature 544. doi: 10.1038/nature22065.

Hollick, A. 1930. The Upper Cretaceous floras of Alaska. U.S. Geological Survey Professional Paper 159. Washington, DC.

Hovers, E. 2017. Unexpectedly early signs of Americans. Nature 544: 420–21.

Hultén, E. 1968. Flora of Alaska and Neighboring Territories: a Manual of the Vascular Plants. Stanford University Press, Stanford.

Jenkins, D. L., et al. (+ 19 authors). 2012. Clovis age western stemmed projectile points and human coprolites at the Paisley Caves. Science 337: 223–28.

Jicha, B. R., D. W. Scoll, B. S. Singer, and G. M. Yodozinski. 2006. Revised age of Aleutian Island Arc formation implies high rate of magma production. Geology 34: 661–64.

Kesling, R. V., and A. Graham. 1962. *Ischadites* is a dasycladacian alga. Journal of Paleontology 36: 943–52.

Knies, J., P. Cabedo-Sanz, S. T. Belt, S. Baranwal, S. Fietz, and A. Rosell-Melé. 2014. The emergence of modern sea ice cover in the Arctic Ocean. Nature Communications 5, article number 5608. doi: 10.1038/ncomms6608.

Lane, L. S. 1997. Canada Basin, Arctic Ocean: evidence against a rotational origin. Tectonics 16: 363–87.

Largent, F. B. 2005. Early humans south of the border. New finds from the Yucatán Peninsula. Mammoth Trumpet 20: 8–11.

Lawver, L. A., R. D. Müller, S. P. Srivastava, and W. Roest. 1990. The opening of the Arctic Ocean. In U. Bleil and J. Thiede (eds.), Geological History of the Polar Oceans: Arctic versus Antarctic. Kluwer Academic Publishers, Dordrecht. Pp. 29–62.

Lebedev, Y. E. 1992. The Cretaceous floras of northeastern Asia. International Geology Review 34: 794–805.

Lister, A. M., and A. V. Sher. 2015. Evolution and dispersal of mammoths across the Northern Hemisphere. Science 350: 805–9.

Maldonado, C., et al. (+ 8 authors). 2015. Estimating species diversity and distribution in the era of Big Data: to what extent can we trust public databases? Global Ecology and Biogeography 24: 973–84.

Matthews, J. V., Jr., and J. G. Fyles. 2000. Late Tertiary plant and arthropod fossils from the high-terrace sediments on the Fosheim Peninsula of Ellesmere Island (Northwest Territories, District of Franklin). Geological Survey of Canada Bulletin 529: 295–317.

McPhee, J. 1976. Coming into the Country. Farrar Straus & Giroux, New York.

Meltzer, D. J. 2006. Folsom: New Archaeological Investigations of a Classic Paleoindian Bison Kill. University of California Press, Berkeley.

———. 2009. First Peoples in a New World: Colonizing Ice Age America. University of California Press, Berkeley.

———. 2015a. The Great Paleolithic War: How Science Forged an Understanding of America's Ice Age Past. University of Chicago Press, Chicago.

———. 2015b. Pleistocene overkill and North American mammalian extinctions. Annual Review of Anthropology 44: 33–53.

———. 2015c. Kennewick Man: coming to closure. Antiquity 89: 1485–93.

Mickey, M. B., A. P. Byrnes, and H. Haga. 2002. Biostratigraphic evidence for the predrift position of the North Slope, Alaska, and Arctic Islands, Canada, and Sinemurian incipient rifting of the Canada Basin. In E. L. Miller, A. Grantz, and S. L. Klemperer (eds.), Tectonic Evolution of the Bering Shelf–Chukchi Sea–Arctic Margin and Adjacent Landmasses. Geological Society of America Special Paper 360. Geological Society of America, Boulder. Pp. 67–75.

Miller, E. L., A. Grantz, and S. L. Klemperer (eds.). 2002. Tectonic Evolution of the Bering Shelf–Chukchi Sea–Arctic Margin and Adjacent Landmasses. Geological Society of America Special Paper 360. Geological Society of America, Boulder.

Nelson, F. E., and K. M. Hinkel. 2002. The far north: a geographical perspective on permafrost environments. In A. R. Orme (ed.), The Physical Geography of North America. Oxford University Press, Oxford. Pp. 249–69.

Norris, Mary. 2015. Between You and Me, Confessions of a Comma Queen. W. W. Norton & Company, New York.

Oldow, J. S., A. W. Bally, H. G. Avé Lallemant, and W. P. Leeman. 1989. Phanerozoic evolution of the North American Cordillera: United States and Canada. In A. W. Bally and A. R. Palmer (eds.), The Geology of North America, Vol. A: The Geology of North America: An Overview. Geological Society of America, Boulder. Pp. 139–232.

Parker, C. L. 2006. Vascular plant inventory of Alaska's Arctic national parklands: Bering Land Bridge National Preserve, Cape Kruensten National Monument, Gates of the Arctic National Park and Preserve, Kobuk Valley National Park, and Noatak National Preserve. National Park Service, Arctic Network Inventory and Monitoring Program Report NPS.AKRARCN.NRTR-2006.01. [Note that this inventory includes one of the same areas (Gates of the Arctic National Park and Preserve) described by McPhee in *Coming into the Country* (1976).]

PBS (Public Broadcasting System). 2015. First Peoples: Americas. April.

Pedersen, M. W., et al. (+ 16 authors). 2016. Postglacial viability and colonization in North America's ice-free corridor. Nature. doi: 10.1038/nature19085.

Raghavan, M., et al. (+ list of affiliated authors). 2015. Genomic evidence for the Pleistocene and recent population history of Native Americans. Science 359. doi: 10.1126/science.aab3884.

Ridgway, K. D., and L. M. Flesch. 2007. Cenozoic tectonic processes along the southern Alaska convergent margin. Geology 35: 1055–56.

Rybczynski, N., J. C. Gosse, C. R. Harington, R. A. Wogelius, A. J. Hidy, and M. Buckley. 2013. Mid-Pliocene warm-period deposits in the High Arctic yield insight into camel evolution. Nature Communications 4, article number 1550. doi: 10.1038/ncomms2516.

Saarnisto, M., and J. P. Lunkka. 2004. Climate variability during the last interglacial-glacial cycle in NW Eurasia. In R. W. Battarbee, F. Gasse, and C. Stickley (eds.), Past Climate Variability through Europe and Africa. Developments in Paleoenvironmental Research 6. Pp. 443–64.

Shephard, G. E., R. D. Müller, and M. Seton. 2013. The tectonic evolution of the Arctic since Pangea breakup: integrating constraints from surface geology and geophysics with mantle structure. Earth-Science Reviews 124: 148–83.

Spielhagen R. F., K.-H. Baumann, H. Erlenkeuser, N. R. Nowaczyk, N. Nørgaard-Pedersen,

C. Vogt, and D. Weiel. 2004. Arctic ocean deep-sea record of northern Eurasia ice sheet history. Quaternary Science Reviews 23: 1455–83.

Tackney, J. C., et al. (+ 8 authors). 2015. Two contemporaneous mitogenomes from terminal Pleistocene burials in eastern Beringia. Proceedings of the National Academy of Sciences USA. 112: 13833–38.

Taylor, T. N., M. Krings, and E. L. Taylor. 2014. Fossil Fungi. Academic Press, London.

Taylor, T. N., E. L. Taylor, and M. Krings. 2009. Paleobotany: The Biology and Evolution of Fossil Plants. Academic Press, London.

Timmerman, A., and T. Friedrich. 2016. Late Pleistocene climate drivers of early human migration. Nature. doi: 10.1038/nature19365.

Timmerman, A., M. Krings, and E. L. Taylor. 2014. Fossil Fungi. Academic Press, London.

Tutin, T. G., V. H. Heywood, N. A. Burges, D. M. Moore, D. E. Valentine, S. M. Walters, and D. A. Webb. (With the assistance of A. O. Chater and I. B. K. Richardson.) 1980. Flora Europea Vol. 5. Pp. 403–40 (index); pp. 441–52 (Index to Families and Genera of Volumes 1–5).

Viereck, L. A., C. T. Dyrness, A. R. Batten, and K. J. Wenzlick. 1992. The Alaska Vegetation Classification. General Technical Report PNW-GTR-286. U.S. Department of Agriculture, Forest Service, Pacific Northwest Research Station, Portland, OR.

Waters, C. N., et al. (+ 23 authors). 2016. The Anthropocene is functionally and stratigraphically distinct from the Holocene. Science. doi: 10.1126/science.aad2622.

Waters, M. R., et al. (+ 10 authors). 2011a. Pre-Clovis mastodon hunting 13,800 years ago at the Manis Site, Washington. Science 334: 351–53.

Waters, M. R., et al. (+ 12 authors). 2011b. The Buttermilk Creek Complex and the origins of Clovis at the Debra L. Friedkin site, Texas. Science 331: 1599–1603 (with online supporting material).

Welsh, S. L. 1974. Anderson's Flora of Alaska and Adjacent Parts of Canada. Brigham Young University Press, Provo.

Wolf, L. W., R. C. McCaleb, D. B. Stone, T. M. Brocher, K. Fujita, and S. L. Klemperer. 2002. Crustal structure across the Bering Strait, Alaska: offshore recordings of a marine seismic survey. In E. L. Miller, A. Grantz, and S. L. Klemperer (eds.), Tectonic Evolution of the Bering Shelf–Chukchi Sea–Arctic Margin and Adjacent Landmasses. Geological Society of America Special Paper 360. Geological Society of America, Boulder. Pp. 25–37.

Zimmer, C. 2017. Humans lived in North America 130,000 years ago, study claims. New York Times. 26 April.

Additional References

Ager, T. A. 2003. Late Quaternary vegetation and climate history of the central Bering land bridge from St. Michael Island, western Alaska. Quaternary Research 60: 19–32.

Bailey, R. G. 1995. Description of the Ecoregions of the United States. U.S. Department of Agriculture, Forest Service, Miscellaneous Publication Number 1391. Washington, DC.

Balter, M. 2015. Ancient infants buried together in Alaska suggest long journey to the Americas. Science 349. doi: 10.1126/science.aad4773.

Barbour, M. G., and W. D. Billings (eds.). 2000. North American Terrestrial Vegetation. 2d ed. Cambridge University Press, Cambridge.

Bliss, L. C. 2000. Arctic tundra and polar desert biome. In M. G. Barbour and W. D. Billings (eds.), North American Terrestrial Vegetation, 2d ed. Cambridge University Press, Cambridge. Pp. 1–40.

Bouchal, J. M., R. Zetter, and T. Denk. 2016. Pollen and spores of the uppermost Eocene Florissant Formation, Colorado: a combined light and scanning electron microscopy study. Grana 55: 179–245. [The site is beyond the southern margin of the eastern BLB as defined here, but the alterations in identifications, ages, and distributions are important biogeographically.]

Clottes, J. (Translated by O. Y. Martin and R. D. Martin.) 2016. What Is Palaeolithic Art? Cave Paintings and the Dawn of Human Creativity. University of Chicago Press, Chicago. [Review: J. Cook. 2016. Old Masters, early cultures. Nature 532: 310–11.]

The Economist. 2015. Canada's Inuit, easier said than written. Economist 417: 30–32. 7 November.

———. 2017a. Skating on thin ice. 29 April.

———. 2017b. Polar bare. 29 April.

Elliott-Fisk, D. L. 2000. The taiga and boreal forest. In M. G. Barbour and W. D. Billings (eds.), North American Terrestrial Vegetation, 2d ed. Cambridge University Press, Cambridge. Pp. 41–73.

Enslin, R. 2014. Geologists shed light on formation of Alaska Range. Syracuse University News, 19 November. http://news.sur.edu/geologists-shed-light-on-formation-of-alaska-range-64688.

Fiorillo, A. R., and P. J. McCarthy. 2010. Preface: Ancient polar ecosystems and environments: proxies for understanding climate change and global warming—an introduction. Palaeogeography, Palaeoclimatology, Palaeoecology 295: 345–47.

Garnett, S. T., and L. Christidis. 2017. Taxonomy anarchy hampers conservation. Nature 546: 25–27.

Fowell, S., and D. Scholl. 2005. The Bering Strait, rapid climate change and land bridge paleoecology. Final Report of the Joint Oceanographic Institutions/International Arctic Research Center Workshop, Fairbanks, Alaska, 20–22 June 2005.

Graham, A. 1999. Late Cretaceous and Cenozoic History of North American Vegetation, North of Mexico. Oxford University Press, Oxford. [See chap. 1, "Setting the Goal: Modern Vegetation of North America, Composition and Arrangement of Principal Plant Communities," pp. 3–27.]

Guthrie, R. D. 1990. Frozen Fauna of the Mammoth Steppe. University of Chicago Press, Chicago.

———. 2013. Ice Age Forensics: Reconstructing the Death of a Wooly Bison. University of Chicago Press, Chicago.

Hamilton, C. 2016. Define the Anthropocene in terms of the whole Earth. Nature 536. doi: 10.1038/536251a.

Harris, AJ., C. Walker, J. R. Dee, and M. W. Palmer. 2016. Latitudinal trends in genus richness of vascular plants in the Eocene and Oligocene of North America. Plant Diversity 38: 133–41.

Herman, A. B. 1994a. A review of Late Cretaceous floras and climates of Arctic Russia. In M. C. Boulter and H. C. Fisher (eds.), Cenozoic Plants and Climates of the Arctic. NATO ASI Series. Series 1: Global Environmental Change, Vol. 27. Pp. 127–49.

———. 1994b. Late Cretaceous Arctic platanoids and high latitude climate. In M. C. Boulter and H. C. Fisher (eds.), Cenozoic Plants and Climates of the Arctic. NATO ASI Series. Series 1: Global Environmental Change, Vol. 27. Pp. 151–59.

Hill, C. A., V. J. Polyak, D. J. Nash, Y. Asmerom, and P. P. Provencio. 2017. The West Water

Formation (Hualapai Plateau, Arizona, USA) as a calcrete-paleosol sequence, and its implications for the Paleogene–Neogene evolution of the southwestern Colorado Plateau. Palaeogeography, Palaeoclimatology, Palaeoecology 479: 146–63.

Holbrook, W. S., D. Lizarralde, S. McGeary, N. Bangs, and J. Diebold. 1999. Structure and composition of the Aleutian Island Arc and implications for continental crustal growth. Geology 27: 31–34.

Jackson, F. G., and F. L. McClintock. 2011. A Thousand Days in the Arctic. Cambridge Library Collection: Polar Exploration, Vol. 2. Cambridge University Press, Cambridge.

Leslie, A. 2005. The Arctic Voyages of Adolf Erik Nordenskiöld 1858–1879. Cambridge University Press, Cambridge.

Levere, T. H. 2004. Science and the Canadian Arctic: A Century of Exploration, 1818–1918. Cambridge University Press, Cambridge.

London, Jack. 1903. The Call of the Wild. Macmillan, New York. [Set in the Yukon Territory of the 1890s during the Klondike Gold Rush.]

MacDonald, G. M. 2002. The boreal forest. In A. R. Orme (ed.), The Physical Geography of North America. Oxford University Press, Oxford. Pp. 270–90.

Matthews, J. V., Jr., and L. E. Ovenden. 1990. Late Tertiary plant macrofossils from localities in Arctic/Subarctic North America: a review of the data. Arctic 43: 364–92.

McCune, J. L., and M. G. Pellatt. 2013. Phytoliths of southeastern Vancouver Island, Canada, and their potential use to reconstruct shifting boundaries between Douglas-fir forest and oak savannah. Palaeogeography, Palaeoclimatology, Palaeoecology 383–84: 59–71.

Natal'in, B. A. 2004. Phanerozoic tectonic evolution of the Chukotka-Arctic Alaska Block: problems of the rotational model. Smithsonian/NASA Astrophysics Data System. http://adsabs.harvard.edu/abs/2004AGUFMGP43C . . . 04N; also available at https://pangea.stanford.edu/research/.

Nature Editorial. 2017. Natural-history collections face fight for survival. Nature 544: 137–38.

Orme, A. R. (ed.). 2002. The Physical Geography of North America. Oxford University Press, Oxford.

PBS (Public Broadcasting System). 2016. The Story of Cats, II: Into the Americas. November. [Includes discussion of exchanges across BLB and CALB.]

Perron, J. T., and J. G. Vanditti. 2016. Megafloods downsized. Nature 538: 174–75.

Qian, H., Y. Jin, and R. E. Ricklefs. 2017. Patterns of phylogenetic relatedness of angiosperm woody plants across biomes and life-history stages. Journal of Biogeography 44: 1383–92.

Reich, P. B., K. M. Sendall, A. Stefanski, X. Wei, R. L. Rich, and R. A. Montgomery. 2016. Boreal and temperate trees show stong acclimation of respiration to warming. Nature 531: 633–36 (plus supporting online material). [The species examined were *Abies balsamea, Picea glauca, Pinus banksiana, P. strobus, Acer rubrum, A. saccharum, Betula papyrifera, Populus tremuloides, Quercus macroocarpa,* and *Q. rubra,* and generically all part of the Cenozoic forests of Beringia.]

Reinhard, K. J. 2006. A coprological view of ancestral Pueblo cannibalism. American Scientist 94: 254–61. [Deals with people who were living on the Colorado Plateau from 1200 BCE onward known as the Anasazi.]

Rivas-Martínez, S., D. Sánchez-Mata, and M. Costa. 1999. North American Boreal and Western Temperate Forest Vegetation. Syntaxonomical Synopsis of the Potential Natural Plant Communities of North America, II. Itinera Geobotanica 12: 5–316.

Scholl, D. W. 2007. Viewing the tectonic evolution of the Kamchatka-Aleutian (KAT) connection with an Alaska crustal extrusion perspective. In J. C. Eichelberger et al. (eds.), Volcanism and Subduction: The Kamchatka Region. Geophysical Monograph 172. American Geophysical Union, Washington, DC. Pp. 3–35.

Sutikna, T., et al. (+ 20 authors). 2016. Revised stratigraphy and chronology for *Homo floresiensis* at Liang Bua in Indonesia. Nature 532: 366–69 (plus additional supplementary information).

Tieszen, L. L. (eds.). 1978. Vegetation and Production Ecology of an Alaskan Arctic Tundra. Springer, Berlin.

Uetz, P., and A. Garg. 2017. Species disconnected from DNA sequences. Nature 545: 412.

Ukraintseva, V. V. 1992. (Edited by L. D. Agenbroad, J. I. Mead, and R. H. Hevly.) Vegetation Cover and Environment of the "Mammoth Epoch" in Siberia. Mammoth Site of Hot Springs, SD.

USDA Office of Communications. 2016. Open data powers new conservation mapping tool for USDA, partners. usda@public.govdelivery.com. ["The University of Montana and other partners have used Google Earth Engine to build a new interactive online map tool that, for the first time, combines layers of data to better target invasive species that are damaging habitat and rangeland. The tool . . . presents geospatial data covering a 100 million acre landscape in eight western states."]

Vallier, T. L., D. W. Scholl, M. A. Fisher, T. R. Bruns, F. H Wilson, R. von Huene, and J. Stevenson. 1994. Geologic framework of the Aleutian Arc, Alaska. In A. W. Bally and A. R. Palmer (eds.), The Geology of North America, Vol. G-1: The Geology of Alaska. Geological Society of America, Boulder. Pp. 367–88.

Warren, W. F. 1885. Paradise Found: The Cradle of the Human Race at the North Pole. Houghton Mifflin, Boston. [Warren was president of Boston University (1873–1903) and a founder of Wellesley College (1870), and also a member of the Mystical Seven and a believer that the human race originated at the North Pole where he also placed Atlantis and the Garden of Eden.]

Williams, C. J. 2006. Paleoenvironmental reconstruction of polar Miocene and Pliocene forests from the western Canadian Arctic. 19th Annual Keck Symposium. http://keck.wooster.edu/publications.

Wilson, K. F., and J. Richardson (eds.). 2008. The Aleutian Islands of Alaska, Living on the Edge. University of Alaska Press, Fairbanks.

Witze, A. 2016. Speedier Arctic data as warm winter shrinks sea ice. Nature 531: 15–16.

Wolf, A., N. B. Zimmerman, W. R. L. Andregg, P. E. Busby, and J. Christensen. 2016. Altitudinal shifts of the native and introduced flora of California in the context of 20th-century warming. Global Ecology and Biogeography 25: 418–29.

Wolfe, J. A. 1994. An analysis of Neogene climates in Beringia. Palaeogeography, Palaeoclimatology, Palaeoecology 108: 207–16.

THREE

North Atlantic Land Bridge: Northeastern North America, Greenland, Iceland, Arctic Islands, Northwestern Europe

In contrast to Beringia, which presents a Late Cretaceous and Tertiary history of construction, continuity, and migration, the record for the North Atlantic Land Bridge is mostly one of destruction, separation, and increasing barriers to migration. That history and its consequence as it relates to biogeography has been reviewed by Tiffney (1985; see also Graham, 1993; and for the late Mesozoic, Brikiatis, 2016). As defined here, the North American portion of the NALB begins at about the Potomac River along the Atlantic Coastal Plain of Virginia and extends northward through the New England states (Fig. 3.1). This allows the relatively few late Mesozoic and Tertiary plant fossil localities of northeastern North America to be included, such as the mid-Cretaceous Potomac Group of Maryland and New Jersey, the late Miocene Brandywine flora of southern coastal Maryland, and the early Miocene Brandon Lignite flora of Vermont. The western part of the NALB continues through eastern Canada, Greenland, and the Arctic Islands of Svalbard (Spitsbergen), Axel Heiberg, Devon, Ellesmere, Queen Elizabeth Islands, and others. These islands are separated by the Beaufort Sea from those of Beringia and northern Russia, such as Franz Josef Land, Novaya Zemlya (= New Land, a nuclear test and storage site since 1954; compare p. 60 above), Severnaya Zemlya (= Northern Land, consisting of October Revolution, Bolshevik, and others), and the New Siberian and Lyakhov Islands (Fig. 3.1). Iceland sits astride the dividing line between the east and west NALB as an emergent peak of the Mid-Atlantic Ridge. The eastern portion extends to Great Britain, including such fossil assemblages as the Late Cretaceous Wealden and the early Eocene London Clay floras, and continues into Western Europe. The NALB and the BLB were the principal means for organisms moving in and out of the New World through the northern latitudes, especially those lacking means of long-distance dispersal. From the Late Cretaceous through the early Eocene, these connections operated in concert, climates were warm, and opportunity for migration of subtropical and warm-temperate organisms was extensive. Later, by about the middle Eocene, the North Atlantic lands had separated substantially, global climates were moderating, and exchanges during the late Eocene and Oligocene to the end of the MMCO along both routes favored cool-temperate organisms. The Arctic Islands, Greenland, and Iceland are mostly north of the BLB, and the Gulf Stream would not reach full force until closure of the Central American Seaway at ca. 3.5 Ma. After the onset of glaciation in the late Pliocene, migrations were rare along both routes, as land free of ice diminished and exchange was limited mostly to chance long-distance dispersal. Physical filtering due to narrowing was not extensive in the Paleogene, because both bridges were wide and opened onto lands of continental extent, but increased in the Neogene as land surfaces were reduced during time of high sea level.

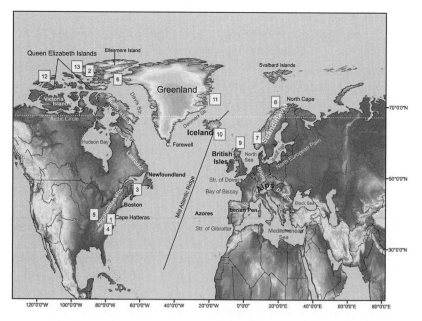

3.1. Index map of place names and physiographic features for the North Atlantic region (North Atlantic Land Bridge). 1 = Potomac R.; 2 = Axel Heiberg Is.; 3 = Prince Edward Is.; 4 = Blue Ridge Mtns.; 5 = Allegheny Mtns.; 6 = Nares Strait; 7 = Denmark Strait; 8 = Norway; 9 = Faeroe Islands; 10 = Shetland Islands; 11 = Reykjavik; 12 = Scoresby Sund; 13 = Banks Is.

Geographic Setting and Climate

The mid-Atlantic Coastal Plain of Virginia at latitude 38°N grades inland into the Piedmont and then into the uplands of the Blue Ridge/Allegheny provinces of the Appalachian Mountains (Shankman and James, 2002). The plain with its numerous marshes and swamps (e.g., the Dismal Swamp near Virginia Beach) extends westward from sea level to the fall line at an elevation of about 900 m where numerous rivers begin a steep descent into the Atlantic Ocean. Virginia Beach at an elevation of 6 m has a MAP of 1200 mm and a MAT of 13.3°C. Prince Edward Island in northeastern North America (46.2°N) has an average elevation of 142 m, a MAP of 890 mm, and a MAT of 9.9°C. It is the extremes in weather that are most important for plant and animal life, and the January low in Virginia Beach is ca. 12°C and the July high is 30°C for a range of 18°C, while on Prince Edward Island the January low is −12.6°C, the July high is 23.2°C, and the range is 35.8°C. In the geologic past the warmest climate for the interval considered here (Late Cretaceous through the Cenozoic) was at the EECO, ca. 55 Ma, when ocean surface MATs in the northern latitudes were 5°C to 8°C warmer than at present (see Fig. 3.12) and both land bridges were in full operation. During this time the Coastal Plain was expanding from deposition of erosion sediments, uplift from compression forces generated by spreading along the Mid-Atlantic Ridge, and lower sea levels as glaciers developed first in the uplands of Antarctica beginning in the late Eocene, then on Greenland in the late Miocene, and later elsewhere in the late Pliocene. The climates and expanded land of the Paleogene provided habitats for an increasing number of moisture- and warmth-requiring organisms occupying lowland to mid-elevations.

Inland from the Coastal Plain the Piedmont extends along the eastern side of the Appalachian Mountains from Alabama to New Jersey at elevations between 50 m and 300 m. It is over 475 km wide in the south and narrows to the north. Farther inland from Georgia to Pennsylvania are the Blue Ridge Mountains. Mt. Mitchell (2037 m), Mt. Craig (2026 m), and Clingmans Dome (2025 m) in North Carolina are the highest peaks in the eastern United States.

In the Canadian Arctic the largest lands in the east are Baffin Island (ca. 69°N; 507,451 km²), Ellesmere Island (79°N; 196,235 km²), and Labrador (47°–60°N; 294,330 km²)/Newfoundland (46°N; 108,160 km²). Ellesmere Island is fully 15° north of the Arctic Circle (66.3°N), the MAT is −18.°C, the MAP is 79.1 mm (a polar desert), and there is near-darkness from mid-October to late February. The highest point is Barbeau Peak at

2616 m, and Cape Columbia (83.4°N) is the northernmost point of land in North America. Kaffeklubben (that is, appropriately, Coffee Club Island, named by the Danes for the Coffee Club in the Copenhagen Geological Museum), off Greenland, is the northernmost island in the world at 83.6°N. The two flowering plants on the island are *Saxifraga oppositifolia* and *Papaver radicum*. With the exception of Barneo, which is on the Arctic ice sheet (88°11′N), the northernmost inhabited land is Alert on Ellesmere Island (82°15′N) with a population of 74.

Among these statistics the diversity of habitats, along with past geologic and climatic change, are of particular interest, for they have combined to produce in the southeastern United States the most species-rich sector of the NALB. Between Virginia Beach and Ellesmere Island altitudes extend from sea level to 2616 m, MAP from 79.1 mm to 1200 mm, and MAT from −18°C to 13.3°C across a latitudinal gradient from 36.8°N to 83.6°N.

In addition to topography and climate, another feature contributing to the exceptional diversity of the biota in the southeastern United States is the north-south orientation of the mountains. In the past when climates changed, as in the Quaternary, the biota was able to migrate south in cooling times, return north during warmer intervals, and be accommodated along the way by a diversity of altitudinally zoned habitats. Another effect of the topography is that cold air (the polar vortex) penetrates far to the south. Wolfe (1979) believes this is a reason for the scarcity of broad-leaved evergreen trees in the eastern forests of North America, as opposed to the deciduous forests of China, where cold fronts from Siberia are deflected by the east-west trending mountain systems. The Appalachian Mountains, like the Urals, are Pennsylvanian in age, so they were even higher in the past, having now undergone nearly 300 Ma years of erosion. Nonetheless, they provided a variety of habitats in the Late Cretaceous and Cenozoic, and as a result the relictual flora and fauna of the southeastern United States persists as one of the most species-rich temperate biotas in the world. New studies are now incorporating paleobotanical data into more precise analyses of the modern vegetation to better explain its history and diversity (see Baskin and Baskin, 2016; Manos and Meireles, 2015; and the section on the modern vegetation, below).

The central North Atlantic presents several features important to the movement of organisms through the region. Most obvious is the extent and central location of Greenland (59°–83°N; Figs. 3.2, 3.3). It is the largest noncontinent island in the world, with an area of 2,166,086 km². Along the northwest coast Greenland is separated from Ellesmere Island by only 30–40 km across the Nares Strait and in the south from Labrador

3.2. Greenland showing extent of ice cap. NASA image, Visible Earth.

3.3. Approaching Greenland from the east along Scoresby Sund.

by 400 km across the Davis Strait. It is 900 km from Norway but with the intervening lands of Iceland, the Faeroe Islands, and the Shetland Islands. The annual temperature range is from $-8\,°C$ to $7\,°C$, and the highest point is Gunnbjørn Fjeld at ca. 3694 m in east-central Greenland. Another notable feature, in addition to its size and position, is that Greenland has the second-largest ice cap in the world (next to Antarctica), covering all but the fringeing coastline. The surface is dome-shaped, but the weight of the ice has depressed the underlying land to a depth of 300 m below sea level. The ice has provided a unique opportunity to study atmospheric chemistry (CO_2, nitrogen, methane), temperature variation, and sea-level change through a core 3028 m long covering the past 100,000 years (European Greenland Ice Core Project [GRIP] and the American Greenland Ice Sheet Program [GISP]; I, 40). Among the findings is that average winter temperatures have risen by $5–6\,°C$ since the LGM at 26,500 BP primarily in concert with increases in atmospheric CO_2 concentration. CO_2 is now established as one of the two principal factors driving climate change (Anagnostou et al., 2016). Ocean cores show that longer term trends also reflect the Milankovitch Variations (changes in the position of the Earth relative to the Sun) on cycles of ca. 100,000, 41,000, and 23,000 years with subcycles. The major glaciations last 100,000 years, and the interglacials have a wider range but have averaged ca. 20,000 years for the last 10 interglacials (Imbrie, 1985; MIS = marine isotope stages):

MIS 1 (last 11.5 cal yr BP): 11.5 kyr

MIS 5.5 (110–128 ka): 18 kyr

MIS 7 (186–245 ka): 59 kyr, but if restricted to the maxima ca. 20 kyr

MIS 9 (303–339 ka): 36 kyr, maxima ca. 20 kyr

MIS 11 (362–423 ka): 61 kyr, maxima ca. 35 kyr

MIS 13 (362–423 ka): 46 kyr, maxima ca. 20 kyr

MIS 15 (565–620 ka): 55 kyr, maxima ca. 30 kyr

MIS 17 (659–689 ka): 30 kyr, maxima ca. 20 kyr

MIS 19 (726–736 ka): ca. 10 kyr

MIS 21 (763–790 ka): 27 kyr, maxima ca. 15 kyr

We are ca. 11,000–12,000 years into the present interglacial, which presumably will be prolonged by the increasing atmospheric CO_2 of the late Anthropocene. It is estimated from the GRIP and GISP cores, and from the European Project for Ice Coring in Antarctica (EPICA) and the US West Antarctica Ice Sheet Divide Project (WAIS), and other sources that atmospheric CO_2 concentration was 180–240 parts per million by volume (ppmv) at the LGM and 280 ppmv just before the Industrial Revolution (Neftel et al., 1982, 1985). Currently it is ca. 400 ppmv. If the Greenland ice cap were to melt completely, sea levels would rise by about 6 m, partially flooding most coastal cities and infiltrating aquifers with brackish water. For Antarctica the rise would be ca. 60 m plus some additional increase from thermal expansion of the water. A caption from the Anchorage Dispatch/Associated Press coverage of global warming ("Alaska feared to be melting away" [31 August 2015]) notes that "The Portage Glacier [about 50 mi south of Anchorage], which is a major Alaska tourist destination, has retreated so far it no longer can be seen from a multimillion-dollar visitors center built in 1986." Study of the Greenland cores revealed that climate changes leading to the 18–20 previous glacial cycles of the Quaternary built up gradually then tumbled rapidly into the interglacials. For example, the transition from the Younger Dryas glacial cold interval into the present interglacial Holocene took place in just a few decades (http://www/esd.ornl.giv/projects/qen/transit.html; Severinghaus et al., 1998). The taiga and temperate deciduous forests expanded briefly during each of the relatively short warm interglacial intervals of the Quaternary, but across the Bering and North Atlantic Land Bridges generally expansion and migration was temporally more limited because the cold glacials last 5x as long as the average interglacials (100,000 vs. 20,000 yrs). This dynamic cycle of climate and vegetation response began in the late Pliocene 3.5 Ma.

During a trip to Iceland and Greenland in September 2007 aboard the Soviet Union's *Aleksey Maryshev*, a former Russian Academy of Science research vessel, the ship went through the Denmark Strait to Scoresby Sund in East Greenland, providing a view of the interglacial landscape at these latitudes (Fig. 3.4). This was the site of the laboratory where Alfred Wegener, early proponent of the concept of continental drift, spent long sessions studying the meteorology of Arctic lands (Fig. 3.5). Logistically, it was a difficult arrangement for the research. In 1930 base camps had been set up at Eismitte (mid-ice) and West Camp:

> On September 24 [1930], although route markers were by now largely buried under snow, Wegener set out [from West Camp to Eismitte] with thirteen Greenlanders and his meteorologist Fritz Loewe to supply the camp by dog sled. During the journey the temperature reached −60° C. Twelve of the Greenlanders returned to West Camp. On October 19, the remaining three members of the expedition reached Eismitte. There being only enough supplies for three at Eismitte, Wegener and Rasmus Villumsen took two dog sleds and made for West Camp. They took no food for the dogs and killed

3.4. Debarking at Scoresby Sund, Iceland.

3.5. Greenland laboratory where Alfred Wegener worked
while making meteorological observations.

them one by one to feed the rest until they could run only one sled. While Villumsen rode the sled, Wegener had to use skis. They never reached the camp. Six months later, on May 12, 1931, Wegener's body was found half-way between Eismitte and West Camp. It had been buried (by Villumsen) with great care and a pair of skis marked the grave site. Wegener had been fifty years of age and a heavy smoker and it was believed that he had died of heart failure brought on by overexertion. His body was reburied in the same spot by the team that found him and the grave was marked with a large cross. After burying Wegener, Villumsen had resumed his journey to West Camp but he was never seen again. Villumsen was twenty three when he died and it is estimated that his body, and Wegener's diary, now lie under more than 100 metres of accumulated ice and snow. (https://en.wikipedia .org/wiki/Alfred_Wegener)

Wegener's theory about past continental movements, based in part on the matching configuration of coastal South America and Africa, was given a plausible physical basis through photographic sea-floor maps prepared in the 1960s by Marie Tharp and Bruce Heezen (posted by Mervine, 2010; http://blogs.agu.org/georneys/2010/12/24/a-famous-ocean-floor-map/; Economist, 2016a,b). The Defense Department was interested in knowing the configuration of the ocean floor to locate potential hiding places for US and enemy submarines and detection devices. The result was discovery of the Mid-Atlantic Ridge and the transform faults associated with spreading. Seismological studies as summarized by Isacks, Oliver, and Sykes (1968; Oliver, 1996) provided a sound geophysical mechanism through the con-centration of earthquake epicenters at plate boundaries (see, e.g., Fig. 4.2). Among the results was an explanation for disruption of the North Atlantic Land Bridge.

Iceland is just across the Denmark Strait ca. 300 km to the east of Green-land. It is a volcanically active place with a maximum altitude of 2119 m on lava peaks near the southeast coast (Figs. 3.6, 3.7). There is deep snow, but only three permanent glaciers cover about 11% of the island. The average January low temperature at Akureyi in the north is ca. $-5.5\,°C$ and the July high is $14.5\,°C$ for a range of $20\,°C$ (along the coast at Reykjavik it is $-3\,°C$ to $13\,°C$; range $16\,°C$). This is compared to the larger continental area of Greenland, where the January low at Kangerlussuaq in the west is $-19\,°C$ and the July high is $10.7\,°C$. Winter storms are frequent (Fig. 3.8). The beds in our Soviet ship were equipped with seatbelts so passengers could take refuge during the day and not be thrown out during the night. The *Aleksey Maryshev* was to return to Reykjavik, but a storm forced taking shelter off

3.6. Lava flow on Iceland, an emergent volcanic peak of the Mid-Atlantic Ridge.

3.7. The inland aspect of volcanic Iceland.

3.8. Storm in the North Atlantic along the northwest coast of Iceland.

northwest Iceland for three days, further documenting the extreme weather conditions of the region.

Even so, the MAT of Iceland is comparatively mild, considering it is virtually on the Arctic Circle (Fig. 3.1). The vegetation is principally *Calluna* heath marsh and *Betula tortuusa* scrub. Where extensive continentality would enhance temperature extremes, the small size and insular environment moderates them. An additional factor is the Gulf Stream. This wind-driven surface current is occasionally envisioned as a serene oceanic river of water bringing heat to the frigid North Atlantic. It is a conveyor of warmth, but the temperature contrasts especially along the margins and toward the north generate highly unstable weather:

The quantities of gulf-weed floating about, and a bank of clouds lying directly before us, showed that we were on the border of the Gulf Stream. This remarkable current is almost constantly shrouded in clouds and is the region of storms and heavy seas. Vessels often run from a clear sky and light wind, with all sail, at once into a heavy sea and cloudy sky, with double-reefed topsails. A sailor once told me that, on a passage from Gibraltar to Boston, his vessel neared the Gulf Stream with a light breeze, clear sky, and studding-sails out, alow and aloft; while before it was a long line of heavy, black clouds, lying like a bank upon the water, and a vessel coming out of it, under double-reefed topsails, and with royal yards sent down. As they drew near, they began to take in sail after sail, until they were reduced to the same condition; and, after twelve or fourteen hours of rolling and pitching in a heavy sea, before a smart gale, they ran out of the bank on the other side, and were in fine weather again, and under their royals and skysails. (Dana, *Two Years before the Mast* [Harper, 1840; reprint, Seven Treasures Publications, 2008], 161)

The current transports 30 million m³ of warm water per second (or 30 Sverdrups) from the south. For comparison, temperatures (surface waters, mid-position) for major New World currents are:

Gulf Stream (off coast of Florida)—ca. 27 °C displacing climate zones 5° to 10° northward
Japan Current—ca. 24 °C with cold continental-side upwelling
cold Humboldt Current—19 °C

The north-south circulation of the Gulf Stream depends on a significant temperature difference between the tropical and temperate Atlantic Ocean whereby equatorial surface waters flow north, cool, sink, and return south at depth where they warm and rise back to the surface (thermohaline circulation). In the meantime, "westerly winds blowing across these waters are warmed; by the time they reach the coast of England [average cold-month temperature of 5.5 °C (41 °F)], this sea to air temperature exchange produces a maritime climate very different from the continental climate in Labrador [average cold-month temperature of −25 °C (−13 °F)], which is at the same latitude (51.5 °N)" (I, 34). The origin of the Gulf Stream, in the sense of an important climate-altering factor for the North Atlantic, was in the middle to late Miocene as the MMCO ended, producing greater temperature contrasts between the surface waters of the equatorial and boreal Atlantic Ocean. This is shown by the initial appearance of ice-rafted quartz grains in the Arctic Ocean: "The data show that the most frequent distribution of glacial grains were deposited at the middle-late Miocene and during the Pliocene and early Pleistocene. Especially, the middle Miocene climatic cooling and the presence of glacial ice are well indicated at ~13.5 Ma" (Immonen, 2013).

Svalbard or Spitsbergen is located about midway between Norway and the North Pole and is north of the Arctic Circle at 74°–81°N. There is some temperature benefit from its proximity to the Gulf Stream, and the average winter temperature is nearly 20°C higher than on lands to the west. Still, the average January temperature is −16°C and the average July high is 6°C. MAP is 1000 mm on the east side of the island and 400 mm to the west. Svalbard is 60% glaciated and another 30% is barren rock, but in the Tertiary there was open land and paleobotanical studies show it was densely vegetated (Budantsev and Golvneva, 2009; McIver and Basinger, 1999).

The eastern end of the NALB opens onto the lowlands of the British Isles and the European continent between the Scandinavian Mountains to the north and Alps-Carpathian Mountains to the south. The Scottish Highlands are relatively low, reaching only 1347 m at the highest point in the British Isles at Ben Nevis, but because of their northerly position at 57°N it is cold and wet (see table 3.1 in online supplementary materials). Much of the vegetation is ericaceous heath with remnants of forest growing in protected sites. The Scandinavian Mountains or Scandes reach 2469 m at Galdhøpiggen Peak in Norway, the Massif Central of France at Puy de Sancy is 1886 m, and the Carpathians of the Czech Republic-Romania are 2600 m in the High Tatras (Gerlachovsky Stit). An aspect of the rainy lowlands south of the Carpathians is captured in the sardonic writings of Paul Theroux (2008, 31), recounting a train ride through the same region 33 years earlier:

> The rain was still falling . . . but this was not life-giving rain, nourishing roots and encouraging growth. It was something like a blight. It spat from the dreary sky, smearing everything it hit, rusting the metal joints of the roof, weakening the station, fouling the tracks. The farming villages could have been illustrations from *Grimm's Fairy Tales*—cottages, huts, outbuildings, barns, all of them aslant, surrounded by fields, no trees. Just at dusk the border, Giurgiu Nord [had] a decayed façade of a station that turned out to have nothing behind it but wasteland.

Steven Pinker (2014, 24–26) refers to the similarly devastated landscapes described by Isabel Wilkerson (2010) in *The Warmth of Other Suns*. The migrations she refers to were also due to environmental deterioration (dust bowls), and more recently to political tyranny (in this case, Nazism), and migration was the natural consequence.

This meagerness of the vegetation in these western European lowlands is also in part a residue of the geologic history. In contrast to North America, the several major mountain systems of Western Europe are oriented east and west. When climates changed in the late Cenozoic, the vegetation was trapped in a squeeze-play between the Scandinavian ice sheet advancing from the north and the Alps-Himalayan Mountains to the south. Van der Hammen et al. (1971, 404) list the following plants present earlier in Western Europe but disappearing between the end of the middle Miocene and the Eemian of the Quaternary (i.e., from ca. 4 Ma through 100 ka):

End of middle Miocene—*Libocedrus, Metasequoia, Pandanus, Castanopsis, Mastixia*

Upper Miocene—*Cinnamomum, Clethra, Engelhardtia, Coriaria*

Brunssumian—*Corylopsis, Cunninghamia, Elaeagnus, Glyptostrobus,* Palmae, *Rhus, Symplocos*

Reuverian—*Aesculus, Diospyros, Liquidambar, Nyssa, Pseudolarix, Styrax, Zelkova* (and probably *Sequoia* and *Taxodium*)

Lower Tiglian—*Fagus, Liriodendron*

Upper Tiglian—*Actinidia, Euryale, Magnolia, Phellodendron*

Waalian—*Carya, Castanea, Juglans, Ostrya, Pterocarya, Tsuga*

Cromer I—*Eucommia, Celtis, Parthenocissus*

Holsteinian—*Azolla*

Eemian—*Brasenia, Dulichium*

Some Western European vegetation persisted in Central Europe (*Abies, Pinus, Corylus, Quercus, Ulmus*) as early documented by Van der Hammen et al. (1971) but now also known from small, isolated "cryptic refugia" in eastern and southern Europe (Bhagwat and Willis, 2008; Birks and Willis, 2008; Leroy and Arpe, 2007; Willis and Niklas, 2004; Willis et al., 2000; Willis and Van Andel, 2004; see Baskin and Baskin, 2016). Milne (2004, 390) lists four species of *Rhododendron* persisting in southwestern Eurasian refugia during the Quaternary (*R. caucasicum, R. ponticum, R. smirnovii, R. ungernii*).

However, when the glacial freeze was over, and these relicts and other plants were beginning to expand into Western Europe during the current interglacial ca. 12,000 BP, early agriculturalists were moving in from the Fertile Crescent between the Tigris and Euphrates rivers, clearing the land with fire and axe and planting crops. The process is clearly captured in the pollen diagrams of Iversen (1941), showing the delayed succession of the natural vegetation and the northwestern movement of crop plants accompanied by weeds associated with cultivation. The result is a comparatively depauperate flora that never fully recovered. These historic events explain some of the differences in vegetation between the western and eastern ends of the NALB. Despite the legacy left to the region by geology, nature, and politics,

Central Europe has a diverse landscape ranging from sea level to 4810 m and until relatively recently (geologically) supported a diverse biota in the Tertiary. The fossil record of that biota (tabulated in appendix 3 in the on-line supplementary materials) reflects in part migrations across the NALB.

The composition of the assemblages suggests some preferential movement from west to east in response to the directional flow of the westerly winds and the establishment of the Gulf Stream in the late Oligocene and Miocene. The question is which organisms actually crossed on the bridge rather than passing over it by long-distance dispersal or around it by alternate routes (see section on utilization of the NALB, below). The answer will be easier than for Beringia because fossil floras are preserved along the numerous emergent remnants of the bridge itself (e.g., Greenland, Iceland, Svalbard/Spitsbergen, Baffin, Victoria, Banks, Queen Elizabeth, and the various Arctic Islands) as opposed to the BLB, where much of the former expanse is submerged (Fig. 2.10).

Geology

The geology of the NALB tells a great deal about the time and nature of its origin, the environments prevailing when it was operational (climatic, topographic, edaphic), usage (through paleobotany and indirectly through distributional, phylogenetic, and statistical/modeling data), and its disruption. Still, little of the information stands alone, and the process was long and complex. Early on, at the far southwestern end, the final phase in the building of the northern Appalachian Mountains took place in the Pennsylvanian Period (Alleghenian Orogeny) at 330–270 Ma, when Africa was pushing against the southeastern edge of the North American plate (Shankman and James, 2002). Farther to the north, some of the oldest rocks on Earth are on the microplate (subplate) of Greenland at 3.7–3.8 billion years old, and the Wegener Fault (Harrison, 2006; Jackson, 1985) running between Ellesmere Island and Greenland, essentially separating North America from Europe, became active ca. 80 Ma. During this time Greenland moved south relative to Ellesmere Island by ca. 320 km (Redfern, 2001). The oldest structural anomaly (evidence for separation) in the Labrador Sea between Greenland and Newfoundland is in the early Late Cretaceous to the south and in the latest Cretaceous to the north. Movement had ceased by the latest Eocene/earliest Oligocene (Trettin, 1989). Near the end of the Eocene a meteorite hit on Devon Island, forming the 16 km wide Haughton Crater (Thorsteinsson and Mayr, 1987), which regionally augmented the trend toward cooler conditions that began after the PETM.

Iceland had a very different history. Rather than being caught between converging tectonic plates, rafted along on a plate, or uplifted above a subduction zone, Iceland is an emergent peak of the Mid-Atlantic Ridge spreading center—Vogt and Tucholke (1989) call the ridge a "crustal factory"—where North America is separating from Europe. Incipient separation began in the far Southern Hemisphere perhaps as early as the Jurassic when South America initially moved away from Africa. Rifting reached the North Atlantic in the Late Cretaceous ca. 80 Ma (Peirce, 1982; Tessensohn et al., 2004) and the latitudes of Iceland in the middle to late Paleocene: "Propagation of the spreading system through the Norwegian-Greenland Sea into the Arctic and massive widespread volcanism, associated with the birth (or great intensification) of the Iceland hot spot, occurred around anomaly 25/24 (~59 Ma)" (Vogt and Tucholke, 1989). After the early Eocene (55 Ma) separation was well advanced. As this was also the time climates were transitioning from the warm conditions of the PETM into the colder Neogene, the two provide causes for the reduced movement of organisms across the NALB after the late Eocene. The Atlantic Ocean connected with the waters of the Arctic Ocean for the first time as Greenland separated from Spitsbergen at ca. 36 Ma, and the Norwegian-Greenland Sea opened at ca. 22 Ma. Iceland's uplift history begins at ca. 25 Ma with elevation of the Reykjanes Ridge above sea level and a reorganization "pulse" as late as 2.5 Ma.

The underlying structural framework of the Arctic Islands also lies in the Precambrian and early Paleozoic (Trettin, 1989). A principal geomorphic feature of the region is the 1000 x 350 km Sverdrup Basin, with Axel Heiberg and Ellesmere Islands, off the northwest coast of Greenland. It is filled with 13 km of eroded Carboniferous and Paleogene sediments and was deformed into approximately its present shape by the Eurekan orogeny and the counterclockwise rotation of Greenland in the Late Cretaceous and Paleogene. On Svalbard/Spitsbergen the Gilsonryggen Formation is dated on the basis of dinoflagellates as Paleocene in age "coinciding with the initial opening of the Norwegian Sea at anomaly 24–25 time," and "an Upper Eocene age is proposed for sediments from Forlandsundet (Sarsbukta)" (Manum and Throndsen, 1986). Fossil plants have long been known from the Arctic Islands and vicinity through the early work of Oswald Heer (1809–83; Fig. 3.9). Subaerial habitats and terrestrial vegetation are known from the Arctic Islands throughout the Late Cretaceous and Cenozoic interval considered here and into the Paleozoic. Fossil pollen and spores (mostly reworked from Cretaceous strata, also known as "palynodebris") and dinoflagellates have been retrieved from these and other formations as part

Oswald Heer.

3.9. Oswald Heer, an early student of the fossil plants
of the Arctic region. His collections still serve as the
basis for many investigations of plant lineages and
plant biogeography. From Wikipedia, public domain.

of offshore drilling in the Barents Sea (Throndsen and Bjaerke, 1983; see
Golynchik, 2009, regarding the geologic structure of Cenozoic basins in
the Barents Sea).

To the east in Europe during the Late Cretaceous (89–95 Ma), limestone was being deposited in a shallow sea in the vicinity of the English
Channel. Inland and to the south at ca. 50 Ma low hills were eroding terrestrial sediments to form the London Clay Formation, preserving one of
the most extensive Eocene floras in the world (Reid and Chandler, 1933).
Some 500,000 years ago Britain was connected to continental Europe by
a series of hills through what is now the ca. 33 km wide Straits of Dover.
Glacial rivers drained into the North Sea, and a prominent sill separated
these waters from the Bay of Biscay. Beginning at about 450,000 years ago,
this sill was breached by two megafloods that scoured a deep channel de-
positing sediments called Fleuve Manche into the Sea of Biscay and forming an extensive lake. With the rise of sea level during the interglacials the

lake overflowed, separating Britain from Europe across the English Channel. The most recent connection was toward the end of the last glacial. After ca. 9000 years ago the latest separation was complete (Gibbard, 1995, 2007; Gupta et al., 2007). It was the last in a long series of geologic and climatic events that began in the Late Cretaceous, intensified near the end of the early Eocene, and disrupted the last segment of the NALB 9000 years ago. The process can be visualized using the map in Fig. 3.1. Viewed as a timeline running backward, Norway and vicinity approach the Mid-Atlantic Ridge from the east, Greenland and vicinity approach from the west, the North Atlantic Ocean closes, the ridge retreats southward, Iceland submerges, and the Arctic Ocean becomes an enclosed polar sea (Fig. 3.10; Galloway et al., 2012). Running the reel forward from the Cretaceous, at

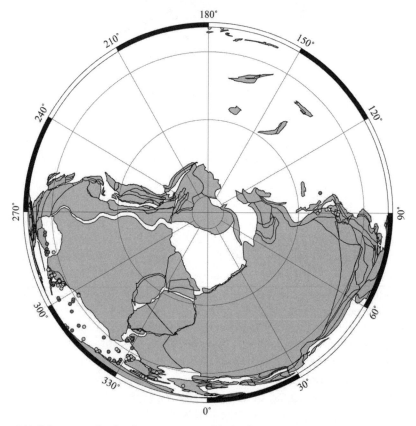

3.10. Polar perspective showing arrangement of the land ca. 100 Ma and a closed Arctic Sea.

ca. 100 Ma the ridge advances northward, the polar lands begin to separate in the Late Cretaceous, climate changes from the hothouse to the icehouse mode at the beginning of the middle Eocene ca. 45 Ma, glaciers form on Greenland at 7–3.2 Ma, Great Britian separates from Europe most recently at 9000 BCE, and the modern scene is essentially set (Fig. 3.1). Several videos are available showing plate movements, including the USGS Educational Videos and Animations (e.g., *Plate Tectonics*). The biotic responses to these geologic and climatic processes are the plant and animal communities presently along the NALB.

Modern Vegetation

The vegetation of North America is most recently and succinctly discussed in the introductory volume of the *Flora of North America North of Mexico* (Flora North America Editorial Committee, 1993). So far 20 of the projected 30 volumes are complete, and the estimated number of species for the United States, Canada, Greenland, and Ellesmere Island is ca. 18,350 (James Zarucchi, editorial director, and Heidi Schmidt, managing editor, pers. comm., 2015). Perhaps one-third, or 6000 species, occur in the region defined as encompassing the western NALB.

An instructive series of maps showing the last deglaciation of North America is presented by Brouillet and Whetstone (1993, and sources cited therein). Those for the LGM at 18,000 yrs BP, early in the present interglacial or Holocene at 10,000 yrs, and essentially modern times (5000 yrs BP) are shown in Figs. 3.11a,b,c. The dynamic nature of the glacial climatic history of North America and the NALB can be envisioned as fluctuating between Fig. 3.11a and 3.11c approximately 20 times during the past 3.5 Ma.

Among the characteristic plants of the polar desert along the northern portions of the NALB are *Draba corymbosa*, *Minuartia rubella*, *Papaver radicatum*, *Puccinellia angustata*, and *Saxifraga oppositifolia*. On less barren sites there is *Dryas integrifolia* grading into dwarf shrub heaths of *Eriophorum varigatum*; low shrub tundra of *Alnus*, *Betula*, and *Salix*, with *Arctostaphylos*, *Empetrum*, and *Vaccinium*; and tall shrub tundra with *Alnus*, *Betula*, and *Salix* reaching 5 m in height. All these genera are widespread and occur throughout the tundra of the boreal region.

The same is true of the plant genera of the boreal forest or taiga, where *Abies balsamea*, *Larix laricina*, *Picea glauca*, *P. mariana*, *Pinus banksiana*, *P. contorta*, and *Thuja occidentalis* are prominent, along with *Alnus crispa*, *Betula papyrifera*, *Populus balsamifera*, *P. tremuloides*, and *Salix* sp. Toward the

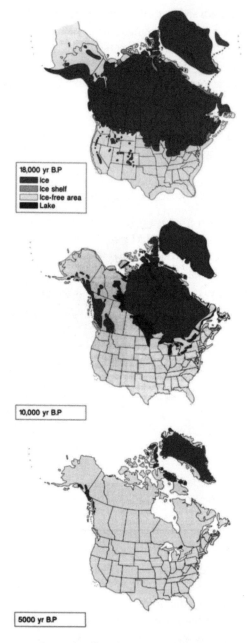

3.11a,b,c. Extent of North American glaciation north of Mexico at 18,000, 10,000, and 5000 yrs BCE. From Brouillet and Whetstone (1993). Used with permission of the Flora of North America Editorial Committee.

south in eastern North America deciduous angiosperms increase, but high elevations are limited and montane gymnosperms are fewer (*Abies* and *Picea*; *Pinus* at mid-elevations). Many genera of the deciduous forest grow far distant from the NALB, eastern North America, and Western Europe. Many are found disjunct in the forests of east-central Asia (e.g., in Japan and the central provinces of China; Donoghue, 2008; Donoghue and Smith, 2004; Donoghue et al., 2001; Milne, 2006; Wen, 1999; Wen and Ickert-Bond, 2009; Wen et al., 2010), fewer in western North America (Wen et al., 2016), and fewer still in Western and Central Europe (see previous discussion). The floristic affinities between eastern Asia and eastern North America in relation to land bridges will be considered later, but a better understanding of the origin of this relationship is currently being gained through studies on the modern vegetation of the eastern United States and its Late Cretaceous and Cenozoic history, mentioned earlier (Baskin and Baskin, 2016; Manos and Meireles, 2015; Soltis et al., 2006). Among new clades more than one-third underwent speciation after arriving in eastern North America; most had affinities with the Old World (that is, came in from the north); arrivals span the long interval of the last 45 Ma, indicating there were likely several periods of introduction; and divergence times in the North American clades suggest that many disjunctions and much of the current diversity were generated in the late Miocene and early Pliocene, that is, during the recent climatic upheavals leading to the Quaternary glaciations (Manos and Meireles, 2015). "The overarching theme of the last 50 million years of Northern Hemisphere plant biogeography is global cooling and repeated opening and closing of land bridges" (Manos, pers. comm., 2014).

Western European vegetation is similar in physiognomy and composition to that of boreal and eastern North America, allowing for historical differences imparted by the east-west orientation of the mountains and the early impact of sedentary agriculturalists toward the end of the last glacial period in Europe. An overview of the vegetation that includes Africa, Southwest Asia, and Europe; Australasia and the Pacific Islands; as well as the Americas (North America, Middle America, the Caribbean islands, and South America) is the three-volume work *Centres of Plant Diversity* (Davis et al., 1994–97). The plant communities pertaining most directly to the NALB are listed below along with some representative species (fide Ozenda, 1979):

Tundra: lowland and desert tundra of Scandinavia and alpine tundra to the south including dwarf *Juniperus communis*, *Betula nana*, *Empetrum nigrum*, *Phyllodoce longipes*, *Salix lanata*, *Vaccinium myrtillus*

Taiga: *Abies alba, Picea abies, Pinus silvestris, Betula pubens*

Deciduous forest: *Acer, Betula pendula, B. pubescens* (wet soils), *Carpinus betulus, Corylus, Crataegus, Fagus silvatica, Fraxinus excelsior, Ilex, Melica, Quercus petra* (dry), *Q. robur, Rhamnus cartharticus, Sambucus nigra, Sorbus aucoparis, Tilia cordata,* and *Ulmus;* including cold-temperate fjord vegetation of *Picea, Betula, Cornus,* and enclaves of possibly human-disturbed coastal heathlands of *Juniperus, Calluna, Crataegus, Genista, Salix, Saxifraga,* sedges (e.g., *Juncus trifidus*), *Ulex, and Vaccinium*

Grassland (steppe): *Juniperus excelsa, Astragalus, Camphora* and *Noaea* (chenoams), and *Quercus anatolica*

Shrubland/chaparral-woodland-savanna (Mediterranean): *Pinus nigra, P. pallasiana, Quercus calliprinos, Q. ilex, Q. ilicis, Q. rotundifolia,* and *Q. subra*

These modern communities along the NALB are the end point of a long process beginning in the Late Cretaceous, and the task is to extract whatever information the paleobotanical and other records will allow regarding their origin, movement, and diversification.

Utilization of the North Atlantic Land Bridge

A list of plant fossils representing the vegetation types listed above is given in appendix 3 in the online supplementary materials, and the Late Cretaceous through Cenozoic paleoenvironmental context (paleotemperature, sea levels, geologic events, fossil floras) is shown in Fig. 3.12. The online appendices document some of the plants actually growing on the North Atlantic Land Bridge because segments of the bridge are still emergent. This is in contrast to Beringia, where vast expanses of the former link are covered by shallow to deeper waters and its use has to be inferred by fossils at either end (Siberia/Kamchatka versus Alaska and adjacent territories) and from analyses of modern distributions.

Another difference between the two boreal connections is that the NALB was not a bridge in the conventional sense—that is, a narrow stretch of land connecting the two continents. Rather, it was the north-central part of Laurasia lying between North America and Europe and the Arctic Ocean and the Tethys Sea. Virtually 100% of the trees and shrubs listed in online appendices 2 and 3 moved across the far Northern Hemisphere sometime during the Cenozoic, consistent with the physiographic reconstructions, expansive extent, and sweeping changes in climate. That being the case, it is plausible to invoke these boreal land bridges as pathways of migration and to explain disjunct occurrences or patterns of diversification for other

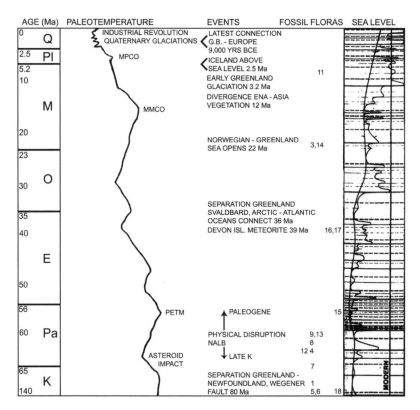

3.12. Paleotemperature and sea-level curves, events, and fossil floras for the North Atlantic Land Bridge and vicinity. See caption to Fig. 2.5 on p. 69 for explanation of abbreviations.

modern taxa, even in the absence of a fossil record, if they had suitable ecologies because supporting context information is available.

A third difference between the two boreal connections is that the NALB was not disrupted by rising seas during the Quaternary as with the BLB; rather, it was torn asunder by promulgation of the Mid-Atlantic Ridge that reached the North Atlantic in the Late Cretaceous and Paleogene. After about the middle Eocene migrations across Beringia were conditioned primarily by the cooling temperatures, while over the NALB it was a combination of progressively cooler climates and fragmentation of the land; that is, environments for migration were comparable for the two routes from the Late Cretaceous through the early Eocene, while afterward they were different. As will be seen later, the equatorial bridges (CALB, ALB) were much more narrow (and the ALB a series of islands throughout its history), so physical filtering was greater while climatic filtering was less. The history of the Antarctica part of the austral land bridge was different yet: some poleward drift; first a cool then a cold isolating circulating current; then extreme and sustained frigid conditions that significantly reduced the higher biota after the MMCO and, with a brief reprieve during the MPCO, eliminated it completely in the Pliocene (see chapter 6).

The scenario for the boreal connections can be examined within the limitations of the paleobotanical record using the extinct genus *Mcclintockia* (Fig. 2.9) as an example. The plant was described by Heer (1868) and assigned by various authors to the Menispermaceae, Proteaceae, and Urticaceae, but its affinities are unknown. Its fossil record is assembled in online appendices 2 and 3 (and references cited therein) and reviewed by Bozukov (2005). In the Cretaceous and Paleocene *Mcclintockia* is known from Russia, Alaska, and Greenland, having also spread into Central Europe where it occurs from the Albian to the middle Miocene. This reflects extensive early use of the boreal land bridges. Along these connections, however, it is not known after the Eocene, consistent with post-EECO cooling along both routes and disruption of the NALB.

The records in online appendix 3 show that ancient Late Paleozoic (Pennsylvanian) and Mesozoic (Triassic and Jurassic) holdovers were still common in the region in the Early Cretaceous. Extinct gymnosperms like *Brachyphyllum* and *Ptilophyllum* were present in Greenland along with Early Cretaceous forms like *Elatocladus* and *Glyptodium*. In the Middle and Late Cretaceous angiosperms begin to appear, with some attributed to modern genera, such as *Ceratophyllum*, *Cinnamomum*, *Hedera*, *Liriodendron*, and *Platanus*. All are present elsewhere, suggesting that the broad land mass or "bridge" *sensu lato* extending across the North Atlantic between the Arctic Ocean and Tethys Sea was functioning from the Cretaceous through the early Eocene. *Abies, Larix, Picea, Pinus,* and *Thuja* are reported for the Paleocene but mixed with *Ginkgo, Glyptostrobus, Sequoia, Taxodium,* and temperate deciduous angiosperms, which diversify in the Eocene through the Miocene. In the Pliocene many of the temperate deciduous gymnosperms disappear. The cold taiga becomes well established along with some cool-temperate deciduous angiosperms, suggesting altitudinal zonation. By the end of the Pliocene and the beginning of the Pleistocene the taiga and cold-temperate angiosperms are prominent. The change toward colder conditions shown by the terrestrial biota (Fig. 3.12) is paralleled in the marine realm (Verhoeven and Louwye, 2013).

These plants present in the Paleocene and Eocene allow the assessment of another facet of biogeography, discussed earlier. Tropical, subtropical, and warm-temperate plants, such as *Anemia, Azolla, Cinnamomum,* and

Iodes, are known from Great Britain, and *Azolla* and *Grewia* are found on Ellesmere and Axel Heiberg islands. These occur in the same deposits as more temperate ones, such as *Abies, Ginkgo, Glyptostrobus, Larix, Picea, Sequoia, Taxodium, Thuja, Acer, Alnus, Betula, Carpinus, Carya, Cornus, Corylus, Fraxinus, Juglans, Liquidambar, Magnolia, Nyssa, Tilia, Ulmus,* and others (see online appendix 3). The same ecological admixture is evident in the early Miocene Brandon Lignite flora of Vermont (*Manikara* and *Carya-Liquidambar-Tilia;* see Figs. 3.13–3.16 for examples of plant microfossils from the Brandon flora). As noted, there was undoubtedly some separation by altitude. Even so, the fossil floras across the northern latitudes suggest an intermingling beyond that expected from zonation alone and reveal that the paleocommunities were not exact analogs of modern ones. This is more consistent with the boreotropical concept of the vegetation as suggested by Wolfe (1975, 1977, 1979) than with the Arcto-Tertiary Geoflora as defined by Chaney (1959; see later section on geofloras).

The shortcomings of the appendices (see online supplementary materials) as presently compiled have been discussed with reference to the BLB (chapter 2; online appendix 2). The same limitations apply to the

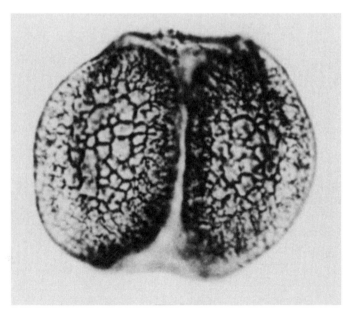

3.13. Plant microfossil from the early Miocene Brandon Lignite, Vermont: *Pinus.* From A. Traverse, 1955, U.S. Bureau of Mines, Washington, DC.

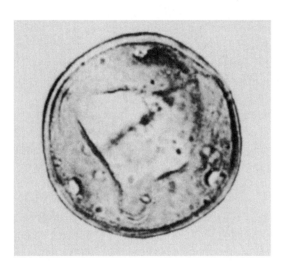

3.14. Plant microfossil from the early Miocene
Brandon Lignite, Vermont: *Carya*. From A. Traverse,
1955, U.S. Bureau of Mines, Washington, DC.

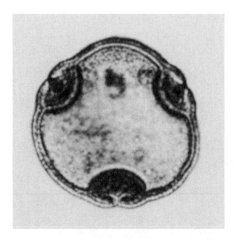

3.15. Plant microfossil from the early Miocene
Brandon Lignite, Vermont: *Tilia*. From A. Traverse,
1955, U.S. Bureau of Mines, Washington, DC.

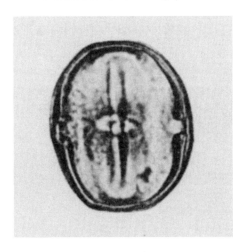

3.16. Plant microfossil from the early
Miocene Brandon Lignite, Vermont:
Manikara. From A. Traverse, 1955, U.S.
Bureau of Mines, Washington, DC.

listings in appendix 3, with some additions. In particular, there are few assemblages of the same age on both sides of the connection that have been studied recently. This requires extended coverage and comparison of Early Cretaceous floras of Great Britain with Middle Cretaceous ones along the North American Coastal Plain. Not all the studies use the same method (macrofossils vs. microfossils) or apply the same nomenclatural system: "We have decided not to use the generic names usually applied to these dispersed pollen taxa," calling them instead Genusbox RETISULC-*Biorecord* RETISULC-MURIVERM (Hughes and McDougall, 1987). After about the early Eocene most plant microfossils can be referred to modern taxa at least to the generic level (e.g., *Carya*) or they can be named by an artificial system (*Triporites psilatus* = a triporate smooth grain) with more casual mention made of some extant plant with similar pollen. Even so, establishing biological affinities involves the tedious and time-consuming task of building, maintaining, and using a herbarium-vouchered reference collection adequate to the task (see Jaramillo, 2008). As a result, microfossils even younger than the early Eocene are often named more expediently by an artificial system, while in other studies they are referred to modern taxa. In spite of this unevenness and these complications, some features of the vegetation across and on either side of the North Atlantic connection through time are discernible from the information available.

Modernization of the Flora

Late Paleozoic and early Mesozoic elements mostly disappear in the Early and Middle Cretaceous, and there is replacement of ancient ferns and gymnosperms by more advanced forms and an increasing prominence of angiosperms. Using the same comparison as with Beringia in chapter 2, the numbers are as follows:

Cretaceous—100 total genera (eastern and western NALB)
    Ferns and allied groups—16 (16%)
    Gymnosperms—48 (48%)
    Angiosperms—36 (36%)
Paleocene—148 genera
    Ferns and allied groups—13 (8.8%)
    Gymnosperms—22 (14.8%)
    Angiosperms—113 (76.4%)
Eocene—356 genera
    Ferns and allied groups—7 (1.9%)
    Gymnosperms—14 (3.8%)
    Angiosperms—335 (94.3%)
Oligocene—56 genera
    Ferns and allied groups—0 (0%)
    Gymnosperms—3 (5.35%)
    Angiosperms—53 (94.7%)
Miocene (beginning of cool conditions in the late Miocene)—116 genera
    Ferns and allied groups—8 (8.9%)
    Gymnosperms (taiga)—12 (10.3%)
    Angiosperms—96 (80.8%)
Pliocene (reduced land surface from expanding ice)—67 genera
    Ferns and allied groups—5 (5.6%)
    Gymnosperms—8 (11.9%)
    Angiosperms—54 (82.5%)

Biodiversity and Vegetation Density

This increased and sustained prominence of the angiosperms from the mid-Cretaceous into the Paleogene is evident in the vicinity of the NALB as it was in Beringia and in other parts of the world. The number of genera recorded in online appendix 3 from the Cretaceous/early Paleogene compared to the late Paleogene/Neogene shows another trend:

Total number of genera, Cretaceous/early Paleogene—607
    Cretaceous—100
    Paleocene—151
    Eocene—356
Total number of genera, late Paleogene/Neogene—225
    Oligocene—52
    Miocene—116
    Pliocene—67

The break chosen for comparison in the above tabulation corresponds to the time of significant physical disruption between the eastern and western North Atlantic and a major shift in climate from tropical to temperate beginning toward the end of the late Eocene. The difference in the number of genera (607 vs. 225) likely reflects, in part, the onset of harsher conditions especially after the MMCO.

Floristic Relationships between Eastern Asia and Eastern North America

Disjunct distributions between widely separated areas are the result of long-distance transport and/or elimination of a former widespread entity from an intervening part of its range. In the case of similarities between the deciduous forests of central China and the eastern United States, the number of genera and the distances are so great that long-distance dispersal alone is implausible as a primary explanation for the many species involved. Further, climate fluctuations and physical changes during the long interval between the Late Cretacous and the end of the Pliocene did not provide a constancy of environments over these great distances sufficient to sustain the migration of each species. Rather, the explanation calls for some disruptive geological event within the former range of the deciduous forest. The paleobotanical record documents widespread distribution of the forest after the middle Eocene cooling event; geological evidence is clear about desertification in the American West after uplift of the Cascades, Coast Ranges, and the Rocky Mountains especially in the late Neogene (late Miocene and Pliocene) and during the glacial phases of the Quaternary; and the destruction of the vegetation in western Europe from glaciations and anthropogenic factors was discussed earlier.The result is that relicts of the once widespread deciduous forest were left in the temperate mountainous regions of eastern Asia and eastern North America.

The literature detailing the origin of this biogeographical relationship is extensive. It begins with the observations of Linnaeus as expressed in the

1750 dissertation of his student Jonas P. Halenius in 1750 (*Plantae Rariores Camschatcenses*; Graham, 1966; Nynäs and Berquist, 2016). Linnaeus had noticed that several plants in the collections of Lerche from Kamchatka were similar to those of Clayton from Virginia, and Sarrazin and Gaulthier from eastern Canada (I, 321–24; Graham, 1972a,b). Linnaeus had no knowledge of the vegetation in the intervening region and was probably unaware these were disjunct distributions. That observation was made by Asa Gray (1840, 1846, 1859). Since then the topic has been reviewed by Boufford (1998), Tiffney (1985), and Wen et al. (2016; western North America and the Bering Land Bridge), and new molecular evidence is available for the timing of the various disjunctions (Xiang et al., 2000; Wen, 1999; Wen et al., 2010). Sequence divergence in the chloroplast gene *rbcl* for 22 species among 11 genera indicates that

> the time of divergence of species pairs examined ranges from 12.56 ± 4.30 Ma to recent with most within the last 10 million years (in the late Miocene and Pliocene). This estimate is closely correlated with paleontological evidence and in agreement with the hypothesis that considers the eastern Asian–eastern North American floristic disjunction to be the result of the range restriction of a once more or less continuously distributed mixed mesophytic forest of the Northern Hemisphere. (Xiang et al., 2000, 462)

and

> that the eastern Asian and eastern North American disjunct distributions are relicts of the maximum development of temperate forests in the northern hemisphere during the Tertiary. Fossil and geologic evidence supports multiple origins of this pattern in the Tertiary, with both the North Atlantic and the Bering land bridges involved. (Wen, 1999, 421)

Information from the boreal land bridges cited in online appendices 2 and 3 is presented in online table 3.2 with reference to the 64 genera with an eastern Asian–eastern North American disjunction listed by Wen (1999, table 1; see online supplementary materials). Taking the identifications and age of the strata in the appendices as stated, and noting that a number of the plants from the Eocene are from the middle and late Eocene (i.e., after the trend toward cooler temperate conditions was under way), the figures for the Cretaceous and Paleogene are Cretaceous, 8 genera; Paleocene, 11 genera; and Eocene, 13 genera. For the late Paleogene and Neogene the figures are Oligocene, 9 genera; Miocene, 16 genera; and Pliocene (after cooling and glaciation had intensified, and one land bridge [NALB] was permanently and the other [BLB] periodically disrupted), 5 genera. In spite of the vagaries in the paleobotanical database, the record is still (barely) consistent with estimates that just before and after the MMCO was the principal time of origin for this disjunct biogeographic pattern.

Geofloras and the Madrean-Tethyan Hypothesis

An update of the lingering concept of geofloras (I, 106–10) can be made based on the inventory of plant fossils in the Arctic realm presented in on-line appendices 2 and 3. Chaney (1959) noted that many Paleogene floras (Paleocene, Eocene, and Oligocene) in the Northern Hemisphere were characterized by a high percentage of tropical and subtropical plants similar to those found today in the equatorial regions. In the Neogene (Miocene and Pliocene) these fossil floras contain more cool- to cold-temperate plants like those of the extant mixed deciduous hardwood forest. Engler (1879/1882) had applied the term *arctotertiary flora* to the modern communities "distinguished by numerous conifers and the numerous genera of [deciduous angiosperm] trees and shrubs that now dominate in North America or extratropical Eastern Asia and in Europe" (quoted in Mai, 1991). Chaney (1959, 12) added a time dimension to this and other communities by referring to them as *geofloras*, which he defined as "a group of plants which has maintained itself with only minor changes in composition for several epochs or periods of earth history, during which time its distribution has been profoundly altered." For North America he recognized two such communities, the Neotropical-Tertiary Geoflora most widespread in the Paleogene and currently in the equatorial regions, and the Arcto-Tertiary Geoflora widespread in the Neogene and presently across much of the boreal zone. Axelrod (1958) added the Madro-Tertiary Geoflora for winter-dry, semidesert to desert vegetation appearing in the late Miocene and Pliocene and at present growing, for example, in the American West and the Mediterranean region. Madro-Tertiary vegetation (as opposed to individual elements) developed in response to the cooling and drying trend of the Late Cenozoic and regionally from mountain uplift. Axelrod (1975) attributed the similarity between the Great Basin/Mohave/Sonoran deserts and the Mediterranean to once greater continuity (the Madrean-Tethyan flora) that later became disrupted by the cold conditions developing across northern Europe, the North Atlan-

tic, and temperate North America. Based on the limited dispersal potential of many of the propagules, it is implausible that the similarity could
be due to long-distance transport. That leaves (1) once greater continuity
of habitats, or (2) parallel evolution in the areas of drying environment.
The first can be assessed from fossil floras found in the intervening area of
the NALB.

There is no geomorphological or sedimentary (edaphic) evidence that
a band of aridity connected or nearly connected the Mediterranean region
across the NALB with the dry regions of the southwestern United States/
northern Mexico. This complicates envisioning a near continuous sclerophyll vegetation through the mid- to high-latitudes of the Northern Hemisphere as suggested by Axelrod (1958, 1975). Isozymes from *Styrax* disjunct between the two regions (Fritsch, 1996) indicate recent divergence,
which is incompatible with the Madrean-Tethyan hypothesis. The result is
consistent with isozyme profiles from an additional 25 taxa, showing that
parallel evolution from common ancestral lineages accounts for the majority of 25 cases of disjunction between western Eurasia and western North
America (Kadereit and Baldwin, 2012). The data presented in online appendices 2 and 3 support the geological and isozyme results in that there
is no evidence for winter-wet or winter-dry arid vegetation in the vicinity of
the NALB or the BLB at any time during the Tertiary.

A hallmark of geofloras is that as the concept developed over time, they
became viewed as ever more closely knit assemblages moving over the landscape as intact blocks in response to physiographic and climatic change.
This was evident from a visit I made to Berkeley in 1961 to get some specimens identified from the Miocene Succor Creek and Trout Creek floras of
southeastern Oregon. The paleovegetation was a rich, deciduous hardwood
forest. Chaney arranged to have Harry MacGinitie present to help: "As I
handed each specimen to Chaney, he gave it a cursory glance and passed it
on to MacGinitie, who provided most of the identifications. The last specimen was a fossil leaf I suggested might be *Arctostaphylos*. Chaney did not
even look at this one and passed it to MacGinitie with the comment, 'It's
not *Arctostaphylos*.' I asked why, and he replied, 'Because *Arctostaphylos* is not
an Arcto-Tertiary element'" (I, 107).

Various shortcomings soon became apparent from such a rigid definition of geofloras (see Wolfe, 1971, e.g., 226; Wolfe 1975, 1977; also
Baskin and Baskin, 2016). For example, the beds Heer (1869) originally
thought to be Miocene in age because of their strong temperate component are now known to be Eocene, thus complicating the sharp distinc-

tion between an Eocene tropical and a Miocene temperate flora, as well as the supposed place of origin (the Arctic) of the Arcto-Tertiary Geoflora. Also, the modern vegetation of the tropics was even less well known then than today, which biased identifications of fossil material toward more familiar temperate plants. This was also prior to Wolfe's (1975, 1977, 1979) characterization of the leaves of distinctly tropical plants as often large, thin-textured, entire-margined, and with drip-tips, compared to those of temperate plants that are frequently smaller, serrate, and lack drip-tips. This observation allows some estimate of the paleoenvironment independent of the identifications. Davis (1981) further showed from the study of Quaternary assemblages that communities seldom migrate as a unit or maintain a constant association of dominants for more than about 2000–3000 years.

Even so, the idea of geofloras, and Madro-Tethyan connections, persists because it is simplistically appealing to envision vegetation types originating in one place and moving as easily recognizable blocks in response to changes in landscape and climate: "For all our palaeophytochronomical studies we confirm the old terms 'Arctotertiary' and 'Paleotropical' and also the concept of geofloras" (Mai, 1991).

Nonetheless, recent evidence indicates that community assemblage, migration, and the origin of affinities between disjunct regions are far more complicated than implied in the original version of geofloras. The complexity is better captured in the boreotropics concept mentioned earlier that envisions the late Cretaceous and Paleogene communities of 100 Myr ago and later as an amalgamation of plants with less sharply defined tropical and temperate ecology. When climates shifted toward cooler temperatures, the most tropical elements were eliminated, or their ranges were altered, or they evolved temperate descendants, while the temperate ones diversified, realigned, and expanded as versions of a mixed evergreen gymnosperm/deciduous angiosperm forest. MacGinitie (1962, 87), a student of Chaney, eventually came to question whether any modern vegetation type originated at one time and place and migrated "with only minor changes in composition for several epochs," and Grímsson et al. (2015) conclude that "Engler's hypothesis of an Arctic origin of the modern temperate woody flora of Eurasia, termed 'Arcto-Tertiary Element,' and later modification by R. W. Chaney and D. H. Mai ('Arcto-Tertiary Geoflora') needs to be modified" (809). A more dynamic view of community evolution is required than implied by geofloras, and the fossil record suggests the boreotropical hypothesis better captures this reality.

Indigenous People

The earliest inhabitants of Greenland were the Saqqaq (2500–800 BCE) and subsequent cultures known as Independence I, Independence II, and Dorset (700–200 BCE). They had migrated eastward from Siberia and the Arctic Islands, and each group ebbed and flowed through the region over the centuries. The indigenous people of far northeastern North America were the Beothuk (Skraelings in the Viking saga [Keneva, 2001]; Icelandic sagas [Sephton, 1880]). They were hunters and fishermen, numbering an estimated few thousand and extinct by 1829 from disease and skirmishes with the Europeans. The name Skraelins sometimes is also applied to the separate Thule people who were ancestors of the Inuits (Inuit Eskimos) and were interacting with the Norse in the 1200s.

Peopling of America (from the East)

At the time of the Viking explorations to Iceland (ca. 874 CE), Greenland (974 CE), and Newfoundland (1003 CE, involving a voyage of 1800 miles, which Leif Erikson made with a crew of 35 men), Iceland and Greenland were uninhabited, although there had been prior cultures. The record of the early introduction of humans into the New World is complicated enough without the added problem of fraudulent data. Such was the case, however, with the discovery that the alleged fifteenth-century Vinland Map was a fake (Skelton et al., 1995; https://en.wikipedia.org/wik/Vinland _map; http://www/econ.ohio-state.edu/jhm/arch/vinland/vinland.htm). The map was bound with a short text referred to as *The Tartar Relation,* dating to 1440, and if genuine would show that a second wave of Norsemen had come to North America 50 years before Columbus. An island was depicted southwest of Greenland named Vinland and now known as Newfoundland. The map was eventually donated to the Yale Library, where suspicions were raised when wormholes on the map did not match those on the *Relation* text. These suspicions were later confirmed when one of the ink components was found to be of modern manufacture. The conclusion is that the map is "unfortunately a fake" (Paul Freedman, 28 November 2011, HIST-210: lecture 22: "Vikings/The European Prospect," Open Yale courses, revised 26 August 2014, http://oyc.yale.edu/transcript/1246/hist -210). As summarized previously (III, 18–19):

> The Vikings were the second group to see the New World [after the Eurasian
> crossing of the BLB], some of them forced to leave Scandinavia because of

obnoxious and downright dangerous behavior even by Viking standards. Thorvald Asvaldsson went to Iceland after being banished from Norway for killing a man. His son, Eric the Red, killed two men in Iceland and settled in Greenland at Midjokull (Middle Glacier) to start a colony with four hundred people in 986 CE. In turn, his son Leif Ericson first landed on Baffin Island, then Labrador, and finally Vinland (Newfoundland) at a site now called L'Anse aux Meadows in about 1000 CE. They were the first Europeans to see the New World.

The temptation to try and the opportunity to succeed in settling Greenland and Newfoundland was in large measure conditioned by climate fluctuations in the postglacial Holocene epoch. It was during the Medieval Warm Period (800–1300 CE) that the Vikings made their extended sea voyages. During the Little Ice Age that followed (1450–1850 CE) they abandoned the sites. Toward the end of that same cold period Washington's troops would suffer a near-fatal winter at Valley Forge in 1777. These climatic changes were on a regional to near-global scale.

As described by Ackroyd, 2005, p. 457:

> On the last day of 1607, Edmund Shakespeare [brother to William] was buried. It was a time of almost unbearable cold. By the middle of December the Thames had frozen solid so that "many persons did walk halfway over the Thames upon the ice, and by the thirtieth of December the multitude . . . passed over the Thames in divers places." A small tent city sprang up on the ice, with wrestling bouts and football matches, barber shops and eating houses, trading upon the novelty of the silent and immobile river.

On 5 October 1492, some 250 years before Vitus Bering's sighting of northwestern North America in 1741, and five hundred years after the Vikings' discovery of the New World in 986 CE, and 18,000 years after the first Eurasians crossed Beringia, a lookout on board the *Pinta* cried out "Tierra" and claimed the 5000-maravedi prize for being the first of his crew to sight land—the [Bahama] island of San Salvador.

The earliest beginnings of human modification of the vegetation in northeastern North America dates from around 1003 CE with harvesting of timber by the Vikings for fuel and transport back to Scandinavia for house and ship construction. The initial alteration was undoubtedly slight and and likely would not affect interpretation of local to regional pollen diagrams, but it was a start and as John Wade said, coining the phrase in 1839, "the rest is history."

References

Ackroyd, P. 2005. Shakespeare: The Biography. Doubleday, London.

Anagnostou, E., et al. (+ 8 authors). 2016. Changing atmospheric CO_2 concentration was the primary driver of early Cenozoic climate. Nature 533: 380–84.

Axelrod, D. I. 1958. Evolution of the Madro-Tertiary Geoflora. Botanical Review 24: 433–509.

———. 1975. Evolution and biogeography of the Madrean-Tethyan sclerophyll vegetation. Annals of the Missouri Botanical Garden 62: 280–334.

Baskin, J. M., and C. C. Baskin. 2016. Origins and relationships of mixed mesophytic forests of Oregon-Idaho, China, and Kentucky: review and synthesis. Annals of the Missouri Botanical Garden 101: 525–52.

Bhagwat, S. A., and K. J. Willis. 2008. Species persistence in northerly glacial refugia of Europe: a matter of chance or biogeographical traits? Journal of Biogeography 35: 464–82.

Birks, H. J. B., and K. J. Willis. 2008. Alpines, trees, and refugia in Europe. Plant Ecology and Diversity 1: 147–60.

Boufford, D. E. 1998. Eastern Asian–North American plant disjunctions: opportunities for further investigations. Korean Journal of Plant Taxonomy 28: 49–61.

Bozukov, V. 2005. *Macclintockia basinervis* (Rossm.) Knobl. in Cenozoic sediments in the Rhodopes Mt. Region (S. Bulgaria). Acta Palaeontologica Romaniae 5: 11–15.

Brikiatis, L. 2016. Late Mesozoic North Atlantic land bridges. Earth-Science Reviews 159: 47–57.

Brouillet, L., and R. D. Whetstone. 1993. Climate and physiography. In Flora of North America Editorial Committee, Flora of North America, Vol. 1: Introduction. Oxford University Press, Oxford. Pp. 15–46.

Budantsev, L. Yu., and L. B. Golovneva. 2009. Fossil Flora of the Arctic. II. Paleogene Flora of Spitsbergen. Russian Academy of Sciences, Komarov Botanical Institute, Moscow.

Chaney, R. W. 1959. Miocene floras of the Columbia Plateau. Part I. Composition and interpretation. Carnegie Institution of Washington Contributions to Paleontology 617: 1–134.

Davis, M. B. 1981. Quaternary history and the stability of forest communities. In D. C. West, H. H. Shugart, and D. B. Botkine (eds.), Forest Succession: Concepts and Application. Springer, Berlin. Pp. 132–53.

Davis, S. D., V. H. Heywood, O. Herrera-MacBryde, J. Villa-Lobos, and A. Hamilton (eds.). 1994–97. Centres of Plant Diversity: A Guide and Strategy for Their Conservation. 3 vols. World Wide Fund for Nature (WWF) and the World Conservation Union, IUCN Publications Unit, Cambridge University, Cambridge.

Donoghue, M. J. 2008. A phylogenetic perspective on the distribution of plant diversity. Proceedings of the National Academy of Sciences USA 105 (supplement 1): 11549–55.

Donoghue, M. J., C. D. Bell, and J. H. Li. 2001. Phylogenetic patterns in Northern Hemisphere plant geography. International Journal of Plant Sciences (supplement) 162: S41–S52.

Donoghue, M. J., and S. A. Smith. 2004. Patterns in the assembly of temperate forests around the Northern Hemisphere. Philosophical Transactions of the Royal Society London B (Biological Sciences) 359: 1633–44.

Doyle, J. A., and L. J. Hickey. 1976. Pollen and leaves from the mid-Cretaceous Potomac Group and their bearing on early angiosperm evolution. In C. B. Beck (ed.), Origin and Early Evolution of Angiosperms. Columbia University Press, New York. Pp. 139–206.

The Economist. 2016a. If the ocean were transparent: the see-through sea. Economist, 16 July. ["The pioneering maps put together by Marie Tharp and Bruce Heezen of Columbia University in the 1950s and 1960s . . . which first identified the structure of the mid-Atlantic ridge . . . let geologists visualize the processes at work in the nascent theory of plate tectonics" (16).]

———. 2016b. In an octopus's garden. Researchers have a plan to chart in detail the depths of the ocean floor. Economist, 29 October.

Engler, A. 1879/1882. Versuch einer Entwicklungsgeschichte der Pflanzenwelt seit der Tertiärperiode I/II. W. Engelmann, Leipzig.

Flora of North America Editorial Committee. 1993. Flora of North America. Vol. 1: Introduction. Oxford University Press, Oxford.

Fritsch, P. W. 1996. Isozyme analysis of intercontinental disjuncts within Styrax (Styracaceae): implications for the Madrean-Tethyan hypothesis. American Journal of Botany 83: 342–55.

Galloway, J., A. R. Sweet, A. Pugh, and A. Embry. 2012. Correlating Middle Cretaceous palynological records from the Canadian High Arctic based on a section from the Sverdrup Basin and samples from the Eclipse Trough. Palynology 36: 277–302.

Gibbard, P. L. 1995. The formation of the Strait of Dover. Geological Society, London, Special Publications 96: 15–26.

———. 2007. Europe cut adrift. Nature 448: 259–60.

Golynchik, P. O. 2009. Geological structure of Cenozoic sedimentary basin in the northwestern margin of the Barents Sea shelf. Moscow University Geology Bulletin 64: 318–23.

Graham, A. 1966. Planate Rariores Camchatcenses: a translation of the dissertation of Jonas P. Halenius, 1750. Brittonia 18: 131–39.

———. 1972a. Floristics and Paleofloristics of Asia and Eastern North America. Elsevier, Amsterdam.

———. 1972b. Outline of the origin and historical recognition of floristic affinities between Asia and eastern North America. In A. Graham (ed.), Floristics and Paleofloristics of Asia and Eastern North America. Elsevier, Amsterdam. Pp. 1–18. [Includes Asa Gray's 1840 and 1846 papers.]

———. 1993. History of the vegetation: Cretaceous (Maastrichtian)—Tertiary. In Flora of North America Editorial Committee, Flora of North America, Vol. 1: Introduction. Oxford University Press, Oxford. Pp. 57–70.

Gray, A. 1840. Dr. Siebold, Flora Japonica. American Journal of Science and Arts 39: 175–76.

———. 1846. Analogy between the flora of Japan and that of the United States. American Journal of Science and Arts 2 (2): 135–36.

———. 1859. Diagnostic characters of new species of phanerogamous plants collected in Japan by Charles Wright, botanist of the U.S. North Pacific Exploring Expedition. (Published by request of Captain James Rodgers, Commander of the Expedition.) With observations upon the relations of the Japanese flora to that of North America, and other parts of the northern temperate zone. Memoir of the American Academy of Sciences 6: 377–452.

Grímsson, F., R. Zetter, G. W. Grimm, G. K. Pedersen, A. K. Pedersen, and T. Denk. 2015. Fagaceae pollen from the early Cenozoic of West Greenland: revisiting Engler's and Chaney's Arcto-Tertiary hypotheses. Plant Systematics and Evolution 301: 809–32.

Gupta, S., J. S. Collier, A. Palmer-Felgate, and G. Potter. 2007. Catastrophic flooding origin of shelf valley systems in the English Channel. Nature 448: 342–45.

Halenius, J. P. 1750. Plantae Rariores Camchatcenses. Thesis, University of Uppsala, Sweden.

Harrison, J. C. 2006. In search of the Wegener Fault: re-evaluation of strike-slip displacements along and bordering Nares Strait. Polarforschung 74: 129–60.

Heer, O. 1868. Fossile Flora von Nordgrönland. In Die fossile Flora der Polarländer. Vol. 1. Zürich. Pp. 78–129.

———. 1869. Flora Fossilis Alaskana. Kongila Svenska Vetenskapsakademienkad Handlingar 8, no. 4. Stockholm.

Hughes, N. F., and A. B. McDougall. 1987. Records of angiospermid pollen entry into the English Early Cretaceous succession. Review of Palaeobotany and Palynology 50: 255–72.

Imbrie, J. 1985. A theoretical framework for the Pleistocene ice ages. Journal of the Geological Society of London 142: 417–32.

Immonen, N. 2013. Surface microtextures of ice-rafted quartz grains revealing glacial ice in the Cenozoic Arctic. Palaeogeography, Palaeoclimatology, Palaeoecology 374: 293–302.

Isacks, B., J. Oliver, and L. R. Sykes. 1968. Seismology and the new global tectonics. Journal of Geophysical Research 73: 5855–99.

Iversen, J. 1941. Land occupation in Denmark's Stone Age. A pollen-analytical study of the influence of farmer culture on the vegetational development. Danmarks Geolgiske Underøgelse 66: 20–65.

Jackson, H. R. 1985. Nares Strait—a suture zone: geophysical and geological implications. Tectonophysics 114: 11–28.

Jaramillo, C. 2008. Donation of the Graham Collection to the Smithsonian Institution. Newsletter, AASP–The Palynological Society 41 (3).

Kadereit, J. W., and B. G. Baldwin. 2012. Western Eurasian-western North American disjunct plant taxa: the dry-adapted ends of formerly widespread north temperate mesic lineages—and examples of long-distance dispersal. Taxon 61: 3–17.

Keneva, K. (translator). 2001. The Saga of Icelanders. Penguin Books, New York.

Leroy, S. A. G., and K. Arpe. 2007. Glacial refugia for summer-green trees in Europe and south-west Asia as proposed by ECHAM3 time-slide atmospheric model simulations. Journal of Biogeography 34: 2115–28.

MacGinitie, H. D. 1962. The Kilgore flora, a late Miocene flora from northern Nebraska. University of California Publications in Geological Sciences 35: 67–158.

Mai, D. H. 1991. Palaeofloristic changes in Europe and the confirmation of the Arctotertiary-Paleotropical geoflora concept. Review of Palaeobotany and Palynology 68: 29–36.

Manos, P. S., and J. E. Meireles. 2015. Biogeographic analysis of the woody plants of the Southern Appalachians: implications for the origins of a regional flora. American Journal of Botany 102: 1–25.

Manum, S. B., and T. Throndsen. 1986. Age of Tertiary formations on Spitsbergen. Polar Research 4: 103–31.

McIver, E. E., and J. F. Basinger. 1999. Early Tertiary floral evolution in the Canadian High Arctic. Annals of the Missouri Botanical Garden 86: 523–45.

Milne, R. I. 2004. Phylogeny and biogeography of *Rhododendron* subsection *pontica*, a group with Tertiary relict distribution. Molecular Phylogenetics and Evolution 33: 389–401.

———. 2006. Northern Hemisphere plant disjunctions: a window on Tertiary land bridges and climate change? Annals of Botany 98: 465–72.

Neftel, A., E. Moor, H. Oeschger, and B. Stauffer. 1985. Evidence from polar ice cores for the increase in atmospheric CO_2 in the past two centuries. Nature 315: 45–47.

Neftel, A., H. Oeschger, J. Schwander, B. Stauffer, and R. Zumbrunn. 1982. Ice core sample measurements give atmospheric CO_2 content during the past 40,000 yr. Nature 295: 220–23.

Nynäs, C., and L. Berquist. 2016. A Linnean Kaleidoscope: Linnaeus and His 186 Dissertations. Hagströmer Medico-Historical Library. Fri Tanke.

Oliver, J. 1996. Shocks and Rocks: Seismology in the Plate Tectonics Revolution. American Geophysical Union, Special Publications 6. Washington, DC.

Ozenda, P. 1979. Vegetation Map (scale 1:3,000,000) of the Council of Europe Member States. European Committee for the Conservation of Nature and Natural Resources, Strasbourg.

Peirce, J. W. 1982. The evolution of the Nares Strait lineament and its relation to the Eurekan Orogeny. In P. R. Dawes and J. W. Kerr (eds.), Nares Strait and the Drift of Greenland: A Conflict in Plate Tectonics. Meddr Grønland Geoscience 8: 237–52. [The Eurekan Orogeny was a Late Cretaceous to Paleogene series of events affecting the Arctic Islands and northern Greenland.]

Pinker, S. 2014. The Sense of Style: The Thinking Person's Guide to Writing in the 21st Century. Penguin, New York.

Redfern, R. 2001. Origins: The Evolution of Continents, Oceans, and Life. University of Oklahoma Press, Norman.

Reid, E. M., and M. E. Chandler. 1933. The London Clay flora. British Museum (Natural History), London.

Sephton, J. (translator). 1880. The Saga of Erik the Red. Icelandic Saga Database. http://sagadb.org.

Severinghaus, J. P., T. Sowers, E. J. Brook, R. B. Alley, and M. L. Bender. 1998. Timing of abrupt climate change at the end of the Younger Dryas interval from thermally fractionated gases in polar ice. Nature 391: 141–46.

Shankman, D., and L. A. James. 2002. Appalachia and the Eastern Cordillera. In A. R. Orme (ed.), The Physical Geography of North America. Oxford University Press, Oxford. Pp. 291–306.

Skelton, R. A., T. E. Marston, and G. D. Painter. (Introduction by A. O. Victor.) 1995. The Vinland Map and the Tartar Relation. Yale University Press, New Haven. [Reprint of the 1965 edition with new prefatory essays by G. D. Painter, W. E. Washburn, T. A. Cahill, B. H. Kusko, and C. Witten II.]

Soltis, D. E., A. B. Morris, J. S. McLachlan, P. S. Manos, and P. S. Soltis. 2006. Comparative phylogeography of unglaciated eastern North America. Molecular Ecology 15: 4261–93.

Tessensohn, F., R. H. Jackson, and I. D. Reid. 2004. The tectonic evolution of Nares Strait: implications of new data. Polarforschung 74: 191–98.

Theroux, P. 2008. Ghost Train to the Eastern Star: On the Tracks of the Great Railway Bazaar. Houghton Mifflin Harcourt, New York.

Thorsteinsson, R., and U. Mayr. 1987. The Sedimentary Rocks of Devon Island, Canadian Arctic Archipelago. Geological Survey of Canada Memoir 411.

Throndsen, T., and T. Bjaerke. 1983. Palynodebris analysis of a shallow core from the Barents Sea. Polar Research 1 (n.s.): 43–47.

Tiffney, B. H. 1985. Perspectives on the origin of the floristic similarity between eastern Asia and eastern North America. Journal of the Arnold Arboretum 66: 73–94.

Traverse, A. 1955. Pollen Analysis of the Brandon Lignite of Vermont. Bureau of Mines, Report of Investigations 5151. Washington, DC.

Trettin, N. 1989. The Arctic Islands. In A. W. Bally and A. R. Palmer (eds.), The Geology of North America, Vol. A: The Geology of North America: An Overview. Geological Society of America, Boulder. Pp. 349—70.

Van der Hammen, T., T. J. Wijmstra, and W. H. Zagwijn. 1971. The fossil record of the late Cenozoic in Europe. In K. K. Turekian (ed.), The Late Cenozoic Glacial Ages. Yale University Press, New Haven. Pp. 391–424.

Verhoeven, K., and S. Louwye. 2013. Palaeoenvironmental reconstruction and biostratigraphy with marine palynomorphs of the Plio-Pleistocene in Tjörnes, northern Iceland. Palaeogeography, Palaeoclimatology, Palaeoecology 376: 224–43.

Vogt, P. R., and B. E. Tucholke. 1989. North Atlantic Ocean Basin: aspects of geologic structure. In A. W. Bally and A. R. Palmer (eds.), The Geology of North America, Vol. A: The Geology of North America: An Overview. Geological Society of America, Boulder. Pp. 53–80.

Wade, J. 1839. British History, Chronologically Arranged. HathiTrust Digital Library (administered by Indiana University and the University of Michigan).

Wen, J. 1999. Evolution of eastern Asian and eastern North American disjunct distributions in flowering plants. Annual Review of Ecology and Systematics 30: 421–55.

Wen, J., and S. Ickert-Bond. 2009. Evolution of the Madrean-Tethyan disjunctions and the North and South American amphitropical disjunctions in plants. Journal of Systematics and Evolution 47: 331–48.

Wen, J., S. Ickert-Bond, Z.-L. Nie, and R. Li. 2010. Timing and modes of evolution of Eastern Asian–North American biogeographic disjunctions in seed plants. In M. Long, H. Gu, and Z. Zhou (eds.), Darwin's Heritage Today: Proceedings of the Darwin 200 Beijing International Conference. Higher Education Press, Beijing. Pp. 252–69.

Wen, J., Z.-L. Nie, and S. M. Ickert-Bond. 2016. Intercontinental disjunctions between eastern Asia and western North America in vascular plants highlight the biogeographic importance of the Bering land bridge from late Cretaceous to Neogene. Journal of Systematics and Evolution 54: 469–90.

Wilkerson, I. 2010. The Warmth of Other Suns. Random House, New York.

Willis, K. J., and K. J. Niklas. 2004. The role of Quaternary environmental change in plant macroevolution: the exception or the rule? Philosophical Transactions of the Royal Society of London B (Biological Sciences) 359: 159–72.

Willis, K. J., E. Ruder, and P. Sümegi. 2000. The full glacial forests of central and southeastern Europe. Quaternary Research 53: 203–13.

Willis, K. J., and T. H. Van Andel. 2004. Trees or no trees? The environments of central Europe during the last glaciation. Quaternary Science Reviews 23: 2369–87.

Wolfe, J. A. 1971. An interpretation of Alaskan Tertiary floras. In A. Graham (ed.), Floristics and Paleofloristics of Asia and Eastern North America. Elsevier, Amsterdam. Pp. 201–55.

———. 1975. Some aspects of plant geography of the Northern Hemisphere during the Late Cretaceous and Tertiary. Annals of the Missouri Botanical Garden 62: 264–79.

———. 1977. Paleogene floras from the Gulf of Alaska region. U.S. Geological Survey Professional Paper 997: 1–108.

————. 1979. Temperature parameters of humid to mesic forests for eastern Asia and relation to forests of other regions of the Northern Hemisphere and Australasia. U.S. Geological Survey Professional Paper 1106: 1–37.

Xiang, Q.-Y., D. E. Soltis, P. S. Soltis, S. R. Manchester, and D. J. Crawford. 2000. Timing the eastern Asian–eastern North American floristic disjunction: molecular clock corroborates paleontological estimates. Molecular Phylogenetics and Evolution 15: 462–72.

Additional References

Ballantyne, A. P., Y. Axford, G. H. Miller, B. L. Otto-Bliesner, N. Rosenbloom, and J. W. C. White. 2013. The amplification of Arctic terrestrial surface temperatures by reduced sea-ice extent during the Pliocene. Palaeogeography, Palaeoclimatology, Palaeoecology 386: 59–67.

Bierman, P. R., L. B. Corbett, J. A. Graly, T. A. Neumann, A. Lini, B. T. Crosby, and D. H. Rood. 2014. Preservation of a periglacial landscape under the center of the Greenland Ice Sheet. Science Express. http://sciencemag.org/content/early/recent. 17 April.

Bohn, U., N. Zazanashvili, and G. Nakhutsrichvili. 2007. The map of the natural vegetation of Europe and its application in the Caucasus ecoregion. Bulletin of the Georgian National Academy of Sciences 175: 112–21.

Couvreur, T. L. P., et al. (+ 6 authors). 2011. Early evolutionary history of the flowering plant family Annonaceae: steady diversification and boreotropical geodispersal. Journal of Biogeography 38: 664–80.

Digital map of European ecological regions (DMEER). s.d. http://www/dmeer-digital-map-of-european-ecological-regions/ eea.europa.eu/data-and-maps/figures.

The Economist. 2016. What if all the ice caps melted? Economist. 16 July.

Eiserhardt, W. L., F. Borchsenius, C. M. Plum, A. Ordonez, and J.-C. Svenning. 2015. Climate-driven extinctions shape the phylogenetic structure of temperate tree floras. Ecology Letters 18: 263–72.

Ellenberg, H. 1988. Vegetation Ecology of Central Europe. Cambridge University Press, Cambridge.

Estes, R., and J. H. Hutchison. 1980. Eocene lower vertebrates from Ellesmere Island, Canadian Arctic Archipelago. Palaeogeography, Palaeoclimatology, Palaeoecology 30: 325–47.

Gilliam, F. S. 2016. Forest ecosystems of temperate climatic regions: from ancient use to climate change. New Phytologist (Tansley Review). doi: 10.1111/nph.14255.

Grímsson, F., G. W. Grimm, B. Meller, J. M. Bouchal, and R. Zetter. 2016. Part IV: Magnoliophyta 2: Fagales to Rosales. Grana 55: 101–63.

Grímsson, F., B. Meller, J. M. Bouchal, and R. Zetter. 2015. Part III: Magnoliophyta 1: Magnoliales to Fabales. Grana 54: 85–128.

Grímsson, F., and R. Zetter. 2015. Part II: Pinophyta (Cupressaceae, Pinaceae and Sciasopityaceae). Grana 50: 262–310.

Grímsson, F., R. Zetter, and C. Baal. 2011. Combined LM and SEM study of the middle Miocene (Sarmatian) palynoflora from the Lavanttal Basin, Austria: Part I: Bryophyta, Lycopodiophyta, Pteridophyta, Ginkgophyta, and Gnetophyta. Grana 50: 102–28.

Guo, Q., H. Qian, R. E. Ricklefs, and W. Xi. 2006. Distributions of exotic plants in eastern Asia and North America. Ecology Letters 9: 827–34.

Harland, W. B. 1997. The geology of Svalbard. Geological Society Memoir No. 17.

Harris, A. J., C. Walker, J. R. Dee, and M. W. Palmer. 2016. Latitudinal trends in genus richness of vascular plants in the Eocene and Oligocene of North America. Plant Diversity 38: 133–41.

Hartley, A., J.-F. Pekel, M. Ledwith, J.-L. Champeaux, E. De Basdts, and S. A. Bartalev. 2007 (last updated 30 November 2009). Vegetation Map of Europe. http://bioval.jrc .ee.europa.eu/products/glc2000/products/europa_poster.jpg. [A jpg version of the vegetation map of Europe.]

Hendriksen, N. 2005. Geological History of Greenland: Four Billion Years of Earth Evolution. Geological Survey of Denmark and Greenland (GEUS), Copenhagen.

Holt, P. C. (ed.). (With the assistance of R. A. Paterson.) 1970. The Distributional History of the Biota of the Southern Appalachians. Part II. Flora. Research Division Monograph 2. Virginia Polytechnic Institute and State University, Blacksburg.

Johnson, W. C., and T. Webb III. 1989. The role of blue jays (*Cyanacitta cristata* L.) in the postglacial dispersal of fagaceous trees in eastern North America. Journal of Biogeography 16: 561–71.

Jones, T. R. 1875. Manual of the Natural History, Geology, and Physics of Greenland and the Neighbouring Regions, Prepared for the Use of the Arctic Expedition of 1875, Under the Direction of the Arctic Committee of the Royal Society. ["Sir George Nares (1831–1915), the expedition's leader, had hoped the North Pole could be reached. Though this proved impossible, a team of his men set a record for the furthest northern latitude attained at the time. Nares' official 1876 report and his 1878 two-volume account of the journey are also reissued in the Cambridge Library Collection."]

Kolbert, E. 2016. Letter from Greenland: a song of ice: what happens when a country starts to melt? New Yorker, 24 October: 51–61.

Kvaček, Z., and S. B. Manum. 1997. A. G. Nathorst's (1850–1921) unpublished plates of Tertiary plants from Spitsbergen. Swedish Museum of Natural History, Stockholm.

Kwok, R., T. Pedersen, P. Gudmandsen, and S. S. Pang. 2010. Large sea ice outflow into the Nares Strait in 2007. Geophysical Research Letters 37: L03502. doi: 1019/2009GL041872.

Lang, G. 1994. Quartäre Vegetationsgeschichte Europas: Methoden und Ergebnisse. Gustav Fischer, Jena.

Mai, D. H. 1995. Tertiäre Vegetationsgeschichte Europas: Methoden und Ergebnisse. Gustav Fischer, Jena. [Mai's interpretation of European Tertiary vegetation history includes acceptance of the geoflora concept as defined by Chaney and Axelrod; see chapter 3.]

Manchester, S. R., Z.-D. Chen, A.-M. Lu, and K. Uemura. 2009. Eastern Asian endemic seed plant genera and their paleogeographic history throughout the Northern Hemisphere. Journal of Systematics and Evolution 47: 1–42.

McKenna, M. C. 1980. Eocene paleolatitude, climate, and mammals of Ellesmere Island. Palaeogeography, Palaeoclimatology, Palaeoecology 30: 349–62.

Nutman, A., et al. (+5 authors). 2016. Rapid emergence of life shown by discovery of 3,700-million-year-old microbial structures. Nature 537: 535–38.

Pedersen, K. R. 1976. Fossil floras of Greenland. Geology of Greenland 1976: 519–38.

Polunin, O., and M. Walters. 1985. A Guide to the Vegetation of Britain and Europe. Oxford University Press, Oxford.

Qian, H., J. J. Wiens, J. Zhang, and Y. Zhang. 2015. Evolutionary and ecological causes of species richness patterns in North American angiosperm trees. Ecography 38: 241–50.

Ritchie, J. C., and G. M. MacDonald. 1986. The patterns of post-glacial spread of white spruce. Journal of Biogeography 13: 527–40.

Rosen, J. 2016. Cold truths at the top of the world. Nature. 19 April. http://www.nature
.com/news/cold-truths-at-the-top-of-the-world-1.19760.

Sánchez-Valásquez, L. R., M. R. Pineda-López, S. G. Vásquez-Morales, and N. L. Avendaño-
Yáñez. 2016. Ecology and conservation of endangered species: the case of magnolias.
In M. Quinn (ed.), Endangered Species. Nova Science Publishers, Hauppauge, NY.
Pp. 64–84.

Sandom, C. J., R. Ejrnaes, M. D. D. Hansen, and J.-C. Svenning. 2014. High herbivore
density associated with vegetation diversity in interglacial ecosystems. Proceedings of
the National Academy of Sciences USA 111: 4162–67.

Schiermeier, Q. 2016. 180,000 forgotten photos reveal the future of Greenland's ice. Nature 535: 480–83.

Schröder-Adams, C. J., J. O. Herrle, A. F. Embry, J. W. Haggart, J. F. Galloway, A. T. Pugh,
and D. M. Harwood. 2014. Aptian to Santonian foraminiferal biostratigraphy and
paleoenvironmental change in the Sverdrup Basin as revealed at Glacier Fiord, Axel
Heiberg Island, Canadian Arctic Archipelago. Palaeogeography, Palaeoclimatology,
Palaeoecology 413: 81–100.

Solgaard, A. M., J. M. Bonow, P. L. Langen, P. Japsen, and C. S. Hvidberg. 2013. Mountain
building and the initiation of the Greenland Ice Sheet. Palaeogeography, Palaeoclimatology, Palaeoecology 392: 161–76.

Svenning, J.-C. 2002. A review of natural vegetation openness in north-western Europe.
Biological Conservation 104: 133–48.

Svenning, J.-C., and F. Skov. 2005. The relative roles of environment and history as controls of tree species composition and richness in Europe. Journal of Biogeography 32:
1019–33.

Wappler, T., et al. (+ 7 authors). 2014. Before the "Big Chill": a preliminary overview of
arthropods from the middle Miocene of Iceland (Insecta, Crustacea). Palaeogeography, Palaeoclimatology, Palaeoecology 401: 1–12.

Wen, J. 2011. Systematics and biogeography of *Aralia* L. (Araliaceae): revision of *Aralia* sects. *Aralia*, *Humiles*, *Nanae*, and *Sciadodendron*. Contributions from the United
States National Herbarium 57: 1–172.

Wen, J., S. T. Bergren, C. Lee, S. Ickert-Bond, T. Yi, K. Yoo, L. Xie, J. Shaw, and D. Potter.
2008. Phylogenetic inferences in *Prunus* (Rosaceae) using chloroplast *ndhF* and nuclear ribosomal ITS sequences. Journal of Systematics and Evolution 46: 322–32.

Wilson, R., D. Greenwood, and J. Basinger. 2013. Fossil remains of the walnut family
(Juglandaceae) from the Eocene fossil forests of Axel Heiberg Island, Canadian High
Arctic. http://people.brandon.ca/greenwood/files/2013/wilsonbasinger-small.pdf.

Witze, A. 2014. Icelandic volcano stuns scientists. Nature 514: 543. [Ca. 250 km from
Reykjavik, spurting 35,000 tons of sulfur dioxide per day.]

PART TWO

Equatorial Land Bridges

FOUR

Antillean Land Bridge

It seems archipelagoes will be well worth examining.

—Charles Darwin, field notes, probably July 1836; Berry, 1984

Stepping Stones or Lost Highway

The configuration of the Antilles reveals their role as a partial land bridge between northern South America, the southeastern United States, and northern Central America/Yucatán Peninsula of Mexico (Fig. 4.1). A map of earthquake epicenters identifying regional subduction zones and plate/subplate contacts (Fig. 4.2) further reveals that the connection extends into deep geological time and has had a dynamic history. The question is whether it was ever continuous, meaning less filtering effect, and could have served as a migration route for organisms with limited means of dispersal, or if it has always been discontinuous so that widespread species had to have some capacity for long-distance dispersal or move by alternate means and routes. In the biogeographic literature both continuous and discontinuous paleolandscapes have been proposed:

> During the Eocene-Oligocene transition, the developing northern Greater Antilles and northwestern South America were briefly connected [for ca. 3 Ma] by a "landspan" (i.e., a subaerial connection between a continent and one or more off-shelf islands) centered on the emergent Aves Ridge [Fig. 4.3]. This structure (Greater Antilles + Aves Ridge) is dubbed GAARlandia. The massive uplift event that apparently permitted these connections was spent by 32 Ma; a general subsidence followed, ending the GAARlandia landspan phase. Thereafter, Caribbean neotectonism resulted in the subdivi-

sion of existing land areas. The GAARlandia hypothesis has great significance for understanding the history of the Antillean biota. Typically, the historical biogeography of the Greater Antilles is discussed in terms of whether the fauna was largely shaped by strict dispersal or strict continent-island vicariance. The GAARlandia hypothesis involves elements of both. (Iturralde-Vinent and MacPhee, 1999, 3)

or

Paleogeographic evidence does not currently support the Aves Ridge land bridge model, in contrast to claims by its strongest advocates Iturralde-Vinent and MacPhee (1999). In their advocacy of that model, with the corresponding need to construct a mid-Cenozoic walkway for land mammals, they have blurred the distinction between paleogeography (past landscapes) and historical biogeography (past distributions of organisms). They refer to their model as "paleogeography" but in reality it is a biogeographic model biased by their desire to create a corridor for mammals when there is no physical evidence to support such an unbroken corridor of land . . . the weight of the evidence supports an origin by over-water dispersal for most West Indian vertebrate fauna. This conclusion stems from (1) the reduced higher-level taxonomic composition of the fauna (now and in the past), (2) the presence of unusually large adaptive radiations, (3) the finding that closest relatives of most Antillean groups are from South America (passive dispersers) or Central and North America (active dispersers), and (4) the finding of divergence time estimates that are not strongly clustered. (Hedges, 2006, 241)

and in response,

Hedges and co-workers have strongly espoused over-water dispersal as the major and perhaps only method of vertebrate faunal formation in the Caribbean region. However, surface-current dispersal of propagules is inadequate as an explanation of observed distribution patterns of terrestrial faunas in the Greater Antilles. Even though there is a general tendency for Caribbean surface currents to flow northward with respect to the South American coastline, experimental evidence indicates that the final depositional sites of passively floating objects is highly unpredictable. Crucially, prior to the Pliocene, regional paleoceanography was such that current-flow patterns from major rivers would have delivered South American waifs to the Central American coast, not to the Greater or Lesser Antilles. Since at least three (capromyid rodents, pitheciine primates, and megalonychid sloths) and possibly four (nesophontid insectivores) lineages of Antillean mammals were already on one or more of the Greater Antilles by the Early Eocene, Hedges' inference as to the primacy of over-water dispersal appears to be at odds with the facts. By contrast, the landspan model is consistent with most aspects of Antillean land-mammal biogeography as currently known; whether it is consistent with the biogeography of other groups remains to be seen. (Iturralde-Vinent and MacPhee, 1999, 3)

In both cases some disjunctions could have arisen through rare and abrupt "sweepstakes" distribution events and/or vicariance.

In addition to changes in the physical continuity of the Antilles, there were simultaneous and interrelated but independent alterations in climate, and new habitats were appearing, ranging from coastal to high altitude, as a result of relative and eustatic changes in sea level and orogenic uplift. All these events and processes had an impact on organisms that were themselves evolving. It was a complex system requiring a multiplicity of approaches generating vast amounts of data best interpreted within the broadest possible context.

The biota of the Antilles serves to emphasize another point. Analyses

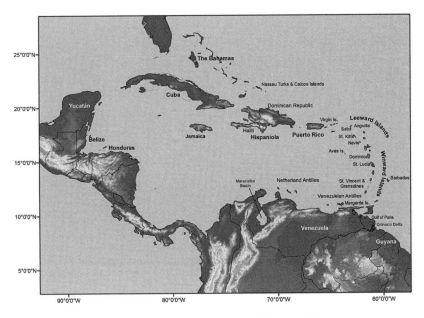

4.1. Index map and place names and physiographic features for the Antillean/Central American region.

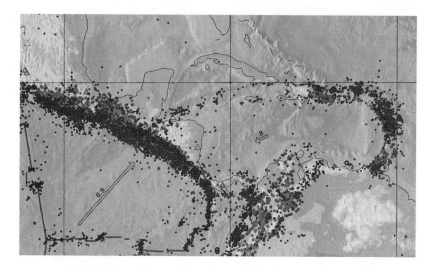

4.2. Distribution of earthquake epicenters in the Caribbean region. From Department of the Interior, U.S. Geological Survey Map prepared in cooperation with the Smithsonian Institution—This Dynamic Planet, Volcanoes Earthquakes, and Plate Tectonics.

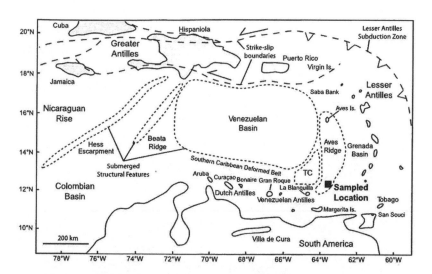

4.3. Structural and tectonic features of the Caribbean Basin. Dashed lines represent submerged escarpments and ridges (note Aves Ridge to the right), V designates subduction zones and the direction of subduction, half-arrows indicate direction of movement along fault contacts. From Neill et al., 2011, Journal of the Geological Society of London. Used with permission.

restricted to patterns, events, and processes operating only within the modern biota, and generalizing from isolated components and approaches, suffer from two shortcomings. First, they sever the connection between interacting factors and tilt the effort toward one of investigating a system that does not exist in the real world. Multiple-factor modeling can be a great asset in these studies and in depicting the results. Second, highly specialized approaches run the risk of subtly shifting from the biota to the approach itself. Understanding modern communities requires both theoretical, ground-truth, and paleo- and neontological approaches as to how they function, and how they got that way. For the Antilles the immediate problem is that fossil plants in existing collections are in serious need of revision, and additional ones are required from stratigraphically and geographically critical sites. The same is true for the other land bridges, but the need is particularly evident in the Antilles.

Geographic Setting and Climate

The islands making up the Antilles are arranged in an arc extending from ca. 10°N to 26°N and from 85°W to 50°W (see table 4.1 in online supplementary materials). They define the Caribbean Basin of 2,754,000 km^3, which, for comparison, is slightly larger than the Mediterranean Sea at 2,505,000 km^2. The greatest depths are 9220 m in a subduction zone called the Puerto Rican Trench along the northern edge and 6800 m in the Cayman Trough running northeast from the junction between Guatemala and Honduras. The greatest heights are Pico Duarte (3098 m) and the adjacent Lorna La Pelona (3094 m) in the Dominican Republic. In size the islands range from Cuba at 110,860 km^2 to the emergent volcanic peak of Mount Scenery, making up most of the island of Saba at 13 km^2. Cuba has a population of 11.3 million people, while Barbuba (161 km^2) is uninhabited. The archipelago includes the Greater Antilles to the north, constituting 90% of both the land and the population. The Lucayan Archipelago (or the Bahamas, Turks and Caicos Islands) extend northwest toward the southeastern United States. They are part of the North American plate and geologically unrelated to the Antilles although otherwise regarded as part of the Caribbean region. The Lesser Antilles trend north-south and are subdivided into the northerly Leeward Islands and the southerly Windward Islands named from the west-blowing trade winds that carried sailing ships to the New World. A third group, bordering the coast of Venezuela, is the Venezuelan Antilles and the Netherland Antilles. All these lands collectively are referred to as the West Indies.

Two geographic features of the Antilles are of particular biogeographic interest. One is the spacing between the islands, which defines the physical barriers to migration for terrestrial organisms. The distances are not great, and considering that land has been present since the Late Cretaceous ca. 100 Ma (II, 67–78) and that early versions of some current islands emerged and have persisted since the middle Eocene ca. 45 Ma, the probability for range extension of almost any organism is high without having to invoke continuous land connection. Regarding the present spacing,

> Cuba is separated from Florida across the Florida Straits by 160 km; from Yucatán across the Yucatán Channel by 210 km; from Haiti across the Windward Passage by 72 km.
>
> The Dominican Republic is separated from Puerto Rico across the Mona Passage by 96 km; from Jamaica by 140 km.
>
> Haiti is separated from Jamaica across the Jamaica Passage by 180 km.
>
> The Virgin Islands are separated from the Lesser Antilles (Anguilla) by 112 km.
>
> Tobago is separated from Trinidad by 34 km.
>
> Trinidad is separated from South America across the Gulf of Paria by 11 km.

The climates on the larger islands are diverse due primarily to altitude. At sea level and up to middle and lower altitudes, the MAT across the Antillean Land Bridge ranges from 22°C to 28°C. Following the average lapse rate of a decrease of 6.4°C per km, the MAT is 12°C at the top of Pico Duarte (3098 m) in the Dominican Republic. The hot months are June through August, and the cold months are December through February. Climates and distances between the islands have fluctuated over Late Cretaceous and Tertiary time, even though at these latitudes changes in MAT were muted (ca. 3°–5°C) compared to the boreal and austral regions (>6°–8°C).

Precipitation varies depending on the location of the site relative to the trade winds; its position along the slopes, with mid-altitude habitats receiving the most rainfall as winds rise and lose moisture upon cooling; and on the extent of adjacent uplands, which, if sufficiently high, cast a rainshadow to the leeward side. At Hill Gardens in Jamaica the MAP is 2367 mm, and at Cienfuegos in Cuba it is 98.3 mm. The wettest months typically are May and June and the dry months December through February. The survival of dispersed propagules depends in part on the climates and soils at the landing site, but the lengthy time available suggests a high probability of eventual success, similar to that in overcoming the relatively short distances.

Geology

Geologic development of the Antilles began with the initial opening of the Caribbean Basin in the Late Cretaceous and subsequent formation of subduction zones along the margins that defined the Caribbean Plate. Superimposed on these processes was the north-to-south jostling between North America and South America to create and modify lands within the basin. It was a complex process that determined the lay of the land and, along with many concurrent climatic and biological factors, influenced the migration, distribution, and geographic affinities of organisms across the region. In the Jurassic to Early Cretaceous, 200–165 Ma, northwestern South America lay appressed to North America in the vicinity of the present Gulf of Mexico (Ross and Scotese, 1989; II, figs. 2.16, 2.17). There were subplates or blocks, such as the Yucatán (Maya) Block that subsequently drifted to its current position along southeastern Mexico and the Chortis Block, which became affixed as the southwestern corner of Mexico. Parts of the ancient sea floor were temporarily emergent (evaporates of Jurassic age 130 Ma are known), but these were subsequently inundated along with many later exposed peaks.

As northwestern South America moved away from southeastern North America, new sea floor was generated and new islands appeared. The Caribbean Basin proper was defined with the development of peripheral subduction zones. As plates dip into these trenches, the leading edge becomes molten at a depth of 100–150 km. The subterranean magma rises to the surface as lava along clusters of fissures called Benioff Zones. The uplifted rock, lava, associated volcanics, and erosion products create new crust inland to the subduction zone. Bordering lands constituting the proto–Greater Antilles formed along the northern edge of the developing Caribbean Basin at the Puerto Rican Trench and along the eastern edge at the Barbados Trench, as North America dipped under the Caribbean Plate. The southern Lesser Antilles appeared along the northern coast of Venezuela as the Caribbean Plate dipped under South America, while the western margin was defined with subduction of the Nazca portion of the Pacific Plate beneath the isthmian region. The time interval for this activity was around the early Eocene:

> The existing Greater Antillean islands, as islands, are no older than middle Eocene. Earlier islands may have existed, but it is not likely that they remained as such (e.g., as subaerial entities) due to repeated transgressions, subsidence, and (not incidentally) the K/T bolide impact and associated

mega-tsunamis. Accordingly, we infer that the on island lineages forming the existing Antillean fauna [and flora] must all be younger than the middle Eocene. (Iturralde-Vinent and MacPhee, 1999, 3)

The location of the Caribbean Basin over time relative to North America and South America is a function of the different rates of plate movement (Bowman et al., s.d.). All three plates are moving westward from the Mid-Atlantic Ridge at ca. 2 cm/yr. This means that since the middle Eocene (55 Ma) they have been displaced ca. 1100 km from the Mid-Atlantic Ridge. However, movement of the Caribbean Plate has ranged from stable to less than ca. 2.0 cm/yr, while the others have drifted between ca. 2.0 and 3.0 cm/yr, which means that the Caribbean Basin slowly attained an easterly position relative to North and South America.

Several models are available describing the extent to which individual islands within the basin have moved since their origin. There is the "fixist" type (Morris et al., 1990; Meschede and Frisch, 1998), suggesting little movement since the middle Eocene, or the "mobilist" type (Anderson et al., 1985; Pindell, 1994; Pindell and Barrett, 1990), proposing greater movement of the islands away from their site of origin. The mobilist view is most easily reconcilable with plate tectonics, and there are two versions. One is based on on-land evidence from along the Polochic-Montagua fault in Guatemala, and this version envisions origin in the vicinity of the Yucatán Peninsula with eastward transport of 130 km. The other utilizes offshore data from the Cayman Trough and proposes transport of 1000 km from near present-day Central America or beyond. As noted by Donnelly et al. (1990a), the problem of distance remains unresolved.

Beginning in the 1980s the complexity of the geology of the Antilles and the Caribbean Basin was balanced by information coming from US and later international mega-oceanographic projects (Emiliani, 1982). The first drilling operation in the deep ocean was on the Nicaraguan Rise. This was later expanded into the DSDP, with its initial site in the Venezuelan Basin, and continued as the ODP until September 2005, when the last cores were retrieved from near Bermuda. The multinational effort is now called the IODP. However, emphasis has shifted geographically, and some investigations and associated websites are no longer actively supported, although existing information is still available (for example, http://www.odsn.de). Further large-scale projects are not likely, and new data will have to come from smaller projects having some economic import and from reanalysis of existing cores. For example, a recent study of the origin of the Aves Ridge and the Netherland Venezuelan Antilles is based on samples collected in

1968 (Neill et al., 2011, 334). The future accessibility of even the existing cores is uncertain due to the expense of storage and maintenance (see chapter 6, below, on Antarctica). These trends leave unresolved several topics of biogeographic interest in the Caribbean (e.g., the existence of continuous versus discontinuous land; extent of island transport). Donnelly (1989b, 110) observes that

> to a geologist it is relatively unimportant if an island arc is broadly emergent or mainly submergent, whereas the biologist is vitally interested in that information. On the other hand, movement of the "Caribbean Plate" eastward from putative Pacific location is a question close to the hearts of geologists, but probably of relatively little interest to biologists. A "proto-Antillean ace" may be of value to biologists regardless of whether it occupies the position of present Central America or the present Antilles as long as connections between North and South America are the major issue.

With regard to the continuity of land surfaces, no tectonic model at present includes an extensive landspan. As suggested by Hedges (2006), it is worthwhile to keep separate the issues of geology and biology. This is what may be called the first principle of historical biogeography: namely, that in the final analysis biological interpretations must adjust to the geological evidence, allowing for some interplay and uncertainties in the initial studies. The same is true for other areas like archaeology, where the age of deposits, artifacts, and splicing together of chronological developments in human history have to be ceded to geology and stratigraphy (e.g., Meltzer, 2015, 11). In my opinion, the geologic evidence is most compatible with the presence of discontinuous land throughout the Antilles during the Late Cretaceous and Cenozoic. Additional evidence for this view will be presented later in the section on the paleobotany of the region.

In terms of the extent or distance of island transport, the lack of conclusive evidence makes only tentative assessments possible. Considering the movement of North and South America away from the Mid-Atlantic Ridge, and of North America away from South America to form the Caribbean Basin; the critical role of subduction in forming the islands; the different rates of movements of the three plates; and the formation of existing islands (e.g., Cuba) by fusion of separate pieces of land, this collectively suggests greatest compatibility with the mobilist model depicting transport from the vicinity of the subduction zone beneath present-day Central America. In other words, for proponents of both a landspan and a "fixist" position of the Antilles, the ball is in their court.

A topic of interest to both geologists and biologists, and relevant to the considerations of movement and distances, is whether a particular island is a coherent whole, a union of nearby fragments, or a composite island that has been pieced together from parts coming in from far afield—the issue of terranes, as in Beringia, but on a smaller scale. Cuba is an example (Cotilla Rodriguez et al., 2007; see Fig. 4.4). Parts of northern Cuba came from the North American Plate with subduction along the Puerto Rican Trench. The Pinar Fault zone extends from the southwest to the northeast across western Cuba and represents two fragments that collided in the late Paleocene–early Eocene (Gordon et al., 1997). In central and eastern Cuba the Canto Fault Zone marks contact in the early to middle Eocene between central Cuba and eastern Cuba with Hispaniola and Puerto Rico attached. When these lands began colliding with the stable Bahamas Platform, the present-day Cuban part slowed first, resulting in the detachment of Hispaniola with Puerto Rico. Puerto Rico separated from Hispaniola between the Oligocene–early Miocene, then western Hispaniola from Cuba in the early to middle Miocene (20–25 Ma). The peninsula of southern Hispan-

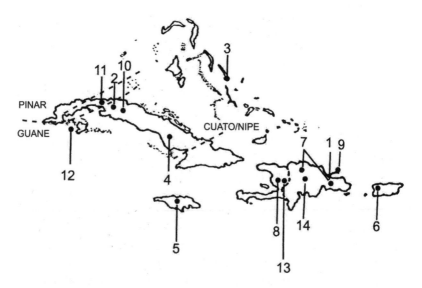

4.4. Location of fossil floras along the Antillean Land Bridge showing the Pinar-Guane and Cuato-Nipe fault zones of Cuba. The peninsula extending from southwestern Haiti was adjoined in the late Miocene. The other islands are mostly of single-unit construction. Cretaceous: 1 = Los Ranchos, 2 = San Andrián, 3 = Cat Island; Eocene: 4 = Saramaguacán, 5 = Guys Hill. Oligocene: 6 = San Sebastian. Miocene: 7 = Dominican Amber, 8 = Artibonite, 9 = Sánchez. Quaternary: 10 = Chapapote, 11 = S. Las Palmas, 12 = Isla De la Juventud, 13 = Lake Miragoâne, 14 = La Nevera Sabana.

iola was added on in the middle Miocene (ca. 15 Ma), and the present configuration was achieved in the late Miocene.

Jamaica's history was different. There is no evidence that it is composed of separate land fragments or that it was ever attached to any other land mass. Also, marine limestone is abundant and terrestrial rocks meager from the middle Eocene to the late Miocene (42–10 Ma), meaning that Jamaica was mostly submerged for this interval and that the modern biota is primarily the result of introductions since that time. To complete the sequence for the Greater Antilles, the Virgin Islands were uplifted principally in the late Eocene.

The subaerial appearance of the Lesser Antilles (as opposed to earliest subduction activity in the Late Cretaceous to early Paleocene; Christeson et al., 2008; Pindell and Kennan, 2009) was in the latest Eocene, but much of that early history is buried beneath younger volcanic material. The islands are located over a subduction zone where ocean crust is dipping under the Caribbean Plate, and it is one of the few active arcs in the Atlantic Ocean. Barbados is geographically part of the Lesser Antilles, but geologically it is separated by the Tobago Trough. The Barbados Ridge began forming ca. 50 Ma, but Barbados as a sustained island appeared only ca. 1 Ma. The southern line of islands along the northern coast of South America are middle Eocene to early Miocene in age (Speed and Keller, 1993; Speed et al., 1993), and some Neogene magmas have erupted as late as ca. 12 Ma. The distance between Trinidad-Togabo across the Gulf of Paria to South America is only 11 km, the average depth is ca. 37 m (maximum 274 m), and the islands were attached to the mainland during each of the 18–20 low sea-level stands of the late Pliocene/Pleistocene.

Subduction of the southern Caribbean Plate margin beneath South America began in the Cretaceous (ca. 88 Ma; Wright and Wyld, 2010; van der Lelij et al., 2010; see review by Neill et al., 2011) with subaerial appearance of the Venezuelan and Netherland Antilles probably in the Eocene during an interval designated as the second deformation phase or F2 (55 Ma; Beardsley and Lallemant, 2007) and corresponding with other uplift events in the Caribbean. The Quaternary history is comparable to that of adjacent Trinidad and Tobago. The revelation that Cuba is an amalgamation of terranes and smaller pieces joined over time must be taken into account when interpreting the vegetation/faunal/climatic history of the island based on fossil floras collected from separate localities. The floras may have been deposited when the terranes were in slightly different zones of longitude (Fig. 4.4). In contrast, formation of most other islands has involved west-to-east transport of single units of land through parallel zones of latitude, be it 130 km or 1000 km. Histories reconstructed from floras at different sites on these latter islands can confidently be read as a sequence of autochthonous events.

Modern Vegetation

Few studies have been made on the geographic affinities of the Antilles vegetation as a whole, and all of them predate modern phylogenetic classifications. The location of the islands on an early trade route between the New World and the Old World requires another cautionary note:

> The voyages of Captain William Bligh for the breadfruit brought plants from Europe to the Southern Hemisphere and from the South Pacific islands to first St. Vincent and then Jamaica. The French used the island of Mauritius as a transplant point for East Indian plants destined for the French colonies and botanic gardens in the Antilles. An interest in establishing plantations of rubber-producing species led to the introduction of such plants from Central America, South America, Africa, Madagascar and the Pacific tropics. Many of the genera and species involved are established and not distinctive in the existing vegetation of the Antilles. (Howard, 1973, 16)

The long history of naturalization of cultivars, other exotics associated with them, and disturbance (Venter et al., 2016) makes it difficult to read the use of the Antillean Land Bridge from the modern vegetation. Another factor that complicates associating the movement of species and vegetation types with geologic and climatic events at specific times, especially in the absence of an adequate fossil record, is long-distance transport that can occur randomly over long intervals and short distances. In the Antilles especially this transport is facilitated by wind and wind-driven ocean currents (Figs. 4.5, 7.1). Gunn and Dennis (1976) recognize 15 currents worldwide (Fig. 4.5) and discuss them in relation to the dispersal of propagules:

> Seed and fruits which drift to beaches have been collected by man for reasons as varied as there are collectors. For the romantic, the disseminules are messengers from exotic lands; for men of the sea, they represent victory over an ancient foe; for the superstitious, they represent gifts from the gods; and for the botanist they are the end product of a plant dispersal mechanism. (3)

The currents most relevant to the Antilles are the Equatorial and Equatorial Counter Currents and the Gulf Stream. This vector is selective, and less

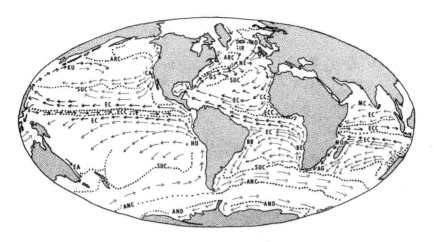

4.5. Ocean currents of the world. Darker arrows indicate velocities greater than 36 nautical miles per day. Currents: ANC, AND = Antarctic Convergence, Divergence; AG = Argulhas; ARC = Arctic Convergence; BE = Benguela; BR = Brazil; CA = California; CC = Canary; EA = East Australia; EC = Equatorial; ECC = Equatorial Counter; GS = Gulf Stream; HU = Humboldt; IR = Irminger; KU = Kuroshio; MC = Monsoon; MO = Mozambique; NE = Northeast Atlantic; NO = Norwegian. From Gunn and Dennis, 1976, Demeter Press. Used with permission.

than 1% of tropical seeds remain viable in sea water for one month; however, the list includes some prominent components of the vegetation (fide Gunn and Dennis, 1976, appendix). In the following list an asterisk (*) indicates the genus has been reported as a fossil (II, appendix 2.2): *Abuta, *Aleurites, Andira, *Annona, *Avicennia, Barringtonia, Bertholletia, Blighia, Caesalpinia, Calatola, Calocarpum, *Clophyllum, Canarium, *Canavalia, Carapa, *Caryocar, *Cassia, Cerbera, *Chrysobalanus, *Chrysophyllum, *Conocarpus, *Crescentia, Crinum, *Crudia, *Dalbergia, Dendrosicus, *Dioclea, Elaeodendron, Enallagma, Entada, *Enterolobium, *Erythrina, Fevillea, *Ficus, Genipa, Grias, *Guazuma, *Hernandia, Hevea, Hippomene, *Hymenaea, *Ipomoea, *Lecythis, *Machaerium, Mamea, Manicaria, Mastichodendron, *Merremia, Mora, *Mucuna, Myristica, Omphalea, Orbignya, Oxyrhynchus, *Pelliciera, Palmae (including Nypa), *Peltophorum, Phytelephas, *Pinus, *Podocarpus, Prioria, *Prunus, *Psidium, *Pterocarpus, *Quercus, *Rhizophora, *Sacoglottis, *Sapindus, *Smilax, *Spondias, *Sterculia, *Swietenia, *Terminalia, *Tournefortia, *Vantanea, and Vinga.

The most recent analysis of the vegetation is by Howard (1973; see also table 4.2 in online supplementary materials). He classifies the plants of the Antilles into 3 formations and 17 subcategories (see online table 4.3). Of particular interest are the geographic affinities within the flora, as this can be an indication of the source of the taxa and the direction of migration. The fossil record of the Antillean flora is given in online appendix 4, and from this data a history of the vegetation can be estimated. Determining whether an organism likely came from the north or south, and the minimum time of its arrival, is where paleobotany can partner with studies on the modern vegetation by providing the oldest known records.

The first pattern is Pan-Caribbean, meaning the plant is found on all the principal islands and in adjacent Mexico, Central America, and northern South America. Among the representative genera are *Podocarpus, Amyris, Anoda, Apteria, Ardesia, Beilschmiedia, *Bumelia, Burmannia, Cedrela, Celtis, Chlorophora, Clidemia, Columnea, Conostegia, Cornutia, Cyrilla, Daphnopsis, Erythroxylum, *Exostema, Graffeneriedia, *Guarea, Guazuma, *Guettarda, Hernandia, Hillia, Hippocratea, Hirtella, Homalium, Jacquinia, Leiphamos, Linociera, Manettia, *Marcgravia, Miconia, *Myrcia, Ocotea, Ossaea, Ouratea, Oxandra, Picramnia, Rondeletia, Salvia, Sapium, Schleglia, Schopfia, Struchium, *Tobouchina, Trichilia, and Weinmannia.

The distribution of the second category suggests some of these may have come from the west at a later date because they are present in the Greater Antilles but have not reached the Lesser Antilles. They are *Cycas (fossils as Cycadopites), Acisanthera, Alchornea, Angelonia, Atriplex (as chenoams, Chenopodiaceae-Amaranthaceae undifferentiated as microfossils), Brunellia, Buchnera, Callicarpa, Cleyera, Dendropemon, Dendrophthora, Drosera, *Eichites, Ehretia, Esenbeckia, Eustoma, Forchhammeria, Forsteronia, Gerardia, Garrya, Gyminda, Helicteres, Hieronima, Hyperbanea, Isocarpha, Juglans, Laetia, Lapacea, Liabum, Licaria, Lunania, Lyonia, Machaonia, Macrocarpaea, Magnolia, Mappea, Marsdenia, Matayba, Meriania, Nama, Neea, Neurolaena, Oocarpon, Ossaea, Oxypetalum, Phaesophaerion, Phylostylon, Portlandia, Pseudolmedia, Rachicallis, Ranunculus, Ravenia, Sagittaria, Salicornia, Salmea, Samolus, Sarcostema, Schaefferia, Scybalium, Solandra, Stegnosperma, Stellaria, Thalia, Valeriana, and Ximenia. As part of this pattern but not extending as far east as Puerto Rico are Alvaradoa, Arenaria, Buddleia, Esenbeckia, Eustoma, Forchhammeria, Garrya, Kosteletzkya, Laetia, Liabum, Lunnania, Mappia, Mascagnia, Neeurolaena, Oocarpon, Phyllostylon, Ranunculus, Samolus, Stegnosperma, Tapura, Valesia, and Zuelania.

Other genera of the West Indies are not known from South America, at least at the time of Howard's (1973) study, suggesting they could have come from the north: Borrichia, Carpodiptera, Dipholis, Ehretia, Erithallis, Ernodea, Exotheca, Gyminda, Krugiodendron, Lyonia, Mastichodendron, Phialanthus, Samyda, Stegnosperma, Tetrazygia, Tropidia, and Urechites. The last category are genera found in the Lesser Antilles that do not extend northward in the Caribbean beyond Puerto Rico and may have migrated from South America: *Amanoa, Aniba, Browallia, Calolisianthus, Centropogon, Chomelia, Chrysochlamys, Codonanthe, Coutoubea, Dacryodes, Drymonia, Dussia, Enicostema, Gonolobus, Ischnosiphon, Licania, Malanea, Nautilocalyx, Norantea, Petra, Prestonia, Pterolepis, Richeria, Rolandra, Ruyschia, Siparuna, Stylogyne, Tovomita,* and *Wittmackia.*

Further afield there are genera in the Antilles known from Africa and/ or Madagascar (and elsewhere). In the following list figures following the names are the number of species in the genus and the number recognized in the Antilles at the time: *Amonoa* 7/1, *Andira* 75/1, *Bertiera* 30/2, *Brachypteris* 3/2, *Chlorophora* 12/1, *Caperonia* 60/3, *Carpodiptera* 80/4, *Cassipourea* 80/7, *Chrysobalanus* 2/2, *Conocarpus* 2/1, *Copaifera* 25/1, *Eichhornia* 7/5, *Genlisea* 15/1, *Guarea* 150/7, *Hesteria* 50/1, *Heteropteris* 100/9, *Hirtella* 95/3, *Laguncularia* 2/1, *Machaerium* (as *Dendrocarpus*) 150/1, *Malouetia* 25/1, *Microtea* 10/2, *Maraca* 10/1, *Parkinsonia* 2/1, *Paullinia* 180/8, *Piriqueta* 20/5, *Quasia* 40/1, *Renealmia* 75/10, *Sabicea* 130/3, *Salvia* 31/12, *Schultesia* 20/2, *Struchium* (as *Sparganium*) 1/1, *Symphonia* 21/1, *Talinum* 50/2, *Tapura* 20/4, *Thalia* 11/1, *Trichilia* 300/20, and *Voyria* 15/5. Germeraad et al. (1968, 275–76) report the African and Indo-Malaysian genus *Ctenolophon* from the Paleocene of northern South America, and Jaramillo and Dilcher (2001, table 6) provide a list of the fossil sporomorphs of the Gulf Coast/ Caribbean–Central American–Colombian region that are also found in Africa. Assessment of the occurrences listed here requires modern systematic study and, as noted by Howard as far back as 1973 (35), some taxonomic judgment.

Genera in the Antilles and in Asia are *Callicarpa* 140/17, *Capsicum* 50/2, *Cedrela* 6/1, *Dendropanax* (as *Gilibertia*) 60/13, *Helicteres* 60/3, *Icanthus* 26/5, *Meliosma* 100/7, *Mitreola* 6/2, *Nelumbo* 2/1, *Sapidus* 13/1, *Schopfia* 35/7, *Sloanea* 120/13, *Symplocos* 350/21, *Talauma* 5/4, *Turpinia* 40/4, and *Xylosma* 100/15. For the most part there is insufficient information on the present relationships and fossil occurrences of the African and Asian components to explain their presence and movement in the Antilles. Multiple explanations are also available for Antillean plants from elsewhere and for documenting their use of the ALB during the Late Cretaceous and Cenozoic. An additional factor needed to appraise the various possibilities is an awareness of the earliest human presence in the Antilles, as noted by Howard (1973).

Indigenous People

For the nonspecialist, the primary source of information on early people in the Caribbean until recently has been the observations of Spanish explorers, who grouped them into the peaceful Arawak and the warlike and cannibalistic Taino. This simplified stereotype was reinforced in the novel *Caribbean* by James Michener (1988). Until recently results from later research were available principally through papers on individual islands, cultures, and time periods. Then in rapid-fire succession there appeared *The Oxford Handbook of Mesoamerican Archaeology* (Nichols and Pool, 2012), *The Oxford Handbook of Caribbean Archaeology* (Keegan, Hofman, and Rodríguez Ramos, 2013), and the *Encyclopedia of Caribbean Archaeology* (Reid and Gilmore III, 2014). These afforded a new perspective on the nature of pre-Columbian people and their environments. One important contribution has been to identify and partially clarify the almost hopeless confusion around the previous classification of the inhabitants into the Ciboney (a stone-age people); the Arawak from South America, who supposedly introduced tools and agriculture into the Antilles; and the Carib, who arrived last from South America and were in the Lesser Antilles at the time of the Spanish arrival. This tripartite characterization is now regarded as incorrect on almost all counts. The first people broadly designated as Archaic are now considered to have arrived in the Caribbean in the early to mid-Holocene ca. 5000–6000 BP, based on pollen, phytoliths, charcoal, and sediment chemistry (Siegel et al., 2015; 4000 BCE, Cooper, 2013, 48). Then, "migrating eastward from the coast of Belize-Honduras [Yucatán Peninsula] into the Caribbean, [they] discovered and settled the then uninhabited Greater Antilles" (Granberry, 2013, 62). Among these were the Casimiroid people speaking the Ciguayo language who did not use stone tools, make pottery, or practice agriculture. Following the sequence proposed by Granberry (2013; but see Reid and Gilmore III, 2014, xvii), this first migration was followed by Saladoid and post-Saladoid groups including the Macoris people in ca. 2000 BCE. They were marginal users of bone and shell as well as stone tools, makers of pottery, and they practiced limited agriculture. Their language is related to Warao of northern Venezuela, the Orinoco Delta, and the Maracaibo Basin, and they are known from the Greater Antilles and the Leeward Islands, so they presumably migrated from South America, although no record of them is known from the other Lesser Antilles. In the earliest stages of the third migration (400 BCE to 1 CE) there were the pre-Taino or Cedrosan people entering from Trinidad, followed by the Taino who by the time of Columbus occupied all the Antilles except for western Cuba: "Certainly from A.D. 1 until about A.D. 500 Taino was the *lingua franca* understood and used by all throughout both the Greater and Lesser Antilles, regardless of their native tongue" (Granberry, 2013, 64). A second wave arrived from the Orinoco Delta and Trinidad region in 500–1000 CE. These were Arawak-speakers called the Barrancoid people. They settled and remained in the Windward Islands. The last (1450 CE to ca. 1600 CE) were the Carib people from the Guianas speaking the Karina language. According to the summary chart prepared by Keegan et al. (2013, fig. 1.5), the migratory sequence was first from the north and west into western and central Cuba, Haiti, and the Dominican Republic, and then later from the north and south into the Lesser Antilles onto St. Kitts ca. 2500 BCE, followed by Antigua (2000–1000 BCE), Martinique (750–600 BCE), and St. Vincent (500–250 BCE). In summary, with regard to the early migrations, "standard models of initial human colonization of the Caribbean indicate two independent entry routes: the Yucatán Peninsula to the Greater Antilles (c. 5900 cal yr BP) and the Orinoco Valley to Trinidad (c. 8000 cal yr BP)" (Siegel et al., 2015). Keegan (pers. comm., 2016) believes there is still some question as to whether the Archaic peoples came from South America via Trinidad and the Lesser Antilles or were a continuation of the movement from Central America ca. 5000 BCE, because there are no Archaic sites in the Windward Islands. Although turbulent seas occur seasonally throughout the Antilles, Keegan et al. (2013, 4) conclude these did not preclude extensive interisland travel. This means that analysis of the past and present vegetation must take into account the possibility that after ca. 6000–5000 BCE plant distributions and ecosystem composition were modified by humans. They certainly were by ca. 1200 BP, as evident at the Coralie Site, Grand Turk, Bahamas. Study of wood charcoal from the site revealed several trees that are today rare or absent from Grand Turk, including wild lime, ironwood, Celastraceae, and palm trunk wood. Buttonwood, which today grows along the margin of the North Creek, was also present in the charcoal samples (Keegan and Carlson, 1997).

The place of origin and probable patterns of migration for many plant genera can now be estimated from molecular phylogenies, although for tropical plants such analyses are still relatively few. Equally important is the surprising fact that most species have never been collected and identified:

> In spite of 250 years of taxonomic classification and over 1.2 million species already cataloged in a central database, our results suggest that some 80% of existing species on Earth and 91% in the ocean, still await description. Re-

newed interest in further exploration and taxonomy is required if this significant gap in our knowledge of life on Earth is to be closed. (Mora et al., 2011)

That is, 80% of existing species are absent or unrecognized in collections used as a source of comparative reference material for identifying fossils (in addition to the taxa that have gone extinct since the time of the fossil biota), and are not represented in the database used for estimating and calibrating molecular phylogenies. Complicating matters is the 21 years between the collection of a new species and its publication (Fontaine et al., 2012). Hence, the fragmentary nature of both the modern and fossil records and from everywhere, including the Antilles.

Utilization of the Antillean Land Bridge

Reconstruction of the historical biogeography of a region benefits from the presence of widely distributed fossil floras of approximately the same age. From these, the broad regional vegetation and environments for particular points in time can be established. It also helps if, within a region, fossil floras of different ages are available in a dated sequence to show changes through time. The most precise analysis is possible when the age of the floras is such that most of the fossils can be referred to modern genera (for plants this is generally after about the early/middle Eocene); it facilitates comparisons if some floras have been studied for both macrofossils and microfossils to provide an extended inventory; and if they have been studied with the same primary intent—biological, where referral to modern taxa is foremost, versus geological (stratigraphic correlation) when an artificial system is sufficient. That is asking a lot of paleobotanical and paleopalynological investigations, especially for the tropics, and to acquire the information fossil floras often must be spliced together from different islands (Fig. 4.4). For example, the Oligocene San Sebastian flora of Puerto Rico (Fig. 4.6) is the only assemblage of Tertiary age known for the island, but it has been studied for both macrofossils (Hollick, 1928) and microfossils (Graham and Jarzen, 1969, Figs. 4.7–4.12). In contrast, only microfossils are known from the Saramaguacan Formation of Cuba and the Guys Hill Formation of Jamaica, but both are of middle Eocene age, allowing for better regional reconstructions. Context information is essential (see Fig. 4.13). The geological studies mentioned earlier indicate that Cuba is composed of fragments from the southern edge of the North American Plate and the northern part of the Caribbean Plate, with miscellaneous bits

4.6. Site of the Oligocene San Sebastian flora of Puerto Rico. The fossil-bearing horizon is behind the falls and is best accessed during the day season because of Schistosomiasis carried by larvae of water snails infected by parasitic flatworms. Author photograph.

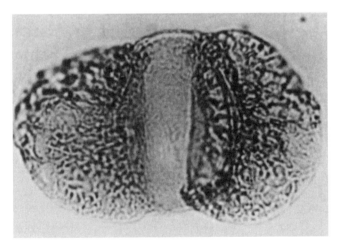

4.7. Plant microfossil from the Oligocene San Sebastian Formation of Puerto Rico: *Podocarpus*. Author photograph.

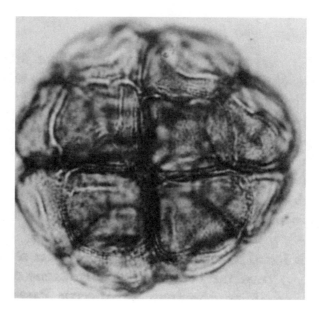

4.8. Plant microfossil from the Oligocene San Sebastian Formation of Puerto Rico: *Acacia*. Author photograph.

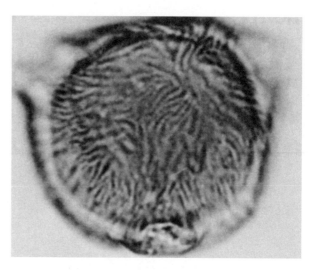

4.9. Plant microfossil from the Oligocene San Sebastian Formation of Puerto Rico: *Bursera*. Author photograph.

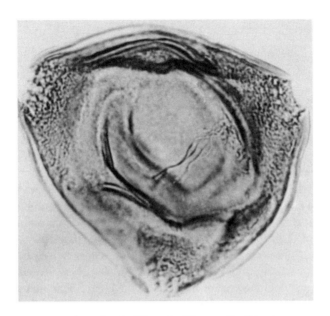

4.10. Plant microfossil from the Oligocene San Sebastian
Formation of Puerto Rico: Onagraceae (*Hauya*). Author photograph.

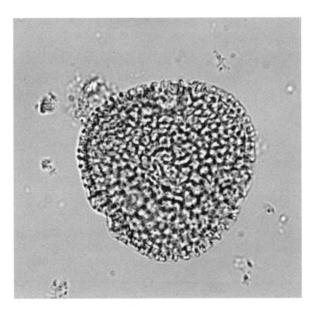

4.11. Plant microfossil from the Oligocene San Sebastian
Formation of Puerto Rico: *Bernoullia*. Author photograph.

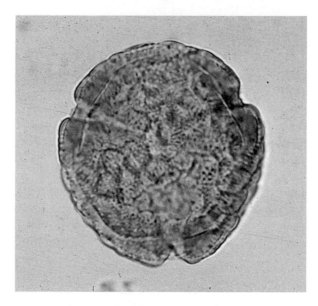

4.12. Plant microfossil from the Oligocene San Sebastian
Formation of Puerto Rico: *Pelliceria*. Author photograph.

added on from movement along the subduction zone of the Puerto Rican
Trench and from transport within the basin. Jamaica originated on the Caribbean Plate, it has never been joined to Cuba, and from the data available
(see appendix 4 in online supplementary materials), few plant genera are
found in the middle Eocene of both islands. Only *Bombacacidites* (a generalized form representing the Bombacaceae) and the unknown *Retimono-
colpites* have been found in common. In contrast, between the two islands
five genera occur only in the Jamaican flora: the fern spore *Deltoidospora*
and pollen of *Corsinipollenites* (Onagraceae), *Cupaneidites* (Sapindaceae),
Mauritidies (Arecaceae/Palmae), and *Psilatricolporites crassus* (*Pelliceria*). And
31 genera are found as fossils only in Cuba. The most parsimonious explanation of the data is that the islands were separate ,and this is consistent
with the geological evidence.

Another comparison that can be made is between the Miocene floras
of Cuba and Hispaniola. Only macrofossils are known for both islands,
the identifications are mostly from early in the century, and age estimates
range from possibly late Oligocene/early Miocene to late Miocene (see II,
305–6). This encompasses the time immediately before to just after the
separation of Hispaniola from Cuba in the early to middle Miocene (25–

20 Ma). The expectation is for some similarity between the floras because they were joined or nearly so at the time. At the generic level 8 genera are reported from the Miocene of both islands: *Bumelia, Ilex, Mimusops, Pisonia, Pithecolobium, Sapindus, Simaruba,* and *Sophora.* Given the poorly constrained information available and the many potential sources of error, this seems significant in light of the comparison between Cuba and Jamaica, which were never connected and have only two poorly defined genera in common. If more paleofloras were available, and if identifications from the older literature were reliable, additional questions of broad biogeographic interest could be addressed. One is whether there is sufficient similarity in the composition of paleofloras from the other islands to suggest they were connected in the past. The geologic evidence previously reviewed says probably not. The following 4 genera of plant microfossils are present in the Cretaceous of Cuba and the Bahamas: *Gleichenidites, Ciatricosisporites*

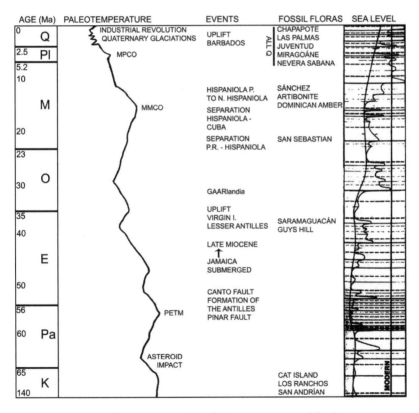

4.13. Paleotemperature and sea-level curves, events, and fossil floras for the Antillean Land Bridge. See legend to Fig. 2.5.

(*Anemia*-type spores), *Araucariaccites* (a generalized pollen type similar to *Araucaria*), and *Eucommiidites* (a gymnosperm pollen grain of unknown affinities). In the Cretaceous of Cuba and Jamaica there is *Bombacacidites* of broad Bombacaceae affinities. In the Miocene there are 8 genera of plant microfossils common to the following islands of the Greater Antilles (see online appendix 4):

	Cuba	Haiti	Dom. Rep.	Puerto Rico
Bumelia	x	x	x	
Inga	x		x	
Mimusops	x	x		
Pisonia	x		x	x
Pithecolobium	x			x
Sapindus	x			x
Simaruba	x		x	
Sophora	x			x

The total number of genera in online appendix 4 is 122, and of those approximately 110 occur on only one island. The tentative conclusion based on the equivocal paleobotanical data is that the composition of the floras argues for separation of the islands except for those (and for the time) where geological evidence indicates union (e.g., Cuba, Hispaniola, Puerto Rico until the Miocene). These are all to the north and west of the proposed Aves landspan.

A related question is whether there is more similarity between the Greater Antilles and South America in the Miocene and afterward (that is, after the proposed landspan) than before the connection. Another way of looking at the problem is that if there was such a connection, notable similarity would be expected throughout the Antilles, and between these islands and South America, and less evident endemism. That is, does explaining the problem of disjunct distributions and similarities for some vertebrates (Iturralde-Vinent and MacPhee, 1999) and plants by invoking a former landspan create a larger problem of why they (and other groups) are so distinct? Sufficient data does not exist for a critical analysis (e.g., there is no paleobotanical information available for the Lesser Antilles). However, the type of data needed can be illustrated by the following comparison of floras of the pre–proposed landspan and post–proposed landspan from northern South America with those from the Greater Antilles (see table 4.4 in the online supplementary materials). Jaramillo and Dilcher (2001) studied the palynology of middle Paleogene deposits from Central Colombia, and Germeraad et al. (1968) have summarized the Late Cretaceous and Paleogene microfossil floras of Colombia and Venezuela (mostly from the Maracaibo Basin, Trinidad, and the Guianas). Because of the age of the deposits, these studies use an artificial system of nomenclature (e.g., *Cicatricosisporites dorogensis*) with suggested affinities (e.g., similar to *Anemia/Mohria* types). Late Cretaceous and Paleogene (Eocene and Oligocene) macrofossil and microfossil floras are also known from the Greater Antilles (II, chap. 5; online appendix 4). There are post-landspan (Neogene) records for South America in Germeraad et al. (1968), and these are compared with those from the Antilles (II, chap. 5; online appendix 4) in table 4.4 in the online supplementary materials. Acknowledging the many ambiguities and shortcomings in the data, it is noted that 8 genera of plant fossils are reported in common between northern South America and the Greater Antilles in the Late Cretaceous and Paleogene: *Anemia/Moria, Alchornea, Catostemma, Chrysophyllum, Irartea, Mauritia, Pelliceria,* and *Rhisophora.* After the proposed landspan there are two, *Alnus* and *Catostemma*—that is, fewer than before the suggested continuity. In summary, the paleobotanical data are equivocal, but it appears that supporting data for a landspan will have to come from other sources.

Regarding paleobotanical evidence for the source and direction of movement for plant populations in the Greater Antilles, the evidence is also suggestive. Using analyses of the modern vegetation by Howard (1973; online table 4.2), and the fossil plant record presented in online appendix 4, there are 11 genera in Tertiary floras belonging to patterns he describes as (a) Pan-Caribbean; (b) coming from the west and extending through the Greater Antilles; (c) coming from the west but not as far as Puerto Rico; and (d) in the Antilles but not known from South America (at the time of Howard's 1973 study). These are *Podocarpus, Alchornea, Brunellia, Bumelia, Echites, Exostema, Guarea, Guettarda, Marcgravia, Pseudolmedia,* and *Trichilia.* This compares to 2 present-day South American genera that extend into the Lesser Antilles and are known in the fossil record of the Greater Antilles: *Norantea* and *Stylogyne.* To the extent these affinities are representative of the overall vegetation, there is some evidence from the modern and paleofloras that population of the Greater Antilles preferentially has been from the west (from Mexico and northern Central America) rather than from South America. This is about as far as analyses of the historical phytogeography of the Antilles can go until additional paleobotanical information becomes available.

References

Anderson, T. H., R. J. Erdlac, and M. A. Sandstrom. 1985. Late Cretaceous allochthons and post-Cretaceous strike-slip displacement along the Cuilco-Chixoy-Polochic fault, Guatemala. Tectonics 4: 453–75.

Beardsley, A. G., and H. G. Ave Lallemant. 2007. Oblique collision and accretion of the Netherlands Leeward Antilles to South America. Tectonics 26: 1–15.

Berry, R. J. 1984. Darwin was astonished. Biological Journal of the Linnean Society 21: 1–4.

Bowman, S., T. Stieglitz, and S. Jagdeo. s.d. A review of the tectonics of the deep water offshore Trinidad/Tobago: implications for hydrocarbon potential. www.spectrumasa.com.

Christeson, G. L., P. Mann, A. Escalona, and T. J. Aitken. 2008. Crustal structure of the Caribbean-northeastern South America arc-continent collision zone. Journal of Geophysical Research 113: B08104. doi: 10.1029/2007/B005373.

Cooper, J. 2013. The climatic context for pre-Columbian archaeology in the Caribbean. In W. F. Keegan, C. L. Hofman, and R. Rodríguez Ramos (eds.), The Oxford Handbook of Caribbean Archaeology. Oxford University Press, Oxford. Pp. 47–58.

Cotilla Rodríguez, M. O., H. J. Franzke, and D. Cordoba Barba. 2007. Seismicity and seismoactive faults of Cuba. Russian Geology and Geophysics 48: 505–22.

Darwin, C. http://darwin-online.org.uk/EditorialIntroductions/Chancellor_fieldNotebooks.html. [See also Darwin's field notebooks, 2009 edition by Chancellor, published by Cambridge.]

Donnelly, T. W. 1989a. Geologic history of the Caribbean and Central America. In A. W. Bally and A. R. Palmer (eds.), The Geology of North America, Vol. A: Overview. Geological Society of America, Boulder. Pp. 299–321.

———. 1989b. History of marine barriers and terrestrial connections: Caribbean paleogeographic inference from pelagic sediment analysis. In C. A. Woods (ed.), Biogeography of the West Indies: Past, Present, and Future. Sandhill Crane Press, Gainesville, FL. Pp. 103–18.

Donnelly, T. W., G. S. Horne, R. C. Finch, and E. López-Ramos. 1990a. Northern Central America: the Maya and Chortis blocks. In G. Dengo and J. E. Case (eds.), The Geology of North America, Vol. H: The Caribbean Region. Geological Society of America, Boulder. Pp. 37–76.

Donnelly, T. W., et al. (+ 9 authors). 1990b. History and tectonic setting of Caribbean magmatism. In G. Dengo and J. E. Case (eds.), The Geology of North America, Vol. H: The Caribbean Region. Geological Society of America, Boulder. Pp. 339–74.

Emiliani, C. 1982. A new global geology. In C. Emiliani (ed.), The Oceanic Lithosphere. John Wiley & Sons, New York. Pp. 1687–728.

Fontaine, B., A. Perrard, and P. Bouchet. 2012. 21 years of shelf life between discovery and description of new species. Current Biology 22 (20 November): R943–44. doi: http://dx.doi.org1016/j.cub.2012.10.029.

Germeraad, J. H., C. A. Hopping, and J. Muller. 1968. Palynology of Tertiary sediments from tropical areas. Review of Paleobotany and Palynology 6: 189–348.

Gordon, M. B., P. Mann, D. Cáceres, and R. Flores. 1997. Cenozoic tectonic history of the North American–Caribbean plate boundary in western Cuba. Journal of Geophysical Research 102: 10055–82.

Graham, A., and D. M. Jarzen. 1969. Studies in neotropical Paleobotany. I. The Oligocene communities of Puerto Rico. Annals of the Missouri Botanical Garden 56: 308–57.

Granberry, J. 2013. Indigenous languages of the Caribbean. In W. F. Keegan, C. L. Hofman, and R. Rodríguez Ramos (eds.), The Oxford Handbook of Caribbean Archaeology. Oxford University Press, Oxford. Pp. 61–69.

Gunn, C. R., and J. V. Dennis. 1976. World Guide to Tropical Drift Seeds and Fruits. Demeter Press, New York.

Hedges, S. B. 2006. Paleogeography of the Antilles and origin of West Indian terrestrial vertebrates. Annals of the Missouri Botanical Garden 93: 231–44.

Hollick, A. 1928. Paleobotany of Porto Rico. Scientific Survey of Porto Rico and the Virgin Islands 7: 177–393.

Howard, R. A. 1973. The vegetation of the Antilles. In A. Graham (ed.), Vegetation and Vegetational History of Northern Latin America. Elsevier, Amsterdam. Pp. 1–38.

Iturralde-Vinent, M. A., and R. D. E. MacPhee. 1999. Paleogeography of the Caribbean region: implications for Cenozoic biogeography. Bulletin of the American Museum of Natural History 238: 1–95.

Jaramillo, C. A., and D. L. Dilcher. 2001. Middle Paleogene palynology of central Colombia, South America: a study of pollen and spores from tropical latitudes. Paleontographica, Abt. B, 258: 87–213.

Keegan, W. E., and B. Carlson. 1997. The Coralie Site, Grand Turk. Caribbean Archaeology , Florida Museum of Natural History. https://www/flmnh.ufl.edu/caribarch/coralie.htm.

Keegan, W. E., C. L. Hofman, and R. Rodríguez Ramos (eds.). 2013. The Oxford Handbook of Caribbean Archaeology. Oxford University Press, Oxford.

Meltzer, D. J. 2015. The Great Paleolithic War: How Science Forged an Understanding of America's Ice Age Past. University of Chicago Press, Chicago.

Meschede, M., and W. Frisch. 1998. A plate-tectonic model for the Mesozoic and early Cenozoic history of the Caribbean plate. Tectonophysics 296: 269–91.

Michener, J. 1988. Caribbean. Knopf, New York.

Mora, C., D. P. Tittensor, S. Adai, A. G. B. Simpson, and B. Worm. 2011. How many species are there on Earth and in the Ocean? PLoS One Biol 9 (8): e1001127. doi: 10.1371/journal.pbio.1001127.

Morris, A. E. L., I. Taner, H. A. Meyerhoff, and A. A. Meyerhoff. 1990. Tectonic evolution of the Caribbean region: alternative hypothesis. In G. Dengo and J. E. Case (eds.), The Geology of North America, Vol. H: The Caribbean Region. Geological Society of America, Boulder. Pp. 433–57.

Neill, I., A. C. Kerr, A. R. Hastie, K.-P. Stanek, and I. L. Millar. 2011. Origin of the Aves Ridge and Dutch-Venezuelan Antilles: interaction of the Cretaceous "Great Arc" and Caribbean-Colombian oceanic plateau. Journal of the Geological Society, London, 168: 333–47.

Nichols, D. L., and C. A. Pool (eds.). 2012. The Oxford Handbook of Mesoamerican Archaeology. Oxford University Press, Oxford.

Pindell, J. L. 1994. Evolution of the Gulf of Mexico and the Caribbean. In S. K. Donovan and T. A. Jackson (eds.), Caribbean Geology: An Introduction. University of the West Indies Publisher's Association (UWIPA), Kingston. Pp. 13–39.

Pindell, J. L., and S. F. Barrett. 1990. Geological evolution of the Caribbean region: a plate tectonic perspective. In G. Dengo and J. E. Case (eds.), The Geology of North America, Vo. H: The Caribbean Region. Geological Society of America, Boulder. Pp. 405–32.

Pindell, J. L., and L. Kennan. 2009. Tectonic evolution of the Gulf of Mexico, Caribbean and northern South America in the mantle reference frame: an update. In K. H.

James, M. A. Lorente, and J. L. Pindell (eds.), The Origin and Evolution of the Caribbean Plate. Geological Society, London. Special Publication 328. Pp. 1–55.

Poinar, G. O., Jr. 2016. The first flower of Bignoniales (Lamiales): *Catalpa hispaniolae* sp. nov. in Dominican amber. Novon 25: 57–63.

Poinar, G. O., Jr., and L. Struwe. 2016. An astrid flower from neotropical mid-Tertiary amber. Nature Plants, Letters. doi: 10.1038/nplants.2016.5.

Reid, B. A., and G. Gilmore III (eds.). 2014. Encyclopedia of Caribbean Archaeology. University Press of Florida, Gainesville.

Ross, M. I., and C. R. Scotese. 1989. A hierarchical tectonic model of the Gulf of Mexico and Caribbean region. In C. R. Scotese and W. W. Sager (eds.), Mesozoic and Cenozoic Plate Reconstructions. Tectonophysics 155: 139–68.

Siegel, P. E., et al. (+ 7 authors). 2015. Paleoenvironmental evidence for first human colonization of the eastern Caribbean. Quaternary Science Reviews 129: 275–95.

Speed, R. C., and C. A. Keller. 1993. Synopsis of the geological evolution of Barbados. Journal of the Barbados Museum of Natural History 41: 113–39.

Speed, R. C., P. L. Smith-Horowitz, K. v. S. Perch-Nielsen, J. B. Saunders, and A. B. Sanfilippo. 1993. Southern Lesser Antilles arc platform: pre–late Miocene stratigraphy, structure, and tectonic evolution. Geological Society of America Special Paper 277: 1–98.

van der Lelij, R., et al. (+ 6 authors). 2010. Thermochronology and tectonics of the Leeward Antilles: evolution of the southern Caribbean plate boundary zone. Tectonics 29, TC6003. doi: 10.1029/TC002654.

Venter, O., et al. (+ 11 authors). 2016. Sixteen years of change in the global terrestrial human footprint and implications for biodiversity conservation. Nature (23 August). doi: 10.1038/ncomms12558.

Wright, J. E., and S. J. Wyld. 2010. Late Cretaceous subduction on the eastern margin of the Caribbean-Colombian oceanic Plateau: one great arc of the Caribbean? Geosphere 7: 468–93.

Additional References

Blanco-Libreros, J. F., E. A. Estrado-Urrea, R. J. Pérez-Montalvo, A. Taborda-Marin, and R. Alvarez-León. 2015. Influencia antrópica en el paisaje de las problaciones de *Pelliciera rhizophorae* (Ericales: Tetrameristaceae) más sureñias del Caribe (Turbo, Colombia). Biologia Tropical 63: 927–42.

Callaghan, R. T. 2010. Crossing the Guadeloupe Passage in the Archaic age. In S. M. Fitzpatrick and A. Ross (eds.), Island Shores, Distant Pasts: Archaeological and Biological Approaches to the pre-Columbian Settlement of the Caribbean. University Press of Florida, Gainesville. Pp. 127–47.

Chambers, K. L., and G. O. Poinar, Jr. 2010. The Dominican amber fossil *Lasiambix* (Fabaceae, Caesalpinioideae?) is a *Licania* (Chrysobalanaceae). Journal of the Research Institute of Texas 4: 217–18.

Graham, A. 1990. New angiosperm records from the Caribbean Tertiary. American Journal of Botany 77: 897–910.

———. 1990. Late Tertiary microfossil flora from the Republic of Haiti. American Journal of Botany 77: 911–26.

———. 1993. Contribution toward a Tertiary palynostratigraphy for Jamaica: the status of Tertiary paleobotanical studies in northern Latin America and preliminary analysis of the Guys Hill Member (Chapelton Formation, middle Eocene) of Jamaica. In R. M. Wright and E. Robinson (eds.), Biostratigraphy of Jamaica. Geological Society of America Memoir 182. Geological Society of America, Boulder. Pp. 443–61.

———. 2003. Historical phytogeography of the Greater Antilles. Brittonia 55: 357–83.

———. 2003. Geohistory models and Cenozoic paleoenvironments of the Caribbean region. Systematic Botany 28: 378–86.

Graham, A., D. Cozadd, A. Areces-Mallea, and N. O. Frederiksen. 2000. Studies in neotropical paleobotany. XIV. A palynoflora from the middle Eocene Saramaguacan Formation of Cuba. American Journal of Botany 87: 1526–39.

Graham, A., and D. M. Jarzen. 1969. Studies in neotropical paleobotany. I. The Oligocene communities of Puerto Rico. Annals of the Missouri Botanical Garden 56: 308–57.

Gregory, B. R. B., M. Peros, E. G. Reinhardt, and J. P. Donnelly. 2015. Middle-late Holocene Caribbean aridity inferred from foraminifera and elemental data in sediment cores from two Cuban lagoons. Palaeogeogrphy, Palaeoclimatology, Palaeoecology 426: 229–41.

Guppy, H. B. 1917. Plants, Seeds, and Currents in the West Indies and Azores. Williams and Norgate, London.

Hearty, P. J., P. Kinder, H. Cheng, and R. L. Edwards. 1999. A +20 m middle Pleistocene sea-level highstand (Bermuda and the Bahamas) due to partial collapse of Antarctic ice. Geology 27: 375–78.

Hofman, C. L., and A. J. Bright. 2010. Towards a pan-Caribbean perspective of pre-colonial mobility and exchange. Journal of Caribbean Archaeology Special Publication No. 3, Mobility and Exchange from a pan-Caribbean Perspective: i–iii (Preface).

Hofman, C. L., A. J. Bright, and R. Rodríguez Ramos. 2010. Crossing the Caribbean Sea: towards a holistic view of pre-Colonial mobility and exchange. Journal of Caribbean Archaeology Special Publication No. 3, Mobility and Exchange from a pan-Caribbean Perspective: 1–18.

Martinelli, J. 2016. As the world warms, how do we decide when a plant is native? Yale e360. Posted 19 April. [This example deals with Magnolia tripetala transported from the south, escaped, naturalized, and long considered native in the deciduous forest of northeastern North America.]

Martínez, M. L., and N. P. Psuty (eds.). 2004. Coastal Dunes: Ecology and Conservation. Springer, Berlin. [Deals with dunes along Lake Erie; Maun (p. 120) mentions examples of plants carried by flotsam and jetsam.]

Rainbird, P. 2007. The Archaeology of Islands. Cambridge University Press, Cambridge.

Ricklefs, R., and E. Bermingham. 2008. The West Indies as a laboratory of biogeography and evolution. Philosophical Transactions Royal Society London B (Biological Sciences) 363: 2393–413.

Rouse, I. 1992. The Tainos: Rise and Decline of the People Who Greeted Columbus. Yale University Press, New Haven.

Salas-Leiva, D. E., et al. (+ 8 authors). 2017. Shifting Quaternary migration patterns in the Bahamian archipelago: evidence from the Zamia pumila complex at the northern limits of the Caribbean island biodiversity hotspot. American Journal of Botany 104: 757–71.

Watts, D. 1987. The West Indies: Patterns of Development, Culture, and Environmental Change since 1492. Vol. 8. Cambridge University Press, Cambridge.

Wilson, S. M. 2007. The Archaeology of the Caribbean. Cambridge University Press, Cambridge.

FIVE

Central American Land Bridge

This avenue between the New World continents in its broadest definition includes the area from northern South America (Colombia and Venezuela), through Central America to Mesoamerica as far north as the Mexican state of Veracruz (latitude 5°N to 20°N; Fig. 5.1), which allows inclusion of important fossil floras from northern Colombia (Cerrejón flora; Jaramillo and Dilcher, 2001), Venezuela, Panama, Costa Rica, and Guatemala to Veracruz (Paraje Solo flora; Graham, 1976a). In a narrower definition, the connection is often referred to as the Panama Land Bridge, but that is too restrictive for present purposes. The limits of the CALB are fluid and even more so, for two reasons, when past history is incorporated: first, many islands now part of the Greater Antilles had their origin in the Isthmian region and were transported eastward during the Late Cretaceous and Cenozoic; second, throughout this interval some land was always available along both routes. The two connections are physically separated, but biogeographically they are related, and they are considered here as the equatorial land bridges.

South and North of the CALB

Immediately south of the CALB in northern South America, there is considerable biotic, topographic, climatic, and habitat diversity. The Northern Andes divide in Colombia into the Cordillera Occidental, Central, and Oriental. The highest peaks are in the Cordillera Central with the Nevado del Huila at 5750 m and Mount Ruiz and Tolima at 5400 m. On the Guajira Peninsula to the northwest of Lago Maracaibo, the Sierra Nevada de Santa Marta is the highest coastal mountain system in the world, reaching an elevation of 5700 m, and in the Cordillera Oriental there is a broad pla-

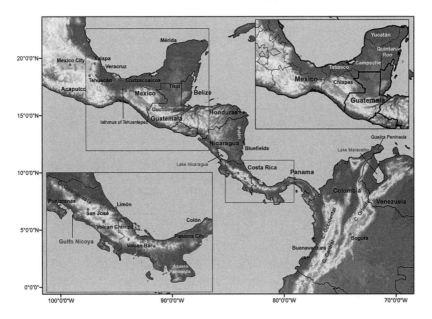

5.1. Index map of place names and physiographic
features for the Central American Land Bridge.

teau called the Sabana de Bogotá with an average elevation of 2550 m. In contrast, in northwestern Colombia there are swamps and marshes at sea level that continue into Panama as the lowlands of the Darién Province. The MAT is ca. 26°C at sea level and decreases to 14.5°C in the highlands around Bogotá. Rainfall is as diverse as the topography and temperature. The MAP in the Chocó on the northwest coast of South America is one of the highest on Earth: "The city of Buenaventura receives 4572 mm of annual rainfall . . . and the isolated swampy region northward from Buenaventura receives nearly perpetual rain (more than 7000 mm/yr to nearly 11,000 mm/yr locally), there are constant clouds, and relative humidity is more than 85%" (II, 36). The driest area is on the Guajira Peninsula with less than 300 mm of rainfall per year.

The vegetation of northern South America reflects this extensive physiographic, climatic, and habitat diversity. The principal plant communities and representative components are given below (an asterisk [*] indicates the genus has been reported in the fossil record of the CALB and vicinity in II [chaps. 4–8, appendices], and additionally from Panama by Jaramillo et al. [2014, tables 1, 4, 6]):

Desert (Guajira Peninsula)—*Cercidium,* *Prosopis,* and cacti such as *Mammillaria* and *Opuntia* are present. *Opuntia* is known from middens in the Quaternary of Chile, and *Prosopis* as pollen in the Quaternary of Argentina/Tierra del Fuego (II, table 7.6).

Shrubland/chaparral-woodland-savanna (neotropical dry forest; DRYFLOR, 2016)—*Agave, Aristida, Bromelia, Bulnesia, Capparis, Castela, Cercidium, Chrysobalanus, Copaifera, Curatella, Lippia,* *Opuntia, Pereskia, Pithecellobium, Platymiscium, Pterocarpus,* and *Tabebuia.* Above ca. 1400 m in the state of Bolivar, southeastern Venezuela, the Gran Sabana consists of scattered trees and shrubs with an understory of grasses and other herbaceous plants. The trees *Maximiliana* and *Vochysia* and the shrubs *Clusia* and *Ouratea* extend northward through Central America, while others, like *Vantanea, Bonneti, Euphronia, Gongylolepis, Thibaudia,* and *Vantanea,* are restricted to South America. Grasses include *Axonopus* and *Trachypogon,* and the prominent herbs are *Bulbostylis, Hypolytrum, Palicourea,* and *Scleria.* All of the herbs are found to the north in Central America. *Clusia* and *Vochysia* have been reported as fossils, but not until the Miocene/Pliocene and both in South America (Brazil) and Trinidad (II, app. 2). Based on these records, the movement of dry upland species appears to have been selective and to have come late, which is consistent with the timing of scattered upland habitats, drying climatic trends associated with cooling temperatures, and continuity of land surfaces. Tracing the history of this community, like the desert, is complicated because clearing of adjacent forests results in expansion of disturbance vegetation that simulates natural desert, shrubland, chaparral, and savannas; dry environments are not suitable for the preservation of most plant fossils; and some of the principal components of dry vegetation, namely, many of the grasses, cannot be differentiated in the fossil record from nonarid genera (II, 186–88). The history of dry communities often must be estimated or inferred from edaphic evidence, the presence of highlands potentially creating dry habitats to the leeward side, the development of cold coastal currents, and within the context of broad cooling temperatures and drying precipitation trends (that is, after the MMCO). Additional data are needed, but generally dry vegetation as a community (as opposed to individual elements) becomes most recognizable and widespread in the Neogene.

Lowland neotropical rainforest—In northern South America this community is found at elevations between ca. 50 and 600 m along the border of Venezuela and Colombia, and in the Amazon Basin of Brazil. It is the most diverse of the New World vegetation, and in the fossil record it is indicative of low elevations; high and relatively uniform MAPs of 3700–5000 mm to as much as 11,000 mm at special places, such as around Buenaventura, Colombia; and a MAT of at least 20°C (average 24°C) with the coldest month 18°C or warmer. The driest three months of the year (25%) receive a minimum of 18% of the rainfall (there is no pronounced dry season), and there is no more than a 3°C difference between the coldest and warmest months. The Bombacaceae is an important family, although reading its fossil record is complicated because *Bombax* is frequently named in the older literature but modern representatives are now treated as an exclusively African genus. Fossil pollen is mostly referred to the artificial genus *Bombacacidites*, which may include more than one genus of the family. Other prominent plants are *Carapa*, *Cedrela*, *Ochroma*, *Parkia*, *Scheelea*, and *Warszewiczia*; and in swampy forests *Bombax*, *Cecropia*, *Ceiba*, *Cordia*, *Inga*, *Macharium*, *Manicaria*, *Montricharida*, *Ochroma*, *Pachira*, *Pterocarpus*, *Symphonia*, and *Terminalia*, with locally dominant palms such as *Euterpe* and *Mauritia* (see II, 597). Bordering the rainforest in coastal brackish-water environments are mangroves of *Avicennia*, *Connocarpus*, *Laguncularia*, *Rhizophora*, and the herbaceous/shrub *Crenea*, which have been reported as fossils in northern South America as far back as the Eocene (*Crenea* to the Oligo-Miocene). The community (as opposed to individual component species) and the habitats and climates implied are ancient and probably developed prominently with onset of the PETM.

Lower to upper montane broad-leaved (deciduous) forest—On adjacent and better drained slopes is a transitional community that at lower elevations grades into either moist rainforest or drier shrubland/chaparral-woodland-savanna and in the highlands into tropical alpine tundra (páramo). It is a highly diverse community (Kreft and Jetz, 2007). In fact, biodiversity on the lower slopes of the east-facing Northern and Central Andes may be as high as, and at places even higher than, in the lowlands. However, this diversity refers only to trees and shrubs, it can vary with plot size, and there is a greater number of habitats along the slopes than in the lowlands. Nonetheless, with these caveats, the models and analyses are important for understanding the vertical and latitudinal patterns of diversity and disjunctions (Jablonski et al., 2013; Renner and Givnish, 2004; Wiens and Donoghue, 2004) and for identifying/predicting/prioritizing sites for collection and preservation (Distler et al., 2009; Jiménez et al., 2009; Tello et al., 2015). Lowland rainforest (and habitats) becomes restricted toward the north, while lower to upper montane vegetation is prominent across Central America and into Mesoamerica.

In northern South America prominent genera of the montane broad-leaved forest include *Podocarpus, Bocconia, Bonnetia, Bravaisia, Brocchinia, Brunellia, *Calophyllum, Chrysophyllum, Clethra,*Clusia, Euterpe, *Ficus, Guettarda, Gyranthera, Hedyosmum, Heliamphora, *Heliconia, Heliocarpus, Laplacea, *Lecythis, Lippia, *Lonchocarpus, Meliosma, *Miconia, Murraya, *Ocotea, *Oreropanax, Palicourea, *Persea, Protium, *Psychotria, Pterocarpus, Saurauia, Sloanea, Stegolepis,*Styrax, *Symplocos, *Tabebuia, Talauma, *Tibouchina, Toxicodendron, *Turpinia, Tyleria, and *Weinmannia. There are also broad-leaved deciduous elements from the north, such as *Alnus, *Ilex, *Myrica,* Prunus, *Quercus, and *Salix. Those of the former group marked with an asterisk (*) are known from the Tertiary of South America and those of the latter group appear later in the Quaternary, with the exception of *Ilex* with its distinctive pollen grains, which are widespread and known back into the Paleogene under the names *Ilex* and *Ilexpollenites*. Pollination in *Ilex* is by bees or flies, and the bright red, fleshy fruits are dispersed by birds (Arrieta and Suárez, 2005; Obeso and Fernández-Calvo, 2002; Stiles, 1980). In *Podocarpus* the reduced seed-bearing cone scale is bright red and berry-like and also eaten by birds (Wilson et al., 1996). Both are examples of the numerous distributions influenced but not determined by the presence of land bridges.

In this connection it is noteworthy that these genera have three parameters useful in reconstructing their historical biogeography and explaining current patterns of distribution: a fossil record, a known means of dispersal, and a relatively accurate picture of the landscapes and climates within their present and past range. More information is needed for other genera, and although the data can be compiled from scattered sources, a more extensive database with the principal dispersal vectors involved would be valuable. This includes whether dispersal capacity is limited (e.g., by gravity, ocean currents, fish, rodents, nonmigratory birds, large terrestrial vertebrates) and continuous land important to essential, or if dispersal capacity is high (e.g., by wind, migratory birds) so that land bridges are less essential. Data on retention time by the vectors, distances traveled, and survivability of the propagules during and after transport are important (Yumoto, 1997). The following migrations are noteworthy (see, e.g., http://nationalzoo.si.edu/scbi/migratorybirds/fact_sheet/fxsht9.pdf):

sooty shearwater (*Puffinus griseus*)—New Zealand, North and South America, Antarctica, Africa, 64,000 km

Arctic tern (*Sterna paradisaea*)—Arctic North America, South America, Europe, Africa, Antarctica, 33,000 km

Pectoral sandpiper (*Calidris melanotos*)—Siberia, North and South America,
  Australia, 28,000 km
short-tailed shearwater (*Puffinus tenuirostris*)—North America, Asia, Australia

Maps of New World bird migrations are available (Fig. 5.2), and the
Cornell Lab of Ornithology has an animated website showing migratory
movement by day of 118 bird species, including routes across the CALB
(cornellbirds@cornell.edu). These are examples of where a complete data-
based life list of bird-transported seeds would be useful (see Carlo et al.,
2013; Fridriksson, 1975; Guttal et al., 2011; Nathan et al., 2008; Viana
et al., 2013, 2016).

Another association in the modern vegetation, but with distributions
likely reflecting events in the past, is seed dispersal anachronisms. As

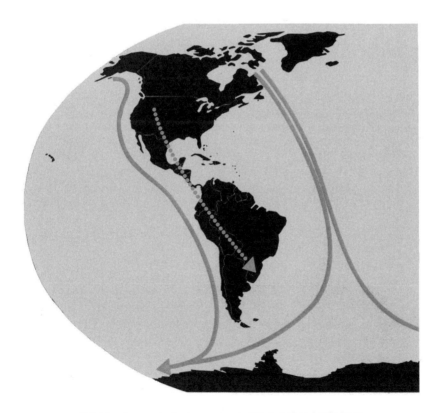

5.2. Migration routes for two neotropical birds. Left and right lines:
Sterna paradisaea (Arctic tern). Center line: *Buteo swainsoni* (Swainson's hawk).
From Wikimedia/commons/migration routes.

defined by Janzen and Martin (1982), these are plants with dispersal syndromes, fruit traits, and phenological patterns best explained as residues of former interactions with extinct animals. They cite species of the following genera as examples from the neotropics (those marked with an asterisk* are also known as fossils; II, app. 2.2): *Acacia, *Acrocomia, *Alibertia, *Andira, *Annona, *Apeiba, *Bactris,* Bromelia,*Brosimum, *Bunchosia, *Byrsonima, *Caesalpinia, *Chlorophora, *Crescentia, *Dioclea,* Diospyros, *Enterolobium, *Ficus, Genipa,*Guettarda, Hippomane, *Hymenaea, Manilkara, Mastichodendron, *Pithecellobium, Prosopis, *Randia, Sapranthus, *Spondias, and *Zizyphus. Guimarães et al. (2008) explore this further by noting that plants with megafaunal dispersal avoid the trade-off between seed size and transport, and that many of the plants have survived the extinction of the megafauna because the seeds are now distributed locally by scatter-hoarding rodents, introduced livestock, runoff and flooding into rivers, gravity, humans, and by frugivores able to carry large seed loads over long distances.

Some examples of genera from the montane broad-leaved forest with a fossil record and studied with reference to dispersal mechanisms are given below. These group into ones emphasizing the vector (wind, birds), vegetation type (subtropical forest understory), or habitat/location (oceanic islands, Barro Colorado Island), and studies focused more on specific plant genera (*Podocarpus, Clusia, Ficus*). A few are from outside the New World:

Vector/Vegetation Type/Habitat

Conceição et al., 2007—58 "soil islands" studied from the Chapada Diamantina of northwestern Brazil; numerous plant taxa including *Clusia* and *Tibouchina*. In *Clusia* there is a seed dispersal syndrome mediated by birds, and the authors note that "the high proportion of dioecious plants found on [analogous] oceanic islands has been associated with seed dispersal by birds (Whittaker, 1998)" (483).

Du et al., 2009—China (see their table 2 for New World countries); numerous taxa including *Quercus* and *Styrax*; principal dispersal mode in subtropical forest understory is by zoochory; dry seasons favor zoochory and wind (5).

Muller-Landau et al., 2008—Barro Colorado Island, Panama; numerous taxa examined, including *Calophyllum* (bats, mammals) and *Tabebuia* (wind); determining factors for distance in wind-dispersed species were the diaspore's terminal velocity, tree height, wind speed; for animal-dispersed species it was seed mass (negative effect) and tree height (positive effect).

Parolin et al., 2013—Amazon floodplain; taxa included *Psychotria* (fish), the riparian fig *Ficus insipida* (fish, bats), *Salix martiana* (and *Ceiba, Pseudobombax*; anemochory= wind), *Calophyllum, Lecythis, Ocotea* (mammaliochory, chiropterochory= bats).

Parrado-Rosselli, 2005—seed dispersal was studied in terra firme rainforests of Colombian Amazonia for *Calophyllum, Clusia, Ficus, Miconia, Ocotea.*

Wilkinson, 1997—The question posed is "Are wind dispersed seeds really dispersed by birds at larger spatial and temporal scales?" Using data on Holocene tree migrations in Europe, "it is suggested that for many 'wind dispersed' seeds the wind dispersal mechanism is adapted to land dispersal (over distances of a few canopy diameters) and larger scale dispersal is due to birds" (61).

Yumoto, 1997—Tropical forest, Colombia; "Although curassows were observed consuming fruits of 13 species belonging to the families Rubiaceae, Meliaceae, Moraceae, Burseraceae, Leguminosae, and Lecythidaceae, only seeds of *Geophila repens* (Rubiaceae) and *Ficus sphenophylla* (Moraceae) were found in their feces. For *G. repens*, the mean and maximum retention times were 1 h 52 min (+/− 1 h 20 min) and 6 h 08 min, and the mean and maximum direct dispersal distances were 245 m (+/− 164 m) and 633 m. More than half the seeds were dispersed in canopy gaps. For *F. sphenophylla*, the mean and maximum retention times were 3 h 15 min (+/− .37 min) and 7 h 08 min, and the mean and maximum direct dispersal distances were 329 m (+/− 46 m) and 451 m. Nearly 60 percent of the estimated seed intake of *G. repens* and 92–94 percent of *F. sphenophylla* were digested or damaged. The retention times and the dispersal distances for *Streptogyna americana*, which has adhesive burs, were also measured. The mean and maximum retention times were 1 h 55 min (+/− 1 h 56 min) and 9 h 11 min, and the mean and maximum direct dispersal distances were 128 m (+/− 68 m) and 280 m."

Genera

Howe and Miriti, 2004 (general)—"A profusion of fruit forms implies that seed dispersal plays a central role in plant ecology, yet the chance that an individual seed will ultimately produce a reproductive adult is low to infinitesimal" (651).

Podocarpus—Wilson et al., 1996; temperate forests, southern Chile; principal agents of dispersal were austral thrushes (*Turdus falcklandii*) "carrying about 18% of seeds at least three tree-crowns away from the parent tree.

Chilean pigeons (*Columba araucana*) virtually always dropped seeds below the parent" (343).

Caesalpinia, Delonix, Helianthus, Lonchocarpus, Spondias—Sánchez-Rojas et al., 2004; tropical dry forest in Mexico; tested postdispersal fruit and seed removal by spiny pocket mice (*Liomys pictus*; fate after dispersal); *Helianthus* and *Delonix* (high, 80%, 70%, respectively), *Spondias* (moderate, 50%), *Lonchocarpus* (no removal).

Calophyllum—Marques and Fischer, 2008; SE Brazil; dispersal of *C. brasiliense* by bats (*Artibeus lituratus, Platyrrhinus lineatus*); germination rates greater for seeds with pulp removed by bats. See also Ribeiro Mello et al., 2005; King, 2003; Kumar et al., 2015.

Clusia—Passos and Oliveira, 2002; rainforest, SE Brazil; studied dispersal system in *C. criuva* producing hundreds of capsules, birds eat 83% of the diaspores on the trees and ants remove 17% on the ground and carry them to their nests.

Casearia—Howe, 1977; Costa Rica; *C. corynbosa* seeds with arils consumed by 22 species of birds (mostly not long-distance-migratory; e.g., masked tityra, *Tityra semifasciata*), many processing the seeds undamaged. See also Howe (1984), *C. corymbosa, Virola sebifera*.

Ficus (numerous studies)—see, e.g., August (1981; llanos of Venezuela; bats), Lomáscolo et al. (2010; Papua New Guinea; birds, bats), Shanahan (2000; Borneo, New Guinea, many vertebrates), Wheelwright et al. (1984; Costa Rica; *Ficus, Hampea, Sapium* attracting more than 20 species of birds).

Heliconia—Uriarte et al. (2011; *H. acuminata*, Amazonia; birds); studied dispersal distances and retention times in relation to forest fragmentation: "Bird dispersers flew farther and faster, and perched longer in primary relative to secondary forests . . . small fragments had reduced densities of *Turdus albicollis*, the largest-bodied disperser . . . to decrease the probability of long-distance dispersal events from small patches." See also Córtes et al., 2013.

Ilex—Obeso and Fernández-Calvo (2002; *I. aquifolium*; northern Spain; birds, migrant thrushes); see also Arrieta and Suárez (2005; Spain); Stiles (1980; numerous genera including *Ilex*, eastern deciduous forest, birds).

Lecythidaceae—Tsou and Mori (2002; Brazil) studied seed coat anatomy in relation to dispersal in *Couroupita guianensis* (exotestal hairs, peccaries); *Cariniana* and *Couratari* (winged seeds, wind); *Allantoma lineata* (papillate exotestal cells, water); *Corythophora* and *Lecythis* (arils, various animals) and *Eschweilera ovalifolia* (sarcotesta, fish). See also Prance and Mori (1978).

Lonchocarpus—Augspurger and Hogan (1983; *L. pentaphyllus*, Barro Colorado Island, Panama); wind-dispersed, 40% fall within radius of tree crown, maximum dispersal distance 70, 66, 44, and 24 m from parent tree for fruits with 0, 1, 2, and 3 seeds.

Miconia—Borges and Melo (2012; *M. theaezans*, palm swamp-gallery forest, central Brazil); 7 bird species, *Tangara cayana* main consumer. See also Christianini and Oliveira (2009; *M. rubiginosa*, neotropical cerrado savanna, ants, birds); Dalling and Wirth (1998; *M. argentea*, leaf-cutting ant *Atta colombica*, Barro Colorado Island, Panama); Loiselle and Blake (1999; 4 shrubs in the Melastomataceae, birds); Magnusson and Sanaiotti (1987; dispersal by the rat *Bolomys lasiurus*).

Ocotea—Beltrán (2015; Monteverde, Costa Rica; *O. floribunda*); birds. See also Wenny and Levey (1998; *O. endresiana*, Monteverde, Costa Rica, bellbirds); Wheelwright (1993; *O. tenera*, Monteverde, Costa Rica, birds).

Persea—Wolstenholme and Whiley (1999; Mexico, *P. americana*, dispersal mostly by ground-dwelling large mammal frugivores after fruit fall (8–10).

Psychotria—Araújo and Cardoso (2006; *P. vellosiana*, cerrado, Brazil); seeds remained viable in the soil until next period of seed dispersal providing for a persistent soil bank. See also Theim et al., 2014.

Styrax—Young (1990; *S. ovatus*, Amazonas, Peru); "it is one of the numerous canopy and subcanopy tree species in Peruvian montane forest that produce fruits apparently attractive to large mammals and birds." Viable seeds were observed in the droppings of the spectacled bear (*Tremarctos ornatus*), "a particularly important seed dispersal agent because of the distances it travels and the rough terrain it crosses." See also Kato and Hiura (1999; *S. obassia*, Japan); Kissmann and Habermann (2013; *S. pohill, S. camporum, S. ferrugineus*, cerrado, Brazil).

Symplocos—Flora Malesiana (s.d.); bats are fond of the hard-fleshed drupes; North America tyrant birds eat fruits of *S. tinctoria*; South America curassow *S. cernua*; fossils numerous at some sites and probably dropped to the soil in situ, abundant and widespread dispersal by birds or bats not considered likely. See Kubitzki (2004, 447) and Kirchheimer (1949) for summary of the fossil record of *Symplocos*.

Tabebuia—Moussa et al. (2014; *T. rosea*, Thailand, winged seeds, wind dispersal, traveled 0.16 m for every meter of height).

Tibouchina—Collevatti et al. (2012; *T. papyrus*, rocky savannas, central Brazil); dispersed by autochory (= active or self-dispersed, e.g., by pressure/explosive mechanism): "Our results support that the disjunct distribution of *T. papyrus* may represent a climatic relict. With an estimated

TMRCA [time to most recent common ancestor] dated from ~836.491 +/− 107.515 kyr BP we hypothesized that the disjunct distribution may be the outcome of bidirectional expansion of the geographical distribution favored by the drier and colder conditions that prevailed in much of Brazil during the Pre-Illinoian glaciation, followed by the retraction as the climate became warmer and moister" (1024).

Turpinia—Tiffney (1979; *T. uliginosa* from the early Miocene Brandon Lignite of Vermont); multiseeded berry with fleshy to woody walls; mode of dispersal not known but assumed to be by birds; current northernmost distribution in the New World is to central Mexico.

These studies are useful in assessing the potential for distribution in individual cases and show that for plants of the montane broad-leaved forest, including many trees and shrubs represented in the fossil record in the vicinity of the equatorial land bridges, there are multiple means of seed dispersal and alternative ways of migrating. In the more distinctive/extreme environments (mangroves, rainforests, páramo) the trend is toward closer association with a particular means of dispersal (ocean currents, primates, migratory birds).

In South America at high elevations above ca. 3600 m where the MAT ranges from 3°C to −6°C, there is a tropical alpine tundra or páramo of shrubs and herbs. Characteristic plants are *Espeletia* and *Puya*. Páramo is scattered along the intermittent highlands of northern South America to the Transvolcanic Belt of Mexico (the "sky islands" of Vásquez et al., 2016). This disrupted occurrence has been true throughout the relatively short history of the páramo. Hence, distribution is primarily by long-distance transport (birds and/or wind), composition at the generic level is mostly distinct, and there are numerous endemic species. The earliest occurrence as a community was probably in the middle to late Pliocene, but older deposits are frequently removed by succeeding glaciations; páramo has been traced back in lacustrine sediments to 2.25 Ma (Torres et al., 2013). As measured by uplift of the Northern Andes, which ended connection between the Pacific and Atlantic oceans across Amazonia, and by flow of the Amazon rivers reversing from west to the north and east as at present (Potter, 1997; Van der Hammen, 1952; Hoorn, 1994a,b; Hoorn et al., 1995, 2010), substantial uplands appeared beginning in the Miocene, and ones sufficient to support elements of páramo in the Mio-Pliocene with the community appearing prominently slightly later, after the MPCO.

North of the CALB in Mexico the lands encountered by organisms transgressing the connection, and the source for many plants and animals moving south from North America, are (1) the saline, brackish, and freshwater habitats along the eastern coastal plain; (2) the physiologically dry porous limestone karst of the Petén in Nicaragua, Guatemala, and the Yucatán Peninsula of Mexico; (3) the hot and climatically dry Isthmus of Tehuantepec; (4) mid-altitude mesic uplands along the eastern slopes of the Sierra Madre Oriental in Veracruz; (5) later in the Neogene the cold alpine highlands of the Transvolcanic Belt across the latitude of about Mexico City (18°N), which includes the highest peaks in Mexico (Pico de Orizaba, 5650 m; Volcán Popocatépetl, 5452 m); and (6) the deserts of northern Mexico and southwestern United States.

The eastern coastal plain of Mexico is composed of alternating layers of erosion sediments, mostly from the Sierra Madre Oriental, and shallow marine and brackish water deposits, including lenses of highly organic lignite with abundant plant microfossils, as in the Pliocene Paraje Solo Formation (Graham, 1976a). Elevations extend from sea level to ca. 200 m, and along the passive margin of eastern Mexico the plain is 100 km wide in Veracruz, extending farther inland to the north at the Texas-Mexico border and to the south across the Yucatán Peninsula. In the past the width varied with orogenic activity and sea-level changes, but it has long provided lagoon, estuary, marsh, and mangrove habitats for organisms exchanging through the lowlands between North and South America. Representative climates in the lowlands along the connection are:

Mérida (20°58′N, 89°37′W)—MAT 26°C; MAP 1036 mm
Veracruz (19°11′N, 96°09′W)—MAT 25.3°C; MAP 1753 mm
Matamoros (25°88′N, 97.5°W)—MAT 23.2°C; MAP 698 mm

In addition to the extensive microfossil assemblage from the Pliocene Paraje Solo lignites in Veracruz, there is a small flora of Eocene age near Laredo/Nuevo Laredo, and microfossil and macrofossil floras of different ages across the Gulf Coastal Plain of Texas, transitioning into the extensive Eocene/late Oligocene Wilcox flora of Kentucky and Tennessee. Throughout the passive-margin province of eastern Mexico, and from similar depositional environments in the Antilles and elsewhere, it is possible to trace the migratory and developmental history of the rainforest (Graham, 1976b) and coastal plants such as *Laguncularia*, *Nypa*, *Pelliceria*, and *Rhizophora* from South America into the southeastern United States (see the section below on utilization of the CALB; chapter 7, case studies [mangroves]). There is also a coastal plain along the active western margin of Mexico, but it is narrow, and few fossil floras of Late Cretaceous and Tertiary age are known from the region.

The Yucatán Peninsula is edaphically dry because of the porous limestone substrate, but in precipitation it is quite moist (1036 mm at Mérida). The average elevation is 200 m. There are extensive dissolution caves called cenotes that preserve artifacts of the early inhabitants. When added to the archaeological record to the north, a history of migration and plant cultivation/domestication is revealed from the Tehuacán Valley with subsequent introduction south into Central and South America, east into the Antilles, and reradiation back north to places like Cahokia Mounds in Missouri and elsewhere. The peninsula is further famous as the site of the terminal-Cretaceous asteroid impact that contributed to the extinction of the nonavian dinosaurs. The Isthmus of Tehuantepec (average elevation 275 m) separates the low mountains of the Sierra Madre del Sur (2438 m) from the Cordillera de los Cuchumantanes of Guatemala (Volcán Tajumulco, 4220 m), which extends into southernmost Mexico in the state of Chiapas.

The 1000 km long Transvolcanic Belt serves to reiterate a principle of historical biogeography noted previously with reference to the Antilles: namely, it is useful to keep geologic events (e.g., the appearance of structural features connecting two regions; whether they were submerged, at sea level, or significantly elevated; permanent or ephemeral) separate from independent albeit related interpretations of the biogeographic consequences of those events (utilization of the connection by plants and animals; Donnelly, 1989; Hedges, 2006). The three events (time of structural continuity, appearance of specific habitats, time of principal biogeographic exchange) may cover a span of millions of years. The Transvolcanic Belt is a case in point. Its origin can be traced back to subduction of the Acapulco section of the Pacific Plate along the western coast of Mexico in the Late Cretaceous. However, the detachment of the mountain system from the southern Sierra Madre Occidental, intensification of volcanic activity, and uplift to high elevations occurred in the middle Miocene between 8 and 11 Ma (Ferrari et al., 1999, 2012). Thus, the Transvolcanic Belt of Cretaceous origin becomes most relevant to the migration of organisms in the Miocene. This distinction will be especially important when considering the uplift history of the southern portion of the CALB in Panama.

The Sierra Madre Oriental is an old Jurassic and Cretaceous mountain system that extends northward into the south-central United States as the Sacramento and Davis Mountains and the Sierra Guadalupe of Texas and

New Mexico near Big Bend National Park. The southern terminus intersects the Transvolcanic Belt to form the high-elevation zone noted earlier and provides habitats ranging from 5650 m east of Mexico City down to sea level at Veracruz. At the highest summits there is *Abies religiosa, Pinus montezumae,* and *P. rudis,* and *Picea* grows at high elevations in the northern Sierra Madre Occidental. As winds rise along the eastern slopes of the Sierra Madre Oriental, moisture is lost and rainfall is high (1587 mm in Xapala, elevation 1425 m). The present-day mid-elevation mesic deciduous forest is important for land bridge history because on the slopes there is a precipitation-related (rather than a temperature-related) deciduous forest that includes trees and shrubs disjunct at the generic level from the broad-leaved temperate woodlands of the eastern United States and separated by the deserts and semideserts of northern Mexico and south Texas: **Liquidambar macrophylla, *Quercus affinis, Q. skinneri, *Ulmus mexicana, Carpinus caroliniana, Cornus florida, Diospyros riojae, *Fagus mexicana,*Ilex vomitoria, *Juglans pyriformis, Magnolia schiedeana, Ostrya virginiana, Prunus brachybotrya, Sambucus mexicana,* and *Viburnum tiliifolium* associated with **Podocarpus matudae, Alfaroa mexicana, Clethra quercifolia, Engelhardia mexicana, Meliosma alba, Persea cinerarascens, Ternstroemia sylvatica,* and *Tournefortia petiolaris.* Intermingled with the deciduous forest is a coniferous forest of pine-oak that at the higher elevations includes *Pinus*(*) *hartwegii, P. oocarpa, P. pseudostrobus, P. teocote, P. rudis, P. strobus* and, at the very high elevations, **Abies religiosa. Pinus, Quercus,* and *Abies* are known as fossils in the Paraje Solo (Pliocene) Formation of Veracruz. Above the treeline there is páramo with grasses, *Draba, Gnaphalium,* and *Senecio.* These zonal communities all shifted downslope with the onset of cold conditions after the MMCO and especially during the glacial periods of the Quaternary. The changes within the Quaternary are beyond the principal focus of this book, but they occurred with muted effect toward the south with the decreasing extent of topographic diversity (altitude) and with the increasing extent and proximity to maritime climates (II, 345–49). Some northern temperate trees are found as fossils in deposits farther south in the late Tertiary, and a few crossed into the uplands of northern South America in the Quaternary (*Alnus, Quercus;* Hooghiemstra, 2006; Torres et al., 2013).

These landscapes and climates of northern South America and southern Mexico provided a variety of changing habitats for accommodating plants dispersed by a variety of vectors. They further give a context for reconstructing the biogeography of plants utilizing (that is, actually living on) the CALB during the Late Cretaceous and Cenozoic.

Geographic Setting and Climate

Panama runs east-west rather than north-south, and the Panama Canal actually goes from the northwest to the southeast. A bit of information given to visitors is that a ship entering the canal from the Atlantic side actually exits farther east than where it went in. The building of the canal was a monumental feat, and the 1915 Panama-Pacific International Exposition World's Fair (including a famous sculpture, *Fountain of Energy*, by Alexander Stirling Calder) was held in San Francisco to commemorate the event. Caledonia Bay on the north coast of Panama was the departure point for Vasco Núñez de Balboa in 1513, as he walked across the isthmus to be the first European to see the Pacific Ocean:

> The original city of Panama had been founded in 1519, or just six years after Balboa's discovery of the Pacific, and by an extraordinarily treacherous individual, Pedro Arias de Ávila, usually referred to as Pedrarias, who had been governor of Castilla del Oro, as the Central American isthmus was known, and who, to solidify his power, had Balboa beheaded on a trumped-up charge of treason. (McCullough, 1977, 111–12)

At the southern terminus of Central America in the Darien Province at the border with Colombia the elevation is virtually at sea level, and at its narrowest point at the Canal Zone it is 48 km wide. The highest elevation in Panama is the Volcán Barú (Volcán Chiriquí) in the northern Cordillera Central (3475 m), but otherwise the topography is moderate to low-lying. The MAT at Colón at sea level on the Atlantic side of the Canal is 27°C and MAP is 1500–3000 mm along the slightly wetter Atlantic Coast and 1140–2290 mm on the Pacific side. Elevations increase in Costa Rica, where the Cerro de Chirripó Grande in the Cordillera de Talamanca is 3819 m. Representative MATs are 25°C at Limón (sea level on the Atlantic coast), 26.6°C at Puntarenas on the Pacific side, and 19.4°C in the capital city of San José at 1200 m elevation. Among other prominent physiographic features of the CALB are Lago de Nicaragua, the largest lake in Central America at 8430 km² (for comparison, Lake Erie is 27,745 km²), and the highlands in and around Bluefields in southern Nicaragua, which is the approximate southern limit for several northern temperate trees like *Juniperus*, *Pinus*, *Carpinus*, *Juglans*, *Liquidambar*, and *Ulmus*.

Forging the Final Link: Geology

The two equatorial connections between North and South America (ALB, CALB) each present unique and some unresolved geologic challenges. For the ALB, the problem of it ever being continuous or nearly so is seemingly resolved in favor of a stepping-stone landscape, while the CALB was continuous to southernmost Panama. For the CALB, the question is the time of the final union between Central and South America, with estimates ranging from 31 Ma, to 15–13 Ma, to 3.5 Ma. Actually the question is more complex, and recognizing that complexity and precisely defining the nature of the paleolandscape is important for formulating views about the past and present biogeography of the New World. As mentioned earlier, it is necessary to know the time, duration, and whether it is a geologically structural union that is being considered or a fully subaerial one accommodating, for example, "walkers" across a variety of altitudinally zoned coastal, upland, and highland habitats, and when each appeared.

The region of the CALB is delineated into Nuclear Central America (Guatemala, Belize, Honduras, El Salvador, northern Nicaragua) and the Isthmian Link (southern Nicaragua, Costa Rica, Panama; Schuchert, 1935/ 1968; Fig. 5.1). The northern boundary of Nuclear Central America is a megafault across the Isthmus of Tehuantepec called the Santa Cruz Suture, and to the south it is defined by the onshore Santa Elena fault zone continuing offshore as the Hess Escarpment just south of Lago de Nicaragua. The Isthmian Link begins at the contact between the North American and Caribbean plates in Costa Rica and extends to the margin of the Caribbean and South American plates across the lowlands of southernmost Panama and adjacent Colombia. The geology of the Isthmian Link is complicated by the tectonic activity of the Cocos and Nazca subplates impinging on the western coast.

The formation of the CALB was a gradual, episodic, and transitional process. Compression and uplift of the Cordillera de los Cuchumatanes (Guatemala) and the Cordillera de Talamanca (Costa Rica) began in the Late Cretaceous, upland slopes were present by the Miocene, major uplift occurred in the Pliocene, and the highest peaks at ca. 3600 m were glaciated in the Quaternary (Anderson, 1968; Anderson et al., 1985; Hastenrath, 1973). Islands and low-elevation peninsulas appeared in the Darien region of southern Panama in the early and middle Cenozoic, with volcanism, uplift, and accretion of the Azuero and other peninsulas in the Late Cenozoic. Important for land bridge considerations is the timing of events that forged the final, lasting connection between Central and South

America in the Late Cenozoic and the nature of the landscape at different stages in its development. The differentiation of Central American marine biotas into Pacific and Atlantic provinces is a function of the rise of the land bridge, and differences appear first with benthic (bottom-dwelling) communities and later among planktic (near-surface and floating) ones. Woodring (1966, 427) found that after a middle Miocene peak, the paci-philes in the Caribbean waters steadily dropped to 2 percent. Citing the literature of the day (e.g., Simpson, 1965), Woodring says, "A few small mammals migrated in both directions before the great interchange, but they doubtless were island-hoppers, to use Simpson's apt term." Bringing the consensus on faunas up to the mid-1980s, Stehli and Webb (1985, 11) state:

> The later Cenozoic interamerican biotic events can be divided into two final episodes. The principal event is the "Great American Biotic Interchange": an episode involving reciprocal passage of numerous land and freshwater taxa between the Americas via the isthmian corridor about 3 million years ago. This major event was preceded by a lesser episode in which a few taxa recip-rocally crossed a filter bridge between 8 and 9 million years ago. The latest biogeographic episodes conform well to Simpson's earlier characterization and still serve well as the "type examples" of a filter bridge and a corridor.

Webb (1985, 210) notes that "the whole land-mammal fauna in North America experienced an extensive turnover between 5 and 3 million years ago. A majority of the immigrants were of Eurasian origin, but a large con-tingent emigrated from South America at about the same time."

Marshall (1985, 76) says that "the earliest known South American mammals to disperse to North America across the Panamanian land bridge occur in rocks dated at 2.8–2.6 Ma," and that "this was the first opportu-nity for taxa to disperse by walking across the land bridge" (78).

The herpetofaunal record is too meager to offer a critical assessment of early continuity, except that, according to Estes and Báez (1985), it seems most consistent with a connection in the Pliocene. Similar views continued into 2004 with the symposium, "Latin American Biogeography—Causes and Effects," at the Missouri Botanical Garden (Graham, 2006; see papers by Webb, 2006; Hooghiemstra et al., 2006).

Donnelly (1985, 112) introduced the important difference between a continuous structural unit and one that was also emergent. He notes that at the end of the Eocene and into later times there was structural continu-ity, but shallow-water pelagic and coastal faunas were similar on both sides (citing Keigwin, 1982), and these conditions surely limited the opportunities for organisms that had to disperse by "walking across."

Broad generalizations about geological and biological history benefit from assessment through these independent lines of evidence. Such was provided by Donnelly (1989) in a chemical analysis of pelagic (open-ocean) sediments as part of the DSDP. Bottom and intermediate-depth waters of the Atlantic and Pacific oceans at present have distinctly different silica content. Differences between the Pacific and the Caribbean part of the Atlantic Ocean first become evident in the middle Miocene (~15 Ma) with the early formation of structural connections across the isthmian region. According to Donnelly (1989) a distinction between surface waters becomes apparent at ca. 4.1 Ma and reaches a maximum between 3.7 and 3.2 Ma, suggesting that was when the final portion of the Central American Land Bridge was in place and the connection continuous and emergent at least to low elevations.

Such was the data available and the interpretation made of it in the literature through ca. 2006. Recently, carbon isotopic variation in planktic foraminifera from ODP cores from the Blake Ridge (NW Atlantic Ocean) indicates a strengthening of the Gulf Stream bringing warm, saline waters northward (Bhaumik et al., 2014). The increased velocity is attributed to final full closure of the Central American Isthmus (latitude ca. 8°N) with the effects felt at the latitude of the Blake Ridge (31°47′N, 75°32′W) at 2.2 Ma. In a review of the literature on the phylogeography and biogeography of Lower Central America, Bagley and Johnson (2014) state that "the most stunning changes occurred over Late Neogene-Quaternary, when gradual emergence of the LCA isthmus cut off the Central American Seaway and permanently linked the Americas for the first time in the Late Pliocene," and that "by ~4.6 Ma, ocean currents and ecosystems became reorganized. The isthmus then became fully closed by at least 3.5 to 3.1 Ma before a permanent Isthmian Link with South America formed 3.1 to 1.8 Ma" (772).

Summarizing opinions through 2014, there was consensus that the final persisting link in the CALB was forged in the Pliocene between ca. 3.5–3.1 Ma with the understanding that (1) this was referring to a complete subaerial connection; (2) some crossings occurred earlier in the Cenozoic depending on the ecological requirements and dispersal capabilities of the organism or its propagules as noted by Woodring (1966) and Stehli and Webb (1985); (3) the initial landscape at ca. 3.5–3.1 Ma was low-lying and allowed interchange of terrestrial organisms of coastal and low-altitude environments; and (4) that scattered higher elevation habitats developed later at an estimated 2.5 Ma (II, 83; see also Leigh et al., 2014).

There is now a new proposal that closure occurred in the middle Miocene ca.15–13 Ma, or 10 Ma earlier than thought (Montes et al., 2012, 2015) and possibly as early as 31–16.3 Ma (Bacon et al., 2013, based on molecular evidence that assumes a constant rate of mitochondrial DNA divergence of 2%/million years). This is at least as significant for New World biogeography as the suggestion that continuous land existed through the Antilles in the Oligocene (Iturralde-Vinent and MacPhee, 1999), with even greater climate implications (ocean water circulation), and deserves careful consideration. If part of the evidence cited in favor of an early connection across the CALB is the presence of comparatively few biotic entities on both sides of the bridge, then there is the question raised by Hoorn and Flantua (2015) about why so many remained separated until the Great American Biotic Interchange at ca. 3.5–3.1 Ma (see e.g., discussion of camelid distributions, chapter 2; also Olmstead, 2013, 81). To my knowledge there is no serious question about the timing of the peak exchange. The basis for the new suggestion is the presence of zircon-containing river deposits considered originating in Panama and found in northern Colombia ca. 15–13 Ma, implying there must have been a land surface over which the river could flow. Dated molecular phylogenies (Bacon et al., 2013) are offered as supporting evidence. Montes et al. (2015) note that at least a segment of the Panama arc had docked by the middle Miocene, and suggest that the delay of 10 Ma before the Great American Biotic Interchange took place "could be unrelated to seaway closure and instead may be linked to Plio-Pleistocene global climatic transitions" (Montes et al., 2015, 228). In the "Perspective" piece written by Hoorn and Flantua (2015), with the declarative subtitle "The land bridge between North and South America formed 10 million years earlier than previously thought," their actual assessment offers more latitude for interpretation. In addition to the question of why there was a 10 Ma delay in peak crossings over the proposed continuous landscape, they state (187) that "a full understanding will require better knowledge of the early land bridge and its environments. Data are also needed on the existence and duration of any intermittent transoceanic connections elsewhere along the narrow strip of land that separates the Atlantic and the Pacific." Among the interchanges of interest are the previously cited presence of northern temperate elements, such as *Alnus* and *Quercus*, present in the Tertiary of Central America but not reported from the uplands of South America until the late Quaternary. *Alnus* first appears in the Northern Andes ca. 1 Ma and *Quercus* at ca. 430 kyr (Torres et al., 2013). The new proposal for the complete closure of the CALB is of considerable interest, the duration and timing are important for interpreting new data, and the nature of the paleolandscape is critical to a range of views that include

> By middle Miocene times (13 to 15 Ma) . . . the Central American seaway was closed. (Montes et al., 2015, 228)

but recently,

> Data from three mid-Pleistocene sites reveal robust evidence of strong seasonal upwelling suggesting that the elevation of the Isthmus must have been sufficiently low to permit wind-jets to form. A low-elevation Isthmus of Panama may have persisted until as recently as the mid-Pleistocene. (O'Dea et al., 2012, 59)

> Molecular divergence estimates suggest that primates arrived in tropical Central America, the southern-most extent of the North American landmass, with several dispersals from South America starting with the emergence of the Isthmus of Panama 3–4 million years ago. (Bloch et al., 2016, 1)

and

> Independent lines of evidence converge upon a cohesive narrative of gradually emerging land and constricting seaways, with formation of the Isthmus of Panama sensu stricto around 2.8 Ma. The evidence used to support an older isthmus is inconclusive, and we caution against the uncritical acceptance of an isthmus before the Pliocene. (O'Dea et al., 2016, 1)

Part of the acceptance is interpreting molecular and other evidence as if early closure were true before it is known to be true. For example, a universal rate of mitochondrial DNA divergence is challenged by O'Dea et al. (2016). Particularly important, they identify 30 Eocene sites in adjacent northern South America with zircon-containing sediments of the same age as those from Panama. If so, this removes the need for a land surface over which a zircon-carrying river could flow. It is an interesting time for contemplating the date of complete and permanent closure of the CALB and the appearance of uplands suitable for the on-land plant migrants and the animal walkers. Following the principle stated earlier, it is primarily a geological problem to which biogeographic interpretation will have to accommodate, and in my opinion and in that sense, the tilt is toward the late Neogene.

Modern Vegetation

The plants of Central America are inventoried in several regional floras including Nicaragua (W. Stevens et al., 2001), Costa Rica (Hammel et al., 2003 et seq.), Panama (D'Arcy, 1987 et seq.; D'Arcy and Correa A., 1985), Barro Colorado Island (Croat, 1978), and Mesoamerica (Davidse et al., 1994 et seq.). There are also databases and websites for monographs, literature, illustrations, and taxonomy/phylogenies involving Central American plants (e.g., Angiosperm Phylogeny Group, 2009; P. Stevens, 2001 et seq.; TROPICOS, http://www.tropicos.org; see also II, 217–42). The following summary is based on the classification of vegetation presented in the introduction, above, and used in the previous volumes (I, II, III) with emphasis on plants and communities represented in the fossil record.

True deserts in the sense of areas receiving ca. 250 mm (10 in) or less of MAP are not present along the CALB. Rather, the driest parts receive 500 mm (20 in) or more of rainfall and are semideserts in edaphically dry habitats or on slopes to the lee of highlands. The narrow width of much of Central America reduces continentality (loss of moisture as winds blow across extensive stretches of flat land), which is a principal cause for aridity in the great interior deserts of the world. Furthermore, there are no cold offshore currents like the Humboldt Current along the western coast of South America, where the winds absorb rather than release moisture upon reaching the land. The driest vegetation in Central America, besides long-disturbed desertified areas (see W. Stevens et al., 2001, xxix–xxxv, for Nicaragua) is the shrubland/chaparral-woodland-savanna (including tropical dry forest, thorn forest, thorn scrub; Fig. 5.3) as found on the karst Petén Plains, and on the south and east-facing slopes of the Sierra de Los Cuchumatanes in Guatemala (II, fig. 3.84). There are rainshadows on the lee side of the mountains in southwestern Nicaragua with MAPs of 290–780 mm and a dry season of 7 months or more. At higher elevations the thorn scrub is locally replaced by pine-oak forest (Fig. 5.4). In Santa Rosa National Park, northwestern Costa Rica around the Gulf of Nicoya, there is dry forest to the lee of the Cordillera de Guanacaste extending into southwestern Nicaragua, and on the drier southern side of Panama to the lee of the Cordillera Central. Among plants in the drier habitats of Central America are *Acacia*(*) *farnesiana* (an asterisk indicates the genus has been reported as a fossil in II, app. 2.2), *Albizia saman, Astronium graveolens, Brosimum alicastrum, Byrsonima crassifolia, Bursera simaruba, Caesalpinia*(*) *eriostachys, Cassia*(*) *grandis, Celtis caudata, Curatella americana, Hymenaea courbaril, Lue-*

5.3. Tropical dry forest, Nicaragua. The plant in the foreground is the cactus *Pilocereus maxonii*. Photograph by Olga Martha Montiel.

5.4. Pine forest, Nicaragua. Photograph by Olga Martha Montiel.

hea candida, Maclura tinctoria, Prosopis juliflora, and *Zanthoxylum setulosum*. In moister areas of the uplands there is the montane forest (Fig. 5.5).

The development of ecosystems typically follows a general three-part history. There is the appearance of lineages with elements or progenitors characteristic of the community (that is, in the Cretaceous). In the case of the desert, for example, these include the Cactaceae and members of the

5.5. Montane forest, Nicaragua. Photograph by Olga Martha Montiel.

Amaranthaceae/Chenopodiaceae, some Fabaceae, Poaceae, and others. There is a period when some of these differentiate into isolated ecotypes occupying scattered dry habitats formed by edaphic, slope/exposure, or other local conditions. Then if a climate trend occurs favoring these pre-adapted types, they coalesce into a recognizable and more broadly distributed ecosystem. For the desert this occurred in the Neogene.

There are no extensive grasslands across the CALB, and those that simulate natural grassy communities and savannas are typically associated with evidence of human disturbance. Beach/strand/dune vegetation is well represented along the coasts, but the environment is not suitable for the preservation of plant fossils, and many represent widespread families in which genera cannot be identified from pollen or vegetative remains (Amaranthaceae/Chenopodiaceae, Cyperaceae, Gramineae) or genera with species occupying a range of different habitats (*Eugenia*). As a result, these communities are sparsely represented and difficult to recognize in the fossil record.

Mangroves extend along both coasts of Central America, and because they grow in lowland brackish-water sites favorable to the preservation of plant microfossils and have distinctive pollen, they are readily identifiable in the fossil record. The widespread plants in the modern communities are *Avicennia germinans, *Conocarpus erecta,* Laguncularia racemosa*, and *Rhizophora mangle* along with the mangrove fern *Acrostichum aureum* and *Hibiscus tiliaceus*. These are all present as fossils along the equatorial land bridges. In addition, pollen of the mangrove *Pelliceria* is found in the Tertiary of Mexico and the Antilles far north of its present distribution along the coasts of Central America and northern South America (see case studies in chapter 7, below), and even the southeast Asian *Nypa* is known from the Eocene of the Americas as far north as south Texas, along with other Old World lineages likely the result of past continental collisions and separations (Figs. 3.10, 5.6, 6.6). Beyond the influence of ocean tides there are bog/marsh/swamps (Fig. 5.7) with plants like *Azolla,* *Salvinia,*Cabomba, Caladium, Calathea, *Ceratophyllum, Drosera,* grasses (*Phragmites*), *Lemna, Ludwigia, Myriophyllum,* *Potamogeton, Sagittaria,* sedges (*Cyperus, Eleocharis*), *Thalia, Typha,* *Utricularia,* and water lilies (*Nuphar, Nymphaea;* *Nym-

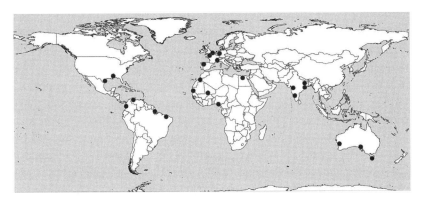

5.6. Current distribution of the mangrove palm *Nypa* (shaded areas, southeastern Asia) and fossil occurrences (dots). From P. Stevens, 2001 et seq., Angiosperm Phylogeny Website.

5.7. Aquatic vegetation in inundated pine savanna, Laguna de Perlas, Nicaragua. Photograph by Olga Martha Montiel.

phaeaceae). Fossils of others are present elsewhere in Latin America, and with the exception of the grasses and sedges, microfossils can be identified to genus.

The lowland neotropical rainforest is a prominent community of the equatorial regions. It reaches its northern limits as a community in southern Mexico near Veracruz and its greatest expanse in the Amazon Basin of South America. In coastal areas it grows inland to the mangroves immediately adjacent to the basins of deposition, and being a diverse, tree-dominated community producing an abundance of pollen/spores, leaves, fruits, and seeds, it is widely represented in the fossil record. One of the unexpected discoveries in tropical biology in the past several decades is that rather than being a stable and unchanging assemblage, as long described in the older literature, the tropical rainforest is revealed as a fragile, delicately balanced community that has undergone considerable alteration in range and composition. This had been suspected from observations on the modern biota (Vuillemier, 1971), such as relict species geographically isolated from the principal community; exceptionally high diversity because generation of novel genotypes and subsequent phenotypes is favored by fluctuating rather than unchanging environments; and later recognition that climatic changes did affect equatorial regions and were interhemispherically connected (Markgraf, 2001). The paleobotanical evidence from Veracruz documented that in an area where rainforest is dominant today, it was absent in the recent geologic past. This discovery and its implications for biodiversity mechanisms and for formulating conservation strategies have been considered elsewhere (II, 499–511; Graham, 1976b, 1977, 1982, 1988, 2015) ,but suffice it to say it altered the way we perceived and modeled speciation processes in the tropical rainforest and developed priorities for its preservation. Among the dominants at its northern limits in Veracruz are *Bernoulia flammea, Brosimum alicastrum, Calophyllum brasiliense, Dialium guianense, *Ficus tecolutensis, Poulsenia armata, Pseudolmedia oxyphyllaria, and Terminalia amazonica.

Near the southern end of the CALB in Panama are *Allophylus occidentalis, Anacardium excelsum, Andira inermis, Astrocaryum standleyanum, *Bombacopsis quintana, Callophyllum candidissmum, *Ceiba pentandra, Chamaedorea wendlandiana, Chomelia recordii, Enterolobium cyclocarpum,* Guarea guidonia, Luehea seemanii, Maeba occidentalis, *Miconia argentea, Neea amplifolia, *Protium panamensis, *Pseudobombax septenatum, *Spondias mombin, Sterculia apetala, Swartzia simplex, Swietenia macrophylla, *Terminalia amazonica, Trophis racemosa, Vitex cymosa, Xylopia aromatica, and Zanthoxylum panamense. Farther to the south in the Amazon Basin, biotic diversity is among the highest in the world, and with the exception of some Old World lineages resulting from past continental movement, the fossil record documents migration in both directions although prominently from south to north during the warmer times of the Paleogene and peaking around the EECO (see the section below on utilization).

Indigenous People

Information about the earliest people in Middle America comes primarily from the classic ceremonial sites in northern and central Mexico (Olmec, Toltec, Teotihuacano, Zapotec, Mixtec, Maya, Aztec), Guatemala (Maya at Tikal), and to the south in the Andes (Inca at Machu Picchu, Peru, 1200–1532 CE; Moche, Chibeha, Cañaris). Movement was principally from the north, and periods of development in Middle America are as follows (Evans, 2012):

> Pre-Archaic—8700 BCE earliest appearance of maize (online table 5.1; Piperno and Smith, 2012, table 11.1)
> Late Archaic—4000–2000 BCE
> Initial Formative (Preclassic)—2000–1200 BCE (Olmec 1500–400 BCE, Veracruz and Tabasco)
> Early Formative—ca. 1200–900 BCE
> Middle Formative—ca. 900–600 BCE
> Middle to Late Formative—ca. 600–300 BCE
> Late Formative—ca. 300 BCE–1 CE
> Terminal Formative—ca. 1–250 CE (Maya 200–900 CE, Yucatán, Guatemala)
> Early Classic—ca. 250–600 CE
> Late Classic—ca. 600–900 CE
> Epiclassic and Early Postclassic—ca. 750–1200 CE
> Middle Postclassic—ca. 1200–1400 CE (Aztec 1345–1521 CE, Valley of Mexico)
> Late Postclassic—ca. 1400–1521 CE

Archaeologists now make a convincing case for greater population densities at the major ceremonial sites and in the surrounding areas than previously thought (McKillop, 2004; the case for the very highest maximum densities and impact is presented by Mann, 2012). There is also a developing consensus about the timing of their emergence as agriculturalists and, therefore, notable modifiers of the land and vegetation:

Combining the archaeological and phytoarchaeological evidence, it appears likely that humans began having some impact on vegetation in Latin America by 15,000 years BP; locally more intensive land modification began at least 12,000 years BP; and some morphological changes in seed and phytolith size suggest plant domestication by ~9000–8000 years BP and cultivation by 7000 years BP (Bush et al., 1989). Markgraf (1989) notes that after 6000 years BP [local] paleovegetation changes can no longer be ascribed automatically to climate alone. (II, 272)

A perusal of the distribution of paleo-Indian sites in Mexico and Central America (Acosta Ochoa, 2012; Fig. 5.8), and the earliest occurrence of cultivated and domesticated plants in Mesoamerica (Piperno and Smith, 2012; online table 5.1), conveys the impression that the pre-6000 yr BP impact of humans on the landscape and vegetation along the CALB was not great. *Spondias* may have been present 7000 yr BP at Tehuacán, and some lake cores show earlier modifications by fire probably attributable to humans (Piperno, pers. comm., 2016). Recently the genome sequence of a maize cob 5310 years old from central Mexico reveals it was more similar to modern maize than to its wild counterpart, suggesting it was already being domesticated (Wales and Ramos-Madirgal, 2016). However, most changes in the vegetation before 6000–7000 BP were likely due to natural (i.e., nonhuman) causes.

Utilization of the CALB

The paleobotanical record for the equatorial land bridges is unusual in that for some places and times there is a fairly extensive, accurate, and up-to-date inventory available for Late Cretaceous and Cenozoic plants (Fig. 5.9). This is due to an unusually large number of Tertiary vegetation history studies, especially for tropical regions, made since about the 1960s, and for Panama from the summary by Jaramillo et al. (2014). Rather than having to depend so extensively on the pioneering works of E. W. Berry, Arthur Hollick, and others, there are a surprising number of newer studies available for northern Latin America. Identifications as part of recent reviews and paleomonographs covering broader regions are the most reliable (e.g., Burge and Manchester, 2008; Corbett and Manchester, 2004; Gonzalez et al., 2007; S. Graham, 2013; Harley, 2006; Jacques, 2009; Jarzen and Dettmann, 1989; Manchester 1987, 1989). However, at present these constitute only a small portion of fossils reported in the literature, so the record for the New World equatorial land bridges is somewhat of a luxury. The list

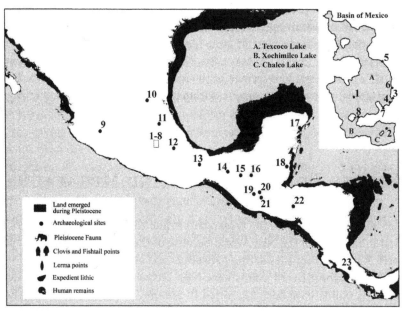

1. Peñon (Romano 1974; González et al. 2003)
2. Tlapacoya (Lorenzo y Mirambell 1986)
3. Los Reyes La paz (García 1973)
4. San Vicente Chicoloapan (Aveleyra 1967)
5. Tepexpan (Arellano 1946)
6. Santa Isabel Iztapan y Tocuila (Aveleyra 1967; Morett et al. 2001)
7. Chimalhuacan (García 1966)
8. Aztahuacan y Atepehuacan (Romano 1974; Aveleyra 1967)
9. Chapala-Zacoalco (Lorenzo 1964)
10. Sierra Gorda (Martz et al. 2000)
11. Oyapa (Cassiano 1990)
12. Valsequillo (Irwin-Williams 1967)

13. Oaxaca Valley (Hole 1986; Marcus and Flannery 1996)
14. Santa Marta (Acosta 2008)
15. Los Grifos (Acosta 2010)
16. Teopisca and Aguacatenango (García-Bárcena 1982)
17. Tulum caves (González et al. 2009)
18. Belize fluted point findings (Lohse et al. 2006)
19. Piedra del Coyote and los tapiales (Gruhn, Bryan and Nance 1977)
20. Valle del Quiché and Sacapulas (Bray 1978; Brown 1980)
21. San Rafael (Coe 1960)
22. La Esperanza (Bullen and Plowden 1968)
23. Turrialba (Snarskis 1979; Pearson 2004)

5.8. Paleo-Indian sites in Mesoamerica. From Acosta Ochoa, 2012, The Oxford Handbook of Mesoamerican Archaeology, Oxford University Press, Oxford. Used with permission.

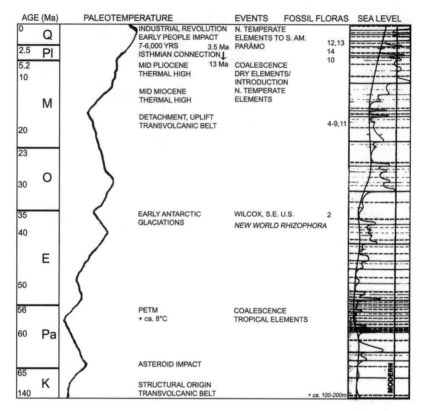

5.9. Paleotemperature and sea-level curves, events, and fossil floras for the Central American Land Bridge. See legend to Fig. 2.5 for explanation of abbreviations. Numbers are fossil floras as listed in II (table 6.1, 323–28): 2 = Gatuncillo Flora, Panama; 4 = Uscari (Panama); 5 = Rio Banano (Costa Rica); 6 = various Miocene formations (Panama); 7 = Culebra (Panama); 8 = Cucaracha (Panama); 9 = La Boca (Panama); 10 = Gatun (Panama); 11 = Padre Miguel (Guatemala); 12 = Herreria (Guatemala); 13, 14 = Rio Banano (Costa Rica).

presented in appendix 5 in the online supplementary materials is based on II (table 6.1) and incorporates these recent revisions and summaries.

In addition to the fossil record, estimates of the existence of a lineage are being made from molecular calculations. For present purposes it is the first appearance of the crown group (modern genus) that is most useful. These estimates are summarized in P. Stevens (2001 et seq.) as part of the APG website, and those pertaining directly to the equatorial land bridges are listed in online table 5.2. The website also may be consulted for boreal and austral lineages, but those for the equatorial connections are especially useful because the tropics contain by far the greatest diversity of plants for which an estimate of age from whatever source is important.

For example, there are about 9–12 prominent tree genera that define the boreal forest, about 67 for the eastern deciduous forest (I, chap. 1), and an undetermined number for the lowland neotropical rainforest. A recent calculation is that there are 11,678 species of trees in Amazonia (Peter M. Jørgensen, pers. comm., 2016; see also Pimm and Raven, 2017). There is an unsettlingly wide range in estimates for the age of the crown groups based on molecular evidence, comparable to that from the fragmented, poorly dated, and often unreliable fossil record, and restraint is needed when re-writing vegetation history and interpreting biogeographic patterns by ei-ther approach. The murky waters of uncertainty make it infinitely easier to be wrong than right.

A cautious use of fossil plants reported for the CALB reveals that dur-ing the Late Cretaceous and Tertiary, it increasingly became a pathway for organisms, although it has always served as a physical filter because of its tapering configuration and, until recently, partial continuity.

Lowland tundra never existed and highlands were not a prominent feature of the landscape, so it is not surprising that distinctive elements of the páramo are mostly different between northern South America and the mountains of southern Mexico. The similarities that do exist are most readily explained by long-distance dispersal and convergent evolution, and facilitated to some extent by expansion of the habitats during the cold in-tervals of the Quaternary when target areas increased.

Desert elements are not widely present or shared between the Ameri-cas (*Ephedra, Acacia, Larrea, Mimosa, Prosopis* are exceptions), and they are most likely the result of long-distance dispersal and/or evolutionary paral-lelisms than of primarily gradual movement across the Americas.

The fossil record of upland mesic vegetation is relatively good in the Tertiary of the CALB and the immediately adjacent regions. The critical question is when these uplands appeared in southern Central America, in contrast to the long-existing lowland habitats, and based on the geological evidence cited earlier (and in II, 78–83), to the north (Mexico, northern Central America) and south (northwestern South America) this was in the Miocene and in southern Central America it was in the late Pliocene.

A few northern boreal coniferous trees like *Picea* and *Abies* had a bet-ter go of it using the high-elevation northern portions of the CALB as a pathway for migration south of their current distribution. Both are found at present in the mountains of northern Mexico and *Abies* extends as far south as the eastern Transvolcanic Belt of Central Mexico. However, *Picea* was present in the Pliocene of Veracruz, *Abies* has been reported as macro-fossils, and cf. *Abies* and *Picea* as microfossils in the Pliocene of Guatemala. *Pinus* currently extends south to southern Nicaragua (e.g., *P. caribaea, P. oocarpa*), and it is found as fossils in the late Cenozoic of northern Central America, but it is not possible to consistently distinguish pollen of the cold, high-altitude species now mixed with *Abies* and/or *Picea* from that of warmer, lower altitude habitats occurring alone or mixed with *Quercus. Pinus* has been reported from the Quaternary of El Salvador, Costa Rica, and Panama (II, 324–25). Boreal coniferous elements extended across the very northern parts of the CALB and reached this southernmost distribution probably after the MMCO and especially in the Pliocene and Quaternary with the onset of cooling climates.

The mid-altitude temperate mixed forests of evergreen conifers and deciduous angiosperms had an even better time of it as climates cooled, land became continuous to eastern Panama, and moderately high altitudes, albeit scattered, were available along the way. In southern Mexico along the eastern slopes of Sierra Madre Oriental between ca. 700 and 2700 m elevation, there is a cloud forest with many genera disjunct to varying degrees from the deciduous forest of the eastern United States. These include *Lycopodium, Pteridium, Abies, Pinus, Acer, Alnus, Carpinus, Cornus, Diospyros, Fagus, Ilex, Juglans, Liquidambar, Magnolia, Myrica, Ostrya, Platanus, Populus, Prunus, Quercus, Rubus, Salix, Sambucus, Smilax,* and *Ulmus.* Some of these continue south to the prominent biogeographic demarcation point in southern Nicaragua around Bluefields (*Abies, Cupressus, Pinus, Carpinus, Celtis, Diospyros, Ilex, Juglans, Liquidambar, Myrica, Ostrya, Prunus, Rhamnus, Rhododendron, Rhus, Rubus, Salix, Sambucus, Typha, Ulmus,* and *Vaccinium*) with *Alnus, Ilex, Myrica, Prunus, Quercus, Salix,* and *Vaccinium* extending into South America. The paleobotanical record documents several of these already in eastern Mexico by the Pliocene or perhaps earlier (II, 252–53): *Alnus, Celtis, Ilex, Juglans, Liquidambar, Myrica, Populus, Quercus,* and *Ulmus.* In the Mio-Pliocene of Guatemala there were *Juglans, Quercus,* and *Ulmus;* in the Plio-Pleistocene of El Salvador *Celtis,* cf. *Hamamelis, Ilex, Juglans, Myrica, Quercus, Rhus, Salix, Ulmus* (and *Carya* reported[?]); *Ilex* in the Pliocene Costa Rica; and *Diospyros, Ilex,* and *Quercus* in the Miocene of Panama.

In the opposite direction some species of *Podocarpus* were moving north, assuming the plant originated in the Southern Hemisphere (ca. 100 species). It now extends from New Caledonia and Australia (Queensland) east along the western coast of South America as far north as Central America and southern Mexico (ca. 8 species). Along the CALB it was present in the Miocene of Panama and the Pliocene of Veracruz, but much earlier in South America with about 480 records of *Podocarpites* (most bisaccates being referred to *Podocarpites*), spanning the time from 59 Ma to today (Jaramillo, pers. comm., 2016). The 1 or 2 red, fleshy, fertile cone scales are eaten by birds, which then disperse the seeds, so a simple south-to-north migration exclusively across the land bridge is unlikely. Leafy shoots identified as *Podocarpus* place the genus in the Mississippi Embayment region during the Eocene (Dilcher, 1969), even though its movement across Central America is currently documented only for the late Tertiary.

There were other northward migrations suggested by the geographic affinities of tropical plants in the modern vegetation of southern Mexico and northern Central America, and several are supported by the paleobotanical record. Wendt (1993) believes ca. 75% of the canopy trees of the region had South American origins. These are listed in the appendix (1993, 657–71) of his publication under column 2 with the designation 3 (= Columbia, Ecuador, Venezuela, the Guyanas) or 4 (+ Peru, Brazil, or farther south). Of the modern Mexican rainforest canopy trees, those reported from the Eocene of southeastern North America and from the Tertiary of northern Latin America are given in table 5.3 in the online supplementary materials. Both are from Wendt (1993) and are selected lists; the one for the southeastern United States includes only post-1960 publications and that for northern Latin America mostly post-1975 (Graham references in Wendt, 1993, 671–80). Thus, the misidentifications present in the older literature, estimated to be as high as 60% (Dilcher, 1971), are reduced. From the 93 combined taxa listed in online table 5.3 it is clear that elements of the tropical rainforest as expected moved extensively northward across the CALB.

There were undoubtedly migrations of tropical trees and shrubs of North and Central American origin into South America, but both modern affinities and the paleobotanical record indicate these were fewer in number. *Lonchocarpus* (76 species in Mexico, 8 in Brazil) and *Sideroxylon* (diverse in Mexico, Central America, and especially the Antilles, poorly developed in the Amazon Basin) may be examples. However, given that a requirement for this community is consistently warm, wet, lowland habitats, and that these habitats were available for most of the Tertiary especially in the southern portion of the CALB, and that the lowlands were nearly continuous, it is not surprising that extensive migrations of tropical rainforest species took place across the connection. Considering the size of the reservoir and the early presence of rainforests in northern South America, it is also not surprising that migrations were primarily from south to north.

The same habitats and conditions were available for aquatic and freshwater herbaceous bog/marsh/swamp vegetation throughout the Late Cretaceous and Cenozoic. Prominent plants in the modern flora include

*Ceratopteris, *Isöetes, *Salvinia, *Cabomba, Calathea, *Ceratophyllum, Drosera, Eichornia, Heliconia, *Lemna, Myriophyllum, *Nymphoides, *Pachira, Phragmites, Pistia, Pontederia, *Potamogeton, *Sagitaria, *Thalia, *Typha, *Utricularia, and Vallesneria in addition to some *Cyperaceae (e.g., Scirpus?), Gramineae/Poaceae (*Phragmites?), and freshwater algae (*Pediastrum).

The two plant communities most immediately adjacent to the lowland depositional basins and receiving highly organic material eventually converted to lignite, other than beach-strand-dune, are the seaward mangroves, and the inland tropical rainforest just beyond the influence of marine and brackish waters. This means the fossil record derived from these sediments is a reasonably accurate reflection of the presence or absence of the communities. The rainforest was absent from at least one of these sites after the MMCO (Veracruz), as well as fluctuating in range and composition elsewhere (the Amazon Basin). The mangrove community has a more consistent presence in the tropical lowlands of the New World, with representation determined primarily by the south-to-north direction of its migration and the time it has taken to reach its most recent northernmost localities. Fossils representing the four principal genera at present making up the New World mangrove community (Avicennia, Conocarpus, Laguncularia, and Rhizophora) are all present in the Tertiary of the CALB. Pollen of the most prevalent, Rhizophora, first appears in the middle to late Eocene in Australia, then soon thereafter in the late Eocene of northern South America (Germeraad et al., 1968) and Panama (Graham, 2006). It is known from the Oligocene of Puerto Rico, and the Miocene (Chiapas) and Pliocene (Veracruz) of Mexico. The northern limit today is typically at 28°N and the minimum MAT is ca. 0°C, but cold-tolerant populations are now being reported from southern Florida, Louisiana, Texas, and even Georgia and North and South Carolina at 35°N. This raises the question of whether the assumed ecological requirements of existing Rhizophora are an accurate indicator of those tolerances, or if those tolerances are actually broader and the plants have only had time to reach southern Florida. A similar situation pertains to the armadillo, which in the literature of the 1930s was assumed limited to warm-temperate environments as in the southern United States, but now has expanded well into Missouri and beyond. Hummingbirds from South America have been moving north and have reached as far north as southern Canada. Global warming is undoubtedly important in the recent expansion (Taulman and Robbins, 1996), but time has likely been a complementary factor. Along with the four prominent genera of the mangrove community, spores of the associated fern Acrostichum, and pollen of Hibiscus/Hampea (possibly representing Hibiscus tiliaceous), Crenea,

Pachria, and others are known as fossils along or immediately adjacent to the CALB. In addition, *Pelliceria* was present in the Tertiary of Mexico and the Antilles far north of its present distribution along the coasts of Central America and northern South America. As mentioned previously, the southeast Asian mangrove palm *Nypa* is known as fossils from both north and south of the CALB (Fig. 5.6). The less tropical Southern Hemisphere *Nothofagus* is reported from Texas and the Pacific Northwest (Elsik, 1974), and *Podocarpus* was in the southeastern United States in the late Eocene, so their presence undoubtedly involves in part and in places closer past continental positions and cooler but still warm climates immediately after the EECO.

The CALB was an important avenue for the interchange of both lowland tropical and upland, more temperate plants between the Americas during different times of the Tertiary. This has long been assumed from the distribution patterns and taxonomic affinities of many modern taxa. The fossil record adds some names, direction, and a temporal and environmental context. In turn, this provides a more defined framework for estimating the migration of other organisms without an extensive fossil record. While the BLB and the NALB were functioning as potential pathways for some warm-temperate and subtropical entities in the Paleogene, and numerous temperate and cold-temperate ones in the Neogene, the ALB and the CALB were accommodating numerous lowland tropical and subtropical species throughout the Tertiary and more upland ones later in the Miocene and especially after ca. 3.5 Ma. Collectively the four connections, along with other means and modes of transport, afforded extensive routes of transport for New World terrestrial plants and animals. Some additions to the inventory were made possible by happenings at the far southern end of the New World.

References

Acosta Ochoa, G. 2012. Ice age hunter-gatherers and the colonization of Mesoamerica. In D. L. Nichols and C. A. Pool (eds.), The Oxford Handbook of Mesoamerican Archaeology. Oxford University Press, Oxford. Pp. 129–40.

Anderson, T. H. 1968. First evidence for glaciation in the Sierra de Los Cuchumatanes Range, northwestern Guatemala. Abstract, Geological Society of America Special Paper 121: 387.

Anderson, T. H., R. J. Erdlac, and M. A. Sandstrom. 1985. Late Cretaceous allochthons and post-Cretaceous strike-slip displacement along the Cuilco-Chixoy-Polochic fault, Guatemala. Tectonics 4: 453–75.

Angiosperm Phylogeny Group. 2009. An update of the Angiosperm Phylogeny Group classification for the orders and families of flowering plants. APG III. Botanical Journal of the Linnean Society 161: 105–21.

Araújo, C. G., and V. J. M. Cardoso. 2006. Storage in cerrado soil and germination of *Psychotria vellosiana* (Rubiaceae) seeds. Brazilian Journal of Biology 66: 709–17.

Arrieta, S., and F. Suárez. 2005. Spatial dynamics of *Ilex aquifolium* populations seed dispersal and seed bank: understanding the first steps of regeneration. Plant Ecology 177: 237–48.

Augspurger, C. K., and K. P. Hogan. 1983. Wind dispersal of fruits with variable seed number in a tropical tree (*Lonchocarpus pentaphyllus*: Leguminosae). American Journal of Botany 70: 1031–37.

August, P. V. 1981. Fig fruit consumption and seed dispersal by *Artibeus jamaicensis* in the llanos of Venezuela. Biotropica 13: 70–76.

Bacon, C. D., A. Mora, W. L. Wagner, and C. A. Jaramillo. 2013. Testing geological models of evolution of the Isthmus of Panama in a phylogenetic framework. Botanical Journal of the Linnean Society 171: 287–300.

Bagley, J. C., and J. B. Johnson. 2014. Phylogeography and biogeography of the lower Central American Neotropics: diversification between two continents and between two seas. Biological Reviews 89: 767–90.

Bell, C. E., D. E. Soltis, and P. S. Soltis. 2010. The age and diversification of the angiosperms re-visited. American Journal of Botany 97: 1296–303.

Beltrán, L. C. 2015. Post-dispersal Seed Fate of *Ocotea floribunda* (Lauraceae) in Monteverde, Costa Rica. Senior thesis, Lake Forest College. Lake Forest College Publications, Lake Forest, IL.

Bhaumik, A. K., A. K. Gupta, and S. Ray. 2014. Surface and deep-water variability at the Blake Ridge, NW Atlantic during the Plio-Pleistocene is linked to the closing of the Central American Seaway. Palaeogeography, Palaeoclimatology, Palaeoecology 399: 345–51.

Bloch, J. I., et al. (+ 11 authors). 2016. First North American fossil monkey and early Miocene tropical biotic interchange. Nature 533: 243–46.

Borges, M. R., and C. Melo. 2012. Frugivory and seed dispersal of *Miconia theaezans* (Bonpl.) Cogniaux (Melastomataceae) by birds in a transition palm swamp-gallery forest in central Brazil. Brazilian Journal of Botany 72: 25–31.

Burge, D. O., and S. R. Manchester. 2008. Fruit morphology, fossil history, and biogeography of *Paliurus* (Rhamnaceae). International Journal of Plant Sciences 169: 1066–85.

Bush, M. B., D. R. Piperno, and P. A. Colinvaux. 1989. A 6,000 year history of Amazonian maize cultivation. Nature 340: 303–5.

Carlo, T. A., D. Garcia, D. Martínez, J. M. Gleditsch, and J. M. Morales. 2013. Where do seeds go when they go far? Distance and directionality of avian seed dispersal in heterogeneous landscapes. Ecology 94: 301–7.

Carvalho, M. R., F. A. Herrera, C. A. Jaramillo, S. L. Wing, and R. Callejas. 2011. Paleocene Malvaceae from northern South America and the biogeographical implications. American Journal of Botany 98: 1337–55.

Christianini, A. V., and P. S. Oliveira. 2009. The relevance of ants as seed rescuers of a primarily bird-dispersed tree in the neotropical cerrado savanna. Oecologia 160: 735–45.

Collevatti, R. G., T. G. de Castro, J. de Souza Lima, and M. P. de Campos Telles. 2012. Phylogeography of *Tibouchina papyrus* (Pohl) Toledo (Melastomataceae), an endangered tree species from rocky savannas, suggests bidirectional expansion due to climate cooling in the Pleistocene. Ecology and Evolution 2: 1024–35.

Conceição, A. A., L. S. Funch, and J. R. Pirani. 2007. Reproductive phenology, pollination and seed dispersal syndromes on sandstone outcrop vegetation in the "Chapada Diamantina," northeastern Brazil: population and community analyses. Revista Brasileira de Botânica 30: 475–85.

Corbett, S. L., and S. R. Manchester. 2004. Phytogeography and fossil history of *Ailanthus* (Simaroubaceae). International Journal of Plant Sciences 165: 671–90.

Córtes, M. C., et al. (+ 6 authors). 2013. Low plant density enhances gene dispersal in the Amazonian understory herb *Heliconia acuminata*. Molecular Ecology. doi: 10.1111/mec.12495.

Croat, T. B. 1978. Flora of Barro Colorado Island. Stanford University Press, Stanford.

Dalling, J. W., and R. Wirth. 1998. Dispersal of *Miconia argentea* seeds by the leaf-cutting ant *Atta colombica*. Journal of Tropical Ecology 14: 705–10.

D'Arcy, W. G. 1987 et seq. Flora of Panama, Checklist and Index. Part I: The Introduction and Checklist. Part II: Index. Monographs in Systematic Botany from the Missouri Botanical Garden 17. Missouri Botanical Garden, St. Louis.

D'Arcy, W. G., and M. D. Correa A. (eds.). 1985. The Botany and Natural History of Panama. Monographs in Systematic Botany from the Missouri Botanical Garden 10. Missouri Botanical Garden, St. Louis.

Davidse, G., M. Sousa Sánchez, and S. Knapp (general eds.). 1994 et seq. Flora Mesoamericana. Universidad Nacional Autónoma de México, México, D.F., in association with the Missouri Botanical Garden, St. Louis, and the Natural History Museum, London.

Dilcher, D. L. 1969. *Podocarpus* from the Eocene of North America. Science 164: 299–301.

———. 1971. A revision of the Eocene flora of southeastern North America. Palaeobotanist 20: 7–18.

Distler, T., P. M. Jørgensen, A. Graham, G. Davidse, and I. Jiménez. 2009. Determinants and predictions of broad-scale plant richness across the western Neotropics. Annals of the Missouri Botanical Garden 96: 470–91.

Donnelly, T. W. 1985. Mesozoic and Cenozoic plate evolution of the Caribbean region. In F. G. Stehli and S. D. Webb (eds.), The Great American Biotic Interchange. Plenum, New York. Pp. 89–121.

———. 1989. History of marine barriers and terrestrial connections: Caribbean paleogeographic inference from pelagic sediment analysis. In C. A. Woods (ed.), Biogeography of the West Indies: Past, Present, and Future. Sandhill Crane Press, Gainesville, FL. Pp. 103–18.

DRYFLORA. 2016. Plant diversity patterns in neotropical dry forests and their conservation implications. Science 353: 1383–87 (plus supporting online material).

Du, Y., X. Mi, X. Liu, L. Chen, and K. Ma. 2009. Seed dispersal phenology and dispersal syndromes in a subtropical broad-leaved forest of China. Forest Ecology and Management 258: 1147–52.

Elsik, W. C. 1974. *Nothofagus* in North America. Pollen et Spores 16: 285–99.

Estes, R., and A. Báez. 1985. Hepertofaunas of North and South America during the Late Cretaceous and Cenozoic: evidence for interchange? In F. G. Stehli and S. D. Webb (eds.), The Great American Biotic Interchange. Plenum, New York. Pp. 139–97.

European Soil Data Centre (ESDAC). 2014. Soil Atlas of Latin America. European Commission, Brussels.

Evans, S. T. 2012. Time and space boundaries, chronologies and regions in Mesoamerica. In D. L. Nichols and C. A. Pool (eds.), The Oxford Handbook of Mesoamerican Archaeology. Oxford University Press, Oxford. Pp. 114–26.

Ferrari, L., T. Esquivel, V. Manea, and M. Manea. 2012. The dynamic history of the Trans-

Mexican Volcanic Belt and the Mexico subduction zone. Tectonophysics 522-23: 122-49.

Ferrari, L., G. Pasquarè, S. Venegas-Salgado, and F. Romero-Ríos. 1999. Geology of the western Mexican Volcanic Belt and adjacent Sierra Madre Occidental and Jalisco block. Geological Society of America, Special Paper 334: 65-83.

Flora Malesiana. s.d. http://portal.cybertaxonomy.org/flora-malesiana/.

Fridriksson, S. 1975. Surtsey: Evolution of Life on a Volcanic Island. Butterworths, London.

Friis, E. M., P. R. Crane, and K. R. Pedersen. 2011. Early Flowers and Angiosperm Evolution. Cambridge University Press, Cambridge.

Germeraad, J. H., C. A. Hopping, and J. Muller. 1968. Palynology of Tertiary sediments from tropical areas. Review of Palaeobotany and Palynology 6: 189-348.

Gonzalez, C. C., M. A. Gandolfo, M. C. Zamaloa, N. R. Cúneo, P. Wilf, and K. R. Johnson. 2007. Revision of the Proteaceae macrofossil record from Patagonia, Argentina. Botanical Review 73: 235-66.

Graham, A. 1976a. Studies in neotropical paleobotany. II. The Miocene communities of Veracruz, Mexico. Annals of the Missouri Botanical Garden 63: 787-842.

———. 1976b. Late Cenozoic evolution of tropical lowland vegetation in Veracruz, Mexico. Evolution 29: 723-35.

———. 1977. The tropical rain forest near its northern limits in Veracruz, Mexico: recent and ephemeral? Boletin de la Sociedad Botánica de México 36: 13-19.

———. 1982. Diversification beyond the Amazon Basin. In G. T. Prance (ed.), Biological Diversification in the Tropics. Columbia University Press, New York. Pp. 78-90.

———. 1988. Some aspects of Tertiary vegetational history in the Gulf/Caribbean region. Transactions of the 11th Caribbean Geological Conference, Barbados. Pp. 3.1-3.18.

———. 2006. Latin American biogeography: causes and effects: introduction. Annals of the Missouri Botanical Garden 93: 173-77.

———. 2015. Past ecosystem dynamics in fashioning views on conserving extant New World vegetation. Annals of the Missouri Botanical Garden 100: 150-58.

Graham, S. A. 2013. The geologic history of the Lythraceae. Botanical Review 79: 48-145.

Guimarães, P. R., Jr., M. Galetti, and P. Jordano. 2008. Seed dispersal anachronisms: rethinking the fruits extinct megafauna ate. PLoS ONE. doi: 10.1371/journal.pone.0001745.

Guttal, V., F. Bartumeus, G. Hartvigen, and A. L. Neval. 2011. Retention time variability as a mechanism for animal mediated long-distance dispersal. PLoS ONE. doi: 10.1371/journal.pone.0028447.

Hammel, B. E., M. H. Grayum, C. Herrera, and N. Zamora (eds.). 2003 et seq. Manual de Plantas de Costa Rica. Monographs in Systematic Botany from the Missouri Botanical Garden 92 et seq. Missouri Botanical Garden, St. Louis.

Harley, M. M. 2006. A summary of fossil records for Arecaceae. Botanical Journal of the Linnean Society 151: 39-67.

Hastenrath, S. 1973. On the Pleistocene glaciation of the Cordillera de Talamanca, Costa Rica. Zeitschrift für Gletscherkunde und Glaziageologie 9: 105-21.

Hedges, S. B. 2006. Paleogeography of the Antilles and origin of West Indian terrestrial vertebrates. Annals of the Missouri Botanical Garden 93: 231-44.

Herrera, F., S. R. Manchester, and C. Jaramillo. 2012. Permineralized fruits from the late Eocene of Panama give clues to the composition of forests established early in the uplift of Central America. Review of Palaeobotany and Palynology 175: 10-24.

Hooghiemstra, H. 2006. Immigration of oak into northern South America: a paleoecological document. In M. Kappelle (ed.), Ecology and Conservation of Neotropical Montane Oak Forests. Studies in Ecology 185. Springer, Berlin. Pp. 17–28.

Hooghiemstra, H., V. M. Wijninga, and A. M. Cleef. 2006. The paleobotanical record of Colombia: implications for biogeography and biodiversity. Annals of the Missouri Botanical Garden 93: 297–324.

Hoorn, C. 1994a. Fluvial palaeoenvironments in the intracratonic Amazonas Basin (early Miocene–early middle Miocene, Colombia). Palaeogeography, Palaeoclimatology, Palaeoecology 109: 1–55.

———. 1994b. An environmental reconstruction of the palaeo-Amazon River system (middle to late Miocene, NW Amazonia). Palaeogeography, Palaeoclimatology, Palaeoecology 112: 187–238.

Hoorn, C., and S. Flantua. 2015. An early start for the Panama land bridge. Science 348: 186–87.

Hoorn, C., J. Guerrero, G. A. Sarmiento, and M. A. Lorente. 1995. Andean tectonics as a cause for changing drainage patterns in Miocene South America. Geology 23: 237–40.

Hoorn, C., et al. (+ 17 authors). 2010. Amazonia through time: Andean uplift, climate change, landscape evolution and biodiversity. Science 330: 927–31.

Howe, H. F. 1977. Bird activity and seed dispersal of a tropical wet forest tree. Ecology 58: 539–50.

———. 1984. Implications of seed dispersal by animals for tropical reserve management. Biological Conservation 30: 261–81.

Howe, H. F., and M. N. Miriti. 2004. When seed dispersal matters. BioScience 54: 651–60.

Iles, W. J., et al. (+ 8 authors). 2014. Reconstructing the age and historical biogeography of the ancient flowering-plant family Hydatellaceae (Nymphaeales). BMC (BioMed Central) Evolutionary Biology. doi: 10.1186/1471-2148-14-102.

Iturraldi-Vinent, M. A., and R. D. E. MacPhee. 1999. Paleogeography of the Caribbean region: implications for Cenozoic biogeography. Bulletin of the American Museum of Natural History 238: 1–95.

Jablonski, D., et al. (+ 7 authors). 2013. Out of the tropics, but how? fossils, bridge species, and thermal ranges in the dynamics of the marine latitudinal diversity gradient. Proceedings of the National Academy of Sciences USA 110: 10487–94.

Jacques, F. M. B. 2009. Fossil history of the Meninspermaceae (Ranunculales). Annales de Paléontologie 95: 53–69.

Janzen, D. H., and P. S. Martin. 1982. Neotropical anachronisms: the fruits the gomphotheres ate. Science 215: 19–27.

Jaramillo, C. A., and D. L. Dilcher. 2001. Middle Paleogene palynology of central Colombia, South America: a study of pollen and spores from tropical latitudes. Palaeontographica Abt. B, 258: 87–213.

Jaramillo, C. A., et al. (+ 12 authors). 2014. Palynological record of the last 20 million years in Panama. In W. D. Stevens, O. M. Montiel, and P. H. Raven (eds.), Paleobotany and Biogeography, A Festschrift for Alan Graham in His 80th Year. Missouri Botanical Garden Press, St. Louis. Pp. 134–251.

Jarzen, D. M., and M. E. Dettmann. 1989. Taxonomic revision of Tricolpites reticulatus Cookson et Cooper, 1953, with notes on the biogeography of Gunnera L. Pollen et Spores 31: 97–112.

Jiménez, I., T. Distler, and P. M. Jørgensen. 2009. Estimated plant richness pattern across northwest South America provides support for the species-energy and spatial heterogeneity hypotheses. Ecography 32: 433–48.

Jud, N. A., C. W. Nelson, and F. Herrera. 2016. Fruits and seeds of *Parinari* from the early Miocene of Panama and the fossil record of the Chrysobalanaceae. American Journal of Botany 103: 277–89.

Kato, E., and T. Hiura. 1999. Fruit set in *Styrax obassia* (Styracaceae): the effect of light availability, display size, and local floral density. American Journal of Botany 86: 495–501.

Keigwin, L. D., Jr. 1982. Isotropic and paleoceanography of the Caribbean and east Pacific: role of Panama uplift in late Neogene time. Science 217: 350–53.

King, R. T. 2003. Seed dispersal by bats and the population dynamics of a canopy tree *Calophyllum brasiliense* (Clusiaceae), in an Amazonian river meander forest. Dissertations from ProQuest 1957. http://scholarlyrepository.miami.edu/dissertations/1957.

Kirchheimer, F. 1949. Die Symplocacean der erdgesichtlichen vergangenheit. Palaeontograpica, Abt B, 90: 1–52.

Kissmann, C., and G. Habermann. 2013. Seed germination performances of *Styrax* species help understand their distribution in cerrado areas in Brazil. Bragantia. doi: 10.1590/brag.2013.030.

Kreft, H., and W. Jetz. 2007. Global patterns and determinants of vascular plant diversity. Proceedings of the National Academy of Sciences USA 104: 5925–30.

Kubitzki, K. (ed.). 2004. The Families and Genera of Vascular Plants. Springer, Berlin.

Kumar, C. N. P., R. K. Somashekar, B. C. Nagaraja, K. Ramachandra, and D. Shivaprasad. 2015. Seed bank estimation and regeneration studies of *Calophyllum apetalum* Willd., from Western Ghats of Karnataka. Proceedings of the International Academy of Ecology and Environmental Sciences 5: 97–103.

Leigh, E. G., A. O'Dea, and G. J. Vermeij. 2014. Historical biogeography of the Isthmus of Panama. Biological Reviews 89: 148–72.

Loiselle, B. A., and J. G. Blake. 1999. Dispersal of melastome seeds by fruit-eating birds of tropical forest understory. Ecology 80: 330–36.

Lomáscolo, S. B., D. J. Levey, R. T. Kimball, B. M. Bolker, and H. T. Alborn. 2010. Dispersers shape fruit diversity in *Ficus* (Moraceae). Proceedings of the National Academy of Sciences USA 107: 14668–72.

Magallón, S., S. Gómez-Acevedo, L. L. Sánchez-Reys, and T. Hernández-Hernández. 2015. A meta-calibrated time-tree documents the early rise of flowering plant phylogenetic diversity. New Phytologist 207: 437–53.

Magnusson, W. E., and T. M. Sanaiotti. 1987. Dispersal of *Miconia* seeds by the rat *Bolomys lasiurus*. Journal of Tropical Ecology 3: 277–78.

Manchester, S. R. 1987. The fossil history of the Juglandaceae. Missouri Botanical Garden Monographs in Systematic Botany 21: 1–137.

———. 1989. Systematics and fossil history of the Ulmaceae. In P. R. Crane and S. Blackmore (eds.), Evolution, Systematics, and Fossil History of the Hamamelidae. Vol. 2. Pp. 221–51. Clarendon Press, Oxford.

Mann, C. G. 2012. 1491: New Revelations of the Americas before Columbus. Knopf, New York. [See review by Tollefson, 2013.]

Manos, P. S., P. S. Soltis, D. E. Soltis, S. R. Manchester, S.-H. Ho, C. D. Bell, D. L. Dilcher, and D. E. Stone. 2007. Phylogeny of extant and fossil Juglandaceae inferred from the integration of molecular and morphological data sets. Systematic Biology 56: 412–30.

Markgraf, V. 1989. Paleoclimates in Central and South America since 18,000 yr B.P. based on pollen and lake-level records. Quaternary Science Reviews 8: 1–24.

——— (ed.). 2001. Interhemispheric Climate Linkages. Academic Press, New York.

Marques, M. C. M., and E. Fischer. 2008. Effect of bats on seed distribution and germination of *Calophyllum brasiliense* (Clusiaceae). Ecotropica 15: 1–6.

Marshall, L. G. 1985. Geochronology and land-mammal biochronology of the Transamerican faunal interchange. In F. G. Stehli and S. D. Webb (eds.), The Great American Biotic Interchange. Plenum, New York. Pp. 49–85.

McCullough, D. 1977. The Path between the Seas: The Creation of the Panama Canal, 1870–1914. Simon & Schuster, New York.

McKillop, H. 2004. The Ancient Maya: New Perspectives. ABC CLIO, Santa Barbara, CA.

Montes, C., et al. (+ 14 authors). 2012. Evidence for middle Eocene and younger land emergence in central Panama: implications for Isthmus closure. Geological Society of American Bulletin 124: 780–99.

Montes, C., et al. (+ 10 authors). 2015. Middle Miocene closure of the Central American Seaway. Science 348: 226–29.

Moussa, Y., S. Ohashi, and S. Wen. 2014. The effects of height, wing length, and wing symmetry on *Tabebuia rosea* seed dispersal. ISB Journal of Science 8: 1–4. http://www/isjos.org.

Muller-Landau, H. C., S. J. Wright, O. Calderón, R. Condit, and S. P. Hubbell. 2008. Interspecific variation in primary seed dispersal in a tropical forest. Journal of Ecology 96: 653–67.

Nathan, R., F. M. Schurr, O. Spiegel, O. Steinitz, A. Trakhtenbrot, and A. Tsoar. 2008. Mechanisms of long-distance seed dispersal. Trends in Ecology and Evolution 23: 638–47.

Nelson, C. W., and N. A. Jud. 2017. Biogeographic implications of *Mammea paramericana* sp. nov. from the lower Miocene of Panama and the evolution of Calophyllaceae. International Journal of Plant Sciences 178. doi: 10.86/689618.

Obeso, J. R., and I. C. Fernández-Calvo. 2002. Fruit removal, pyrene dispersal, post-dispersal predation and seeding establishment of a bird-dispersed tree. Plant Ecology 165: 223–33.

O'Dea, A., N. Hoyos, F. Rodríguez, B. De Gracia, and C. Degarcia. 2012. History of upwelling in the tropical Eastern Pacific and the paleogeography of the Isthmus of Panama. Palaeogeography, Palaeoclimatology, Palaeoecology 348–49: 59–66.

O'Dea, A., et al. (+ 34 authors). 2016. Formation of the Isthmus of Panama. Science Advances 2016;2:e1600883. http://advances.sciencemag.org.

Olmstead, R. G. 2013. Phylogeny and biogeography in Solonaceae, Verbenaceae and Bignoniaceae: a comparison of continental and intercontinental diversification patterns. Botanical Journal of the Linnean Society 171: 80–102.

Pan, A. D., B. F. Jacobs, J. Dransfield, and W. J. Baker. 2006. The fossil history of palms (Arecaceae) in Africa and new records from the late Oligocene (28–27 Mya) of northwestern Ethiopia. Botanical Journal of the Linnean Society 151: 69–81.

Parolin, P., F. Wittmann, and L. Ferreira. 2013. Fruit and seed dispersal in Amazonian floodplain trees—a review. Ecotropica 19: 19–36.

Parrado-Rosselli, A. 2005. Fruit Availability and Seed Dispersal in Terra Firme Rain Forests of Colombian Amazonia. Ph.D. diss., University of Amsterdam. Tropenbos International, Wageningen.

Passos, L., and P. S. Oliveira. 2002. Ants affect the distribution and performance of seedlings of *Clusia criuva*, a primarily bird-dispersed rain forest tree. Journal of Ecology 90: 517–28.

Pimm, S. L., and P. H. Raven. 2017. The fate of the world's plants. Trends in Ecology and Evolution 32: 317–20.

Piperno, D. R., and B. D. Smith. 2012. The origins of food production in Mesoamerica. In D. L. Nichols and C. A. Pool (eds.), The Oxford Handbook of Mesoamerican Archaeology. Oxford University Press, Oxford. Pp. 151–64.

Potter, P. E. 1997. The Mesozoic and Cenozoic paleodrainage of South America: a natural history. Journal of South American Earth Sciences 10: 331–44.

Prance, G. T., and S. A. Mori. 1978. Observations on the fruits and seeds of neotropical Lecythidaceae. Brittonia 30: 21–33.

Renner, S. A., and T. J. Givnish (organizers). 2004. Tropical Intercontinental Disjunctions. International Journal of Plant Sciences 165 (supplement): S1–S138.

Ribeiro Mello, M. A., N. O. Leiner, P. R. Guimarães, Jr., and P. Jordano. 2005. Size-based fruit selection of *Calophyllum brasiliense* (Clusiaceae) by bats of the genus *Artibeus* (Phyllostomidae) in a restinga area, southeastern Brazil. Acta Chiropterologica 7: 165–88.

Sánchez-Rojas, G., V. Sánchez-Cordero, and M. Briones. 2004. Effect of plant species, fruit density and habitat on post-dispersal fruit and seed removal by spiny pocket mice (*Liomys pictus*, Heteromyidae) in a tropical dry forest in Mexico. Studies on Neotropical Fauna and Environment 39: 1–6.

Schuchert, C. 1935/1968. Historical Geology of the Antillean-Caribbean Region. Facsimile edition. Hafner, New York.

Shanahan, M. J. 2000. *Ficus* seed dispersal guilds: ecology, evolution and conservation implications. PhD. diss., University of Leeds.

Simpson, G. G. 1965. The Geography of Evolution: Collected Essays. Chilton Books, Philadelphia.

Stehli, F. G., and S. D. Webb (eds.). 1985. The Great American Biotic Interchange. Plenum, New York.

Stevens, P. 2001 et seq. Angiosperm Phylogeny Website. http://www.mobot.org/MOBOT/research/APweb.

Stevens, W. D., C. Ulloa Ulloa, A. Pool, and O. M. Montiel (eds.). 2001. Flora of Nicaragua. Monographs in Systematic Botany from the Missouri Botanical Garden 85. Missouri Botanical Garden, St. Louis.

Stiles, E. W. 1980. Patterns of fruit presentation and seed dispersal in bird-disseminated woody plants in the eastern deciduous forest. American Naturalist 116: 670–88.

Tank, D. C., et al. (+ 8 authors). 2015. Nested radiations and the pulse of angiosperm diversification. New Phytologist 207: 454–67.

Taulman, J. F., and L. W. Robbins. 1996. Recent range expansion and distributional limits of the nine-banded armadillo (*Dasypus novemcinctus*) in the United States. Journal of Biogeography 23: 635–48.

Taylor, D. W., and W. L. Crepet. 1987. Fossil floral evidence of Malpighiaceae and an early plant-pollinator relationship. American Journal of Botany 74: 274–86.

Tello, J. S. (+10 authors). 2015. Elevational gradients in ß-diversity reflect variation in the strength of local community assembly mechanisms across spatial scales. PLOS One. doi: 10.1371/journal.pone.0121458.

Theim, T. J., R. Y. Shirk, and T. J. Givnish. 2014. Spatial, genetic structure in four understory *Psychotria* species (Rubiaceae) and implications for tropical forest diversity. American Journal of Botany 101: 1189–99.

Tiffney, B. H. 1979. Fruits and seeds of the Brandon Lignite III. *Turpinia* (Staphyleaceae). Brittonia 31: 39–51.

Tollefson, J. 2013. Footprints in the forest. Nature 502: 160–62.

Torres, V., H. Hooghiemstra, L. J. Lourens, and P. C. Tzedakis. 2013. Astronomical tuning of long pollen records reveals the dynamic history of montane biomes and lake levels in the tropical high Andes during the Quaternary. Quaternary Science Reviews 63: 59–72.

Tsou, C.-H., and S. A. Mori. 2002. Seed coat anatomy and its relationship to seed dispersal in subfamily Lecythidoideae of the Lecythidaceae (the Brazil nut family). Botanical Bulletin Academia Sinica 43: 37–56.

Uriarte, M., M. Anciães, M. T. da Silva, P. Rubim, E. Johnson, and E. M. Bruna. 2011. Disentangling the drivers of reduced long-distance seed dispersal by birds in an experimentally fragmented landscape. Ecology 92: 924–37.

Van der Hammen, T. 1952. Geología del Río Apaporis entre Soratama y Cachivera La Playa. Informe Servicio Geológia Nacional, Bogotá, Colombia, 834.

Vásquez, D. L. A., H. Balslev, M. M. Hansen, P. Skenáᛗ, and K. Romeroux. 2016. Low genetic variation and high differentiation across sky island populations of *Lupinus alopecuroides* (Fabaceae) in the northern Andes. Alpine Botany. http://link.springer.com/article/10.1007%2Fs00035-016-0165-7.

Viana, D. S., L. Gangoso, W. Bouten, and J. Figuerola. 2016. Overseas seed dispersal by migratory birds. Proceedings of the Royal Society B (Biological Sciences). doi: 10.1098/rspb.2015.2406.

Viana, D. S., L. Santamaría, T. C. Michot, and J. Figuerola. 2013. Allometric scaling of long-distance seed dispersal by migratory birds. American Naturalist 181: 649–62.

Vuillemier, B. S. 1971. Pleistocene changes in the flora and fauna of South America. Science 173: 771–80.

Wales, N., and J. Ramos-Madirgal. 2016. Genome sequence of a 5,310-year-old maize cob provides insight into the early stages of maize domestication. Current Biology. www.cell.com/current-biology/fulltext/s0960-9822(16)31120-4. doi: 10.1016/j.cub.2016.09.036.

Wang, Q., S. R. Manchester, C. Li, and B. Geng. 2010. Fruits and leaves of *Ulmus* from the Paleogene of Fushun, northeastern China. International Journal of Plant Sciences 171: 221–26.

Webb, S. D. 1985. Main pathways of Mammalian diversification in North America. In F. G. Stehli and S. D. Webb (eds.), The Great American Biotic Interchange. Plenum, New York. Pp. 201–17.

———. 2006. The Great American Biotic Interchange: patterns and processes. Annals of the Missouri Botanical Garden 93: 245–57.

Wendt, T. 1993. Composition, floristic affinities, and origins of the canopy tree flora of the Mexican Atlantic Slope rain forests. In T. P. Ramamoorthy, R. Bye, A. Lot, and J. Fa (eds.), Biological Diversity of Mexico, Origins and Distributions. Oxford University Press, Oxford. Pp. 595–680.

Wenny, D. G., and D. J. Levey. 1998. Directed seed dispersal by bellbirds in a tropical cloud forest. Proceedings of the National Academy of Sciences USA 95: 6204–7.

Wheeler, E. A., and T. M. Lehmann. 2009. New Late Cretaceous and Paleocene dicot woods of Big Bend National Park, Texas, and review of Cretaceous wood characteristics. IAWA (International Association of Wood Anatomists) Journal 30: 293–318.

Wheeler, E. A., T. M. Lehmann, and P. E. Gasson. 1994. *Javelinoxylon*, an Upper Cretaceous dicotyledonous tree from Big Bend National Park with presumed malvalean affinities. American Journal of Botany 81:703–10.

Wheelwright, N. T. 1993. Fruit size in a tropical tree species: variation, preference by birds, and heritability. Vegetatio 107–8: 163–74.

Wheelwright, N. T., W. A. Haber, K. G. Murray, and C. Guindon. 1984. Tropical fruit-eating birds and their food plants: a survey of a Costa Rica lower montane forest. Biotropica 16: 173–92.

Whittaker, R. J. 1998. Island Biogeography: Ecology, Evolution, and Conservation. Oxford University Press, Oxford.

Wiens, J. J., and M. J. Donoghue. 2004. Historical biogeography, ecology and species richness. Trends in Ecology and Evolution 19: 639–44.

Wikström, N., V. Savolainen, and M. W. Chase. 2001. Evolution of the angiosperms: calibrating the family tree. Proceedings of the Royal Society B (Biological Sciences) 268: 2211–20.

Wilkinson, D. M. 1997. Plant colonization: are wind dispersed seeds really dispersed by birds at larger spatial and temporal scales? Journal of Biogeography 24: 61–65.

Wilson, M. F., C. Sabag, J. Figueroa, and J. J. Armesto. 1996. Frugivory and seed dispersal of *Podocarpus nubigena* in Chiloé, Chile. Revista Chilena de Historia Natural 69: 343–49.

Wolstenholme, B. N., and A. W. Whiley. 1999. Ecophysiology of the avocado (*Persea americana* Mill.) tree as a basis for pre-harvest management. Revista Chapingo Serie Horticultura 5: 77–88.

Woodring, W. P. 1966. The Panama Land Bridge as a sea barrier. Proceedings of the American Philosophical Society 110: 425–433,

Xi, Z., et al. (+10 authors). 2012. Phylogenomics and *a posteriori* partitioning resolve the Cretaceous angiosperm radiation in Malpighiales. Proceedings of the National Academy of Sciences USA 109: 17519–24.

Xiang, X.-G., et al. (+ 7 authors). 2014. Large-scale phylogenetic analyses reveal fagalean diversification promoted by the interplay of diaspores and environments in the Paleogene. Perspectives in Plant Ecology, Evolution, and Systematics 16: 101–10.

Young, K. R. 1990. Dispersal of *Styrax ovatus* seeds by the spectacled bear (*Tremarctos ornatus*). Vida Silvestre Neotropical 2: 68–69.

Yumoto, T. 1997. Seed dispersal by Salvin's curassow, *Mitu salvini* (Cracidae), in a tropical forest of Colombia: direct measurements of dispersal distance. Biotropica 31: 654–60.

Zhou, L.-C., J. B. Bachelier, X.-H. Zang, M.-R. Luo, Z.-Y. Chang, and Y. Rem. 2014. Floral organogenesis in *Dysosma versipellis* and its implications for the systematics and the evolution of petals in Berberidaceae. Abstract, Botany 2014, New Frontiers in Botany.

Additional References

Amazon Conservation Association (ACA) and Conservatión Amazónica (ACCA). 2016. #30, Gold mining invasion of Tambopata National Reserve Intensifies. #35, Confirming Amazon deforestation by United Cacao in 2013. amazonconservation.org. [High-resolution photos.]

Anderson, B. W., A. Hendy, E. H. Johnson, and W. D. Allmon. 2017. Paleoecology and paleoenvironmental implications of turritelline gastropod-dominated assemblages from the Gatun Formation (upper Miocene) of Panama. Palaeogeography, Palaeoclimatology, Palaeoecology 470: 132–46.

Arbogast, B. S., and G. J. Kenagy. 2001. Comparative phylogeography as an integrative approach to historical biogeography. Journal of Biogeography 28: 819–25.

Augliere, B. 2016. Simulating sustainability: Stanford computer model sheds light on how modern interventions can affect tropical forests and indigenous peoples. Stanford Report, 11 March. http://news.stanford.edu/news/2016/march/amazon-model-fragoso-031116.html.

Awe, J. J. 2012. The archaeology of Belize in the twenty first century. In D. L. Nichols and C. Pool (eds.), The Oxford Handbook of Mesoamerican Archaeology. Oxford University Press, Oxford. Pp. 69–82.

Báez, S., et al. (+ 21 authors). 2015. Large-scale patterns of turnover and basal area change in Andean forests. PLoS ONE. doi: 10.1371/journal.pone.0126594.

Baker, R. R. 1982. Migration: Paths through Time and Space. Hodder & Stoughton, London.

Bartlett, A. S., and E. S. Barghoorn. 1973. Phytogeographic history of the Isthmus of Panama during the past 12,000 years (a history of vegetation, climate, and sea-level change). In A. Graham (ed.), Vegetation and Vegetational History of Northern Latin America. Elsevier, Amsterdam. Pp. 203–99.

Bond, W. J. 2016. Ancient grasslands at risk. Science 351: 120–22.

Buchs, D. M., R. J. Arculus, P. O. Baumgartner, and A. Ulianov. 2011. Oceanic intraplate volcanoes exposed: example from seamounts accreted in Panama. Geology 39: 335–38.

Callaway, E. 2017. Collapse of Aztec society linked to catastrophic salmonella outbreak. Nature 542: 404.

Cevallos-Ferriz, S. R. S., G. C. López, and L. A. Flores Rocha. In press. *Laurinoxylon chalatenagensis* sp. nov. from the Miocene Chalatenango Formation, El Salvador. Review of Palaeobotany and Palynology (2016).

Chinchilla Mazariegos, O. 2012. Archaeology in Guatemala, nationalist, colonialist, imperialist. In D. L. Nichols and C. A. Pool (eds.). The Oxford Handbook of Mesoamerican Archaeology. Oxford University Press, Oxford. Pp. 55–68.

Cody, S., J. E. Richardson, V. Rull, C. Ellis, and R. T. Pennington. 2010. The Great American Biotic Interchange revisited. Ecography 33: 326–32.

Crawford, A. J., E. Bermingham, and P. S. Carolina. 2007. The role of tropical dry forest as a long-term barrier to dispersal: a comparative phylogeographical analysis of dry forest tolerant and intolerant frogs. Molecular Ecology 16: 4789–807.

Dick, C. W., and M. Heuretz. 2008. The complex biogeographic history of a widespread tropical tree species. Evolution 62: 2760–74.

Funk, J., P. Mann, K. McIntosh, and J. Stephens. 2009. Cenozoic tectonics of the Nicaraguan depression, Nicaragua, and Median Trough, El Salvador, based on seismic-reflection profiling and remote-sensing data. Geological Society of America Bulletin 121: 1491–521.

Gaston, K. J. 2000. Global patterns in biodiversity. Nature 405: 220–27.

González, C., L. E. Urrego, and J. L. Martínez. 2006. Late Quaternary vegetation and climate change in the Panama Basin: palynological evidence from marine cores ODP 677B and TR 163–38. Palaeogeography, Palaeoclimatology, Palaeoecology 234: 62–80.

González, S., D. Huddart, M. R. Bennett, and A. González-Huesca. 2006. Human footprints in central Mexico older than 40,000 years. Quaternary Science Reviews 25: 201–22.

Hall, J. S., V. Kirn, and E. Y. Fernández (eds.). 2015. Managing watersheds for ecosystem services in the steepland Neotropics. Smithsonian Tropical Research Institute/Inter-American Development Bank, Panama City, Panama. doi: 10.18235/0000163#sthash.hFLgv9dJ.dpuf.

Hall, S. A. 2010. Early maize pollen from Chaco Canyon, New Mexico, USA. Palynology 34: 125–37.

Hauff, F., K. Hoernle, P. van den Bogaard, G. Alvarado, and D. Garbe-Schönberg. 2000. Age and geochemistry of basaltic complexes in western Costa Rica: contributions to the geotectonic evolution of Central America. Geochemistry, Geophysics, Geosystems 1: 1–41.

Hernández, H. M., and C. Gómez-Hinostrosa. 2011, 2015. Mapping the Cacti of Mexico. Part I (2011). Part II (*Mammillaria*) (2015). Succulent Plant Research 9: 1–189. David Hunt, series editor. DH Books, Milborne Port, England.

Hickerson, M. J., et al. (+ 8 authors). 2010. Phylogeography's past, present, and future: 10 years after Avise, 2000. Molecular Phylogenetics and Evolution 54: 291–301.

Hubbell, S. P. 2001. The Unified Theory of Biodiversity and Biogeography. Princeton University Press, Princeton.

Hubbell, S. P., and R. B. Foster. 1986. Biology, chance, and history and the structure of tropical rain forest tree communities. In J. Diamond and T. J. Cawe (eds.), Community Ecology. Harper and Row, New York. Pp. 314–29.

Hughes, L., et al. (+ 6 authors). 1994. Predicting dispersal spectra: a minimal set of hypotheses based on plant attributes. Journal of Ecology 82: 933–50.

Hull, P. M., S. A. F. Darroch, and D. H. Erwin. 2015. Rarity in mass extinctions and the future. Nature 528: 345–51.

Iles, W., S. Y. Smith, M. A. Gandolfo, and S. W. Graham. 2015. Monocot fossils suitable for molecular dating analyses. Botanical Journal of the Linnean Society 178: 346–74.

Jackson, S. T., and L. D. Walls (eds.). 2014. Views of Nature: Alexander von Humboldt. University of Chicago Press, Chicago.

Kappelle, M. 2015. Costa Rican Ecosystems. University of Chicago Press, Chicago.

Kirby, M. X., and B. MacFadden. 2005. Was southern Central America an archipelago or a peninsula in the middle Miocene? A test using land-mammal body size. Palaeogeography, Palaeoclimatology, Palaeoecology 228: 193–202.

Kirby, M. X., D. S. Jones, and B. J. MacFadden. 2008. Lower Miocene stratigraphy along the Panama Canal and its bearing on the Central American Peninsula. PLoS One 3: e2791.

Kreft, H., and W. Jetz. 2007. Global patterns and determinants of vascular plant diversity. Proceedings of the National Academy of Sciences USA 104: 5925–30.

Levine, N. M., et al. (+ 16 authors). 2015. Ecosystem heterogeneity determines the ecological resilience of the Amazon to climate change. Proceedings of the National Academy of Sciences USA. doi: 10.1073.pnas.1511344112.

Losos, J. B., and R. E. Glor. 2003. Phylogenetic comparative methods and the geography of speciation. Trends in Ecology and Evolution 18: 220–27.

Lyons, S. K., et al. (+ 28 authors). 2016. Holocene shifts in the assembly of plant and animal communities implicate human impacts. Nature 529: 80–83 (plus additional supplementary information).

Mann, P., R. D. Rogers, and L. Gahagan. 2007. Overview of plate tectonic history and its unresolved tectonic problems. In J. Bundschuh and G. E. Alvarado (eds.), Central America: Geology, Resources and Hazards. Taylor & Francis, Philadelphia. Pp. 205–41.

Mongabay Weekly Newsletter. 1999–2016 et seq. [See, e.g., Top Vatican official: climate change action is a "moral imperative" (02/04/2016). A railroad that crosses the Amazon could be an infeasible, expensive dream for Peru (02/04/2016). Oil extraction threatens to expand further into Ecuadorean rainforest under new 20-year contract

(02/03/2016). New deforestation hotspot threatens southern Peru's tremendous biodiversity (02/02/2016).]

Morecroft, M. D. 2015. Review: Invasive Species and Global Climate Change (L. H. Ziska and J. S. Dukes). Frontiers of Biogeography 7: 179–80.

Myers, N., R. A. Mittermeier, C. G. Mittermeier, G. A. B. da Fonseca, and J. Kent. 2000. Biodiversity hotspots for conservation priorities. Nature 403: 853–58.

Nature. 2017a. Hydroclimate changes across the Amazon lowlands over the past 45,000 years. Nature 541: 204–7.

———. 2017b. Climate change: save last cloud forests in western Andes. Nature 541: 157.

Olson, S. 2002. Mapping Human History, Genes, Race, and Our Common Origins. Mariner Books, Boston.

Pascual, R. 2006. Evolution and geography: the biogeographic history of South American mammals. Annals of the Missouri Botanical Garden 93: 209–30.

Pascual, R., M. G. Vucetich, G. J. Scillato-Yané, and M. Bond. 1985. Main pathways of mammalian diversification in South America. In F. G. Stehli and S. D. Webb (eds.), The Great American Biotic Interchange. Plenum, New York. Pp 219–47.

Perez, T. M., J. T. Stroud, and K. J. Feeley. 2016. Thermal trouble in the tropics: tropical species may be highly vulnerable to climate change. Science 351: 1392–93.

Pimm, S. L., et al. (+ 8 authors). 2014. The biodiversity of species and their rates of extinction, distribution, and protection. Science. doi: 10:1126/science 1246752.

Piperno, D. L., and K. V. Flannery. 2001. The earliest archaeological maize (Zea mays L) from highland Mexico: new accelerator mass spectrometry dates and their implications. Proceedings of the National Academy of Sciences USA 98: 2101–3.

Piperno, D. L., and D. M. Pearsall. 1998. The Origins of Agriculture in the Lowland Tropics. Academic Press, San Diego.

Posadas, P., J. V. Crisci, and L. Katinas. 2006. Historical biogeography: a review of its basic concepts and critical issues. Journal of Arid Environments 66: 389–403.

Sahu, S. K., R. Singh, and K. Kathiresan. 2016. Multi-gene phylogenetic analysis reveals the multiple origin and evolution of mangrove physiological traits through exaptation. Estuarine, Coastal and Shelf Science 183: 41–51.

Snow, D. W. 1985. Seed dispersal. In B. Campbell and E. Lack (eds.), Dictionary of Birds. Poyser, Calton, Staffordshire, UK. [Reprint, Bloomsbury Publishers, 2013.]

Steadman, D. W. 1985. Fossil birds. In B. Campbell and E. Lack (eds.), Dictionary of Birds. Poyser, Calton, Staffordshire, UK.

Stephens, J. L. 1841. (Illustrations by F. Catherwood.) Incidents of Travel in Central America. 2 vols. Harper, New York.

Sussman, D. R., F. W. Croxen III, H. G. McDonald, and C. A. Shaw. 2016. Fossil porcupine (Mammalia, Rodentia, Erethizontidae) from El Gulfo de Santa Clara, Sonora, Mexico, with a review of the taxonomy of the North American Erethisontids. Contributions in Science 524: 1–29.

Turner, B. L., II, and P. D. Harrison (eds.). 1983. Pulltrouser Swamp: Ancient Maya Habitat, Agriculture and Settlement in Northern Belize. University of Texas Press, Austin.

Volkov, I., J. R. Banavar, S. R. Hubbell, and A. Maritan. 2009. Inferring species interactions in tropical forests. Proceedings of the National Academy of Sciences USA 106: 13854–9.

Wade, L. 2016. Monkey ancestors rafted across the sea to North America. Science. doi: 10.1126/science.aaf4154.

———. 2016. A nation divided: some scientists see Nicaragua's plans for a Grand Canal as a boon for an ailing land; others predict ecological catastrophe. Science 351: 220–23.

Wang, W., et al. (+ 10 authors). 2012. Menispermaceae and the diversification of tropical rainforests near the Cretaceous-Paleogene boundary. New Phytologist 195: 470–78.

Webb, S. L. 1986. Potential role of passenger pigeons and other vertebrates in the rapid Holocene migrations of nut trees. Quaternary Research 26: 367–75.

Weber, S., and K. Buckingham. 2016. How Brazil, Panama and Costa Rica breathed new life into their degraded lands. WRI Restoration Diagnostic, 3 August.

Wiegand, T., et al. (+ 7 authors). 2012. Testing the independent species' arrangement assertion made by theories of stochastic geometry of biodiversity. Proceedings of the Royal Society B (Biological Sciences) 279. doi: 10.1098/rspb.2012.0376.

Wilder, B. T., C. O'Meara, L. Monti, and G. P. Nabhan. 2016. The importance of indigenous knowledge in curbing the loss of language and biodiversity. Biosicence. http://bioscience.oxfordjournals.org.

PART THREE

Austral Land Bridge

SIX

Magellan Land Bridge: Cono del Sur and Antarctica

We arrived at a new land which, for many reasons that are enumerated in what follows, we observed to be a continent.

—Amerigo Vespucci on the recognition of South America as a continent, quoted in Boorstin, 1983, 233

The Magellan Land Bridge extends from Patagonian South America across the Strait of Magellan and the Drake Passage to Antarctica (Figs. 6.1, 6.2). Islands at the tip of South America are separated from the mainland by the Strait of Magellan and make up Tierra del Fuego with the Beagle Channel running along its southern coast. Beyond Antarctica the austral biogeographic province extends to Australasia (Australia, New Zealand, New Guinea, and neighboring islands) and beyond that into southeastern Asia. The past and present distribution of plants such as the gymnosperms *Araucaria* and Podocarpaceae and the angiosperms *Nothofagus* and *Nipa* is consistent with the region once having been part of a continuous and well-traveled route between the New World and Old World of the Southern Hemisphere. The focus here is on that portion of the connection between southern South America and Antarctica.

Cono del Sur

Geographic Setting and Climate

The Cono del Sur consists of Argentina, southern Brazil, Chile, Paraguay, and Uruguay. Patagonia is the southern part, lying mostly in Argentina and partly in Chile. The name Patagonia derives from "Patagón," the name

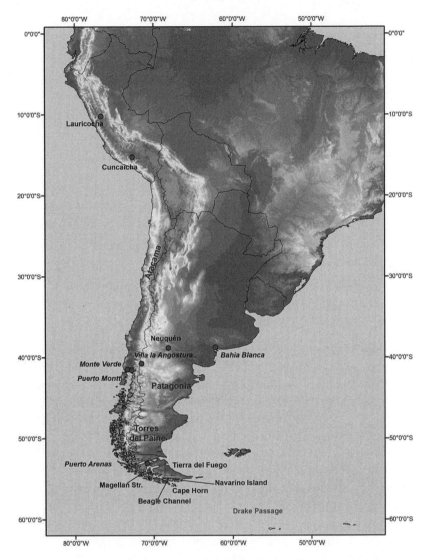

6.1. Index map of place names and physiographic features for southern South America (northern Magellan Land Bridge).

bestowed by Magellan in 1520 on the people the early Spaniards considered giants at around 6 feet tall, or from "Pathagoni" (big feet) for the large, furry shoes worn by the natives. Nequén, Venezuela, in the north has a population of 345,000; Punta Arenas, Chile, in the south has 116,000 people, and the intervening area is sparsely populated.

The Andes Mountains extend along the western edge of South America and divide Chilean Patagonia into dry east-facing slopes, moister uplands, and extensively glaciated highlands (II, figs. 1.49, 1.51). East of the mountains the rainshadow-influenced dry lands are a semidesert or a grassland steppe or pampas. As the name implies, the Cono del Sur tapers to a point and constitutes a physical filter for organisms migrating southward and an expanding target area for those moving north. Winds are from the west (Fig. 4.5), and throughout its long Mesozoic and Cenozoic history, "the flux of the westerlies would be maintained at approximately the same latitude as today" (Compagnucci, 2011, 230) with all that implies for the past and present distribution of wind-disseminated propagules. Once the clockwise (east-flowing) Antarctic Circum-Current (ACC) was established in the early Oligocene, movement through the marine realm was also predominantly in that direction. The Humboldt/Peru Current developed at about this time, transporting ca. 20 Sv of water per second and deflecting propagules, flotsam, and jetsam northward along the western coast. For these reasons distributions in the Cono del Sur from the Oligocene onward were preferentially from the west, coastal, and northward rather than from the east, inland, and southward (Fig. 4.5). Geographic similari-

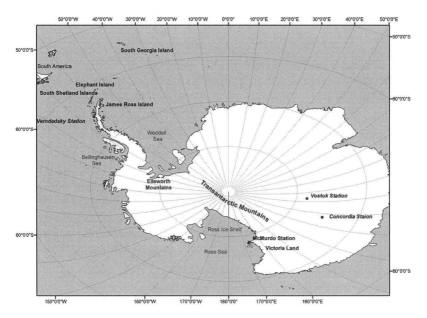

6.2. Index map of place names and physiographic features
for Antarctica (southern Magellan Land Bridge).

ties among the older fossil floras (Triassic, Jurassic, earliest Cretaceous) and in the modern residuals between Australasia and South America, as shown by primitive ferns (*Dicksonia*) and gymnosperms (*Araucaria*), are the consequence of land continuity before the breakup of Gondwana that began in the Jurassic (see Fig. 3.10).

Physiographic and habitat diversity is high in northern Chile, ranging from coastlands to the volcano Ojos del Salado on the border with Argentina at 6893 m, moist east-facing mid-altitude slopes, and the hyper-arid Atacama Desert, which receives virtually no rainfall for extended numbers of years. The presence of commercial quantities of evaporitic minerals like nitrates, copper, halite, and gypsum in northern Chile reveals that local to regional dry conditions are ancient (Clarke, 2006). The origin of extensive arid communities probably corresponds to the time cold waters started being transported north by the Humboldt Current. This was after significant ice had formed on Antarctica and calved into the ocean following the MMCO between 21[17]–15 Ma. The drying effect on vegetation was augmented by rise of the Andes to significant heights in the middle to late Miocene ca. 10 Ma (Gregory-Wodzicki, 2000a,b; Gregory-Wodzicki et al., 1998; Graham, 2009; Graham et al., 2001). The amalgamation of individual elements into dry vegetation dates from this time.

The climates are cool to frigid, and with the exception of mid-altitude zones they are mostly dry. At Puerto Montt in northwestern Chile the MAT is 11 °C with annual extremes from 25.5 °C to −15 °C. To the far southwest at Punta Arenas, it is 6 °C with lesser extremes of 24.5 °C to −2 °C. The MAP at Torres del Paine, Chile, is 4000–7000 mm and drops to 800–200 mm in the eastern hills. In the northeast at Bahia Blanca, Argentina, the MAT is 15 °C (35 °C to −5 °C); the MAP around Villa La Angostura near the Chilean border is 2074 mm, while to the east it is as low as 125 mm (Vizcaíno et al., 2012, 9). The drier part of northwestern Argentina is the pampas and the monte, which has a physiography similar to the Basin and Range Province of the western United States and several dry-habitat plants in common (e.g., *Ephedra*). The similarity in vegetation between these two remote regions is likely the result of long-distance transport by birds (Fig. 5.2).

To the south the Strait of Magellan separates Patagonia from Tierra del Fuego. The strait is a 600 km long treacherous maze of islands, channels, and shoals, dark in the long winter, and often horrendously stormy. It is a passage between the Atlantic and Pacific oceans, although most commercial ships now take the wider and more open but equally stormy Drake Passage to the south. The heroic efforts of Ferdinand Magellan in traversing the strait in 1520 are vividly captured by Laurence Bergreen (2003) in

Over the Edge of the World. Magellan left Sanlucar de Barrameda, Spain, on 20 September 1519 with 5 ships and a crew of 260. The single surviving ship returned on 6 September 1522 with the remaining crew of 18, who were the first to circumnavigate the world. After exiting the strait, Magellan followed the Humboldt Current north and then west seeking the Spice Islands. He was killed on 27 April 1521 by natives on Mactan in the Philippines who believed he was siding with a warring faction against them. The first Englishman to traverse the Strait was Francis Drake in 1578. Upon entering the Pacific Ocean, he was driven south by storms and found the more navigable passage that now bears his name.

For many of these early explorers, their efforts were complicated by the inaccuracy of most maps available at the time. The three known continents of Europe, Asia, and Africa were frequently arranged as a T inside a circle representing the Earth,

> so medieval maps of this genre are referred to as "T in O" maps. To remain consistent with religious traditions, T in O maps located Jerusalem at dead center, with Paradise floating vaguely at the top. (Bergreen, 2003, 75)

Many explorers preferred to use ruttiers. These were accounts written by earlier travelers, often incorporating the myths of indigenous people about rich and fabled lands or conjured to get the intruders to move on, and with sketches charting the course of previous expeditions. At the time of Magellan the Pacific Ocean was thought to be a narrow inlet between the New World and Asia, only a few days' sail away at the most, and food, water, medicine, and other necessities were stored accordingly. With such charts, maps, and provisions they headed out toward what many still believed was the edge of the world.

The climates of Patagonia extend into Tierra del Fuego, named by Magellan for the bonfires set by the natives. The coldest temperatures are ca. $-20°C$ in the northeast interior and $-10°C$ under the more maritime conditions to the southeast. MAP ranges from 800 to 600 mm, and there is heavy snowfall with some snow falling even in the summer months. The largest island of the archipelago is Isla Grande de Tierra del Fuego, administered jointly by Argentina and Chile. The Beagle Channel, 240 km long and 5 km wide, separates Isla Grande from the smaller islands of Tierra del Fuego. When Pringle Stokes, captain of HMS *Beagle*, committed suicide during its first voyage (1827–30), the command passed to Robert Fitz-Roy, who on the second voyage in 1832 took Charles Darwin along as a naturalist.

At the southernmost point of the Tierra del Fuego group is Cape Horn (55°58′S), marking the boundary with the Drake Passage. "Rounding the Cape," as it is called in sailing parlance, is a highly dangerous undertaking, and throughout the centuries over a thousand ships have been lost. It has become a challenge, therefore, to some modern yachtsmen, including Nigel Seymour who in 1987 accomplished the feat in dual "sailing kayaks" each with two sails mountable to any of the four sailing positions, and Howard Rice who sailed and paddled a 15-foot collapsible canoe around the Cape. Puerto Toro on the eastern coast of Isla Navarino, Tierra del Fuego, is the southernmost settlement in the world at 55°05′N, with 36 people—even fewer than on Alert, Ellesmere Island (82°15′N), which has 79 people (see chapter 3, above). This all speaks to the individualism that has compelled adventurers past and present to seek discoveries and fame, and to test the limits of endurance in explorations from Beringia to Antarctica.

South of the Drake Passage is the Scotia Sea, surrounded by the Scotia Ridge. The ridge formed ca. 40 Ma by back-arc extension as the boundary between the South American and Antarctic plates moved eastward at a rate of ca. 50 mm/yr and was completed in the Late Cenozoic (Galindo-Zaldívar et al., 2006a). Beyond the Scotia Sea is the island continent of Antarctica at 82°S, 135°E.

Geology

The physiographic configuration of southern South America is a consequence of its geologic history, which involved (1) separation of the Gondwana continents of Africa and South America (McLoughlin, 2001) beginning ca. 135 Ma in the south during the Early Cretaceous and proceeding northward; (2) pressure along the western margin as this movement compressed South America against the Pacific Plate; and (3) formation of a subduction zone along the contact between these two plates beginning in the Eocene that ultimately gave rise to the Southern Andes. There were two periods of rapid plate convergence, 50–42 Ma in the middle Eocene and 25–10 Ma in the middle to late Miocene (Suárez et al., 2000). This established a major part of the landscape discussed in the previous section and further involved marine transgressions and regressions on the Atlantic Patagonian Platform in the Late Cretaceous, middle Eocene, and late Oligocene–early Miocene. There was uplift of the Taitao and other lava plateaus off Patagonia resulting from back-arc volcanism (Malumían and Náñez, 2011; see also Folguera et al., 2011). Other features of the landscape resulted from islands and island arcs being accreted onto the western coast through subduction. It was a dynamic setting, creating numerous ephemeral terrestrial habitats for migrating biotas and putting environmental stress on ones already there (Nullo and Combina, 2011; Aragón et al., 2011). The Southern Andes terminate at the Isla Grande de Tierra del Fuego at the east-west trending Cordillera Darwin north of the Beagle Channel. The mountains are covered by rapidly retreating ice and have an elevation of 2000 m. Glaciation began early, as evident by tills beneath basalts with an age between the latest Miocene and the Quaternary. In the Pleistocene, "the Patagonian Andes were covered by a continuous mountain ice sheet, from 37°S to Cape Horn (56°S), in at least five major glaciations over more than 15 cold events over the last million years" (Rabassa et al., 2011, 316). Presumably this was the case during each of the 18–20 glaciations of the Pleistocene. Outwash from these glaciers created high-energy stream flows (Martínez and Kutschker, 2011) and leveled the terrain forming the Patagonian plains of the eastern Cono del Sur.

Modern Vegetation

A flora of the Cono del Sur is being compiled as the *Catalogo de las Plantas Vasculares del Cono Sur* by the Instituto de Botanica Darwinion, Buenos Aires, in collaboration with the Missouri Botanical Garden and other institutions in Argentina, southern Brazil, Chile, Paraguay, and Uruguay (Zuloaga et al., 2008). There are 17,693 species listed. Previous inventories for the modern vegetation of Chile are given in Marticorena and Quezada (1985), Marticorena and Rodríguez (1995 et seq., *Flora de Chile*), and Gajardo (1994), and for Argentina by Cabrera (1971, 1976). The biogeographic regions recognized in the *Catalogo* are shown in Fig. 6.3.

Chile

The elongated dimension of Chile, extending from ca. 18°S to 55°S (5200 km) and from sea level to 6893 m at the volcano Ojos del Salado, provides an array of habitats for the existing vegetation. The Southern Andes is the oldest section of the Cordillera, with the uplift pulses previously mentioned at 50–42 and 25–10 Ma, and the Humboldt Current (still moderately warm at ca. 30 Ma) becoming increasingly cold after the MMCO at 21–15 Ma. This means that considerable diversity for accommodating organisms through the northern portion of the Magellan Land Bridge has existed for much of the Cenozoic. The following list enumerates the principal communities according to the classification used here. The asterisk (*) indicates the genus is reported as a fossil from the Cono del

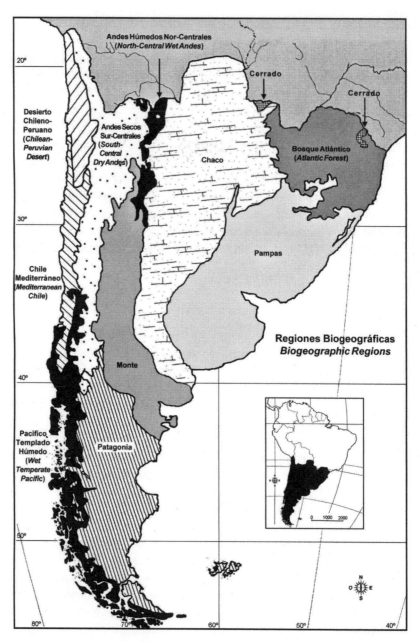

6.3. Biogeographic regions of the Cono del Sur, South America. From Zuloaga et al., 2008, Missouri Botanical Garden Press and CONICET. Used with permission.

Sur in either Chile and/or Argentina (II, app. 2, sometimes using an artificial system of nomenclature, e.g., *Araucariacites, Podocarpidites*). As previously mentioned, some microfossils belonging to stenopalynous families (Cyperaceae, Poaceae [except *Zea*], Asteraceae [because of the large size of the family relative to pollen diversity], Chenopodiaceae/Amaranthaceae [except *Iresine*], Ericaceae, and others) are occasionally named to genus, if for some reason the choices can be limited to regional taxa (e.g., because of the young age of the flora).

Desert (Antafagasto, latitude ca. 24°S)—*Ephedra, Adesmia, Babia, Chuquiraga, Cristaria, Echinopsis, *Euphorbia, Glypothmnium, Heliotropum, Lycium, Nolina, Ophyrosporus, Oxalis, Polyachyrus, Proustia, Salvia.*

Shrubland/chaparral-woodland-savanna (Serena–La Ligua, 30°S; central zone, 32°S–37°S; MAP 50–300 mm)—*Acacia, Adesmai, Alstroemeria, Aristolochia, Fuchsia, Haplopappus, Jubaea, Lapageria, Lithrea, Loasa, Mutisia, *Nothofagus, Peumus, Pouteria, Puya, Quillaja, Schizanthus, Tropaeolum*

Lower to upper montane broad-leaved forest—*Hymenophyllum, Copiapoa, Croton, *Drimys, Griselinia, Mitraria, Peperomia, Salvia, Tillandsia; Los Andes de la Zona Central—Alstroemeria, *Berberis, Calandrinia, Calceolaria, Chaetanthera, Happlopappus, Loasa, Maibuenia, Nassauvia, Nastanthus, *Nothofagus, Oxalis, Schizanthus, Senecio, Viola;* bosque de *Araucaria;* bosque pluvioso siempreverde (temperate rain forest, MAP ~1000 mm)—*Adiantum, *Blechnum, *Hymenophyllum, Hypolepis, *Lophosoria, Megalastrum, *Austrocedrus, *Fitzroya, *Podocarpus, Boquila, Campsidium, Chusquea, Cissus, *Drimys, *Escallonia, Eucryphia, Fascicularia, Fuchsia, Gevuina, *Gunnera, Lapageria, Lardizabala, Laurelia, Laureliopsis, Luzuriaga, *Myrceugenia, *Persea, Pilgerodendron, Proustia, Sarmienta, Saxegothaea, Ugni;* Patagonia—*Austrocedrus, Azorella, Baccharis, *Berberis, Bolax, Chiliotrichum, *Escallonia, Festuca, Gaultheria, *Gunnera, Junellia, Mulinum, *Nothofagus, Pilgerodendron*

Tundra (Chile Antártico)—*Deschampia, Colobanthus*

Alpine tundra (páramo, puna, 1900 to above 4000 m)—*Baccharis, Mentzelia, Trixis, Viguera;* 3600–4000 m, *Baccharis, Fabiana, Parastrephia, Senecio;* above 4000 m, *Azorella, *Polylepis, Pycnophyllum*

Argentina

Shrubland/chaparral-woodland-savanna (chaco, monte; with inundated areas of freshwater herbaceous bog/marsh/swamp, aquatics, MAT 20°C–23°C, MAP 340–1200 mm)—Aquatics: *Cabomba, Ceratophyllum,*

Eichhornia, Myriophyllum, Pondederia, *Potamogeton, Reussia, Sagittaria, *Sparganium, Thalia, Utricularia, Victoria; inundated areas (*Cyperaceae undifferentiated): Cyperus, Eriocaulon, Fuirena, Scirpus, *Typha; dry areas: *Ephedra, *Acacia, Acanthosyris, Aechmea, Allenrolfea, Aloysia, Aspidosperma, *Astronium, Atamisquea, Bothriochloa, Bougainfillea, Bromelia, Bulnesia, *Bumelia, *Caesalpinia, *Capparis, Cattela, *Celtis, Cercidium, Cereus, Chorisia, Cleistocactus, Copernicia, Deinacanthum, Digitaria, Diplokeleba, Dyckia, Elionurus, Eriocereus, *Fagara, Geoffroea, Gleditsia, Gouinia, Heterostachys, Jodina, Leptochloa, Leptocoryphium, Lithraea, *Maytenus, Melica, *Mimosa, Opuntia, Parkinsonia, Paspalum, Patagonula, Pennisetum, *Pereskia, Phorodendron, Phrygilanthus, Phyllostylon, Pisonia, *Pithecellobium, Prosopis, *Psittacanthus, Quiabentia, *Ruprechtia, Salix, Schinopsis, *Schinus, Schizachyrium, Setaria, Sorghastrum, Spartina, *Tabebuia, Tessaria, Trichloris, Trithrinax, Ziziphus; highest altitudes: Polylepis, Festuca, Stipa

Grassland—(pampas; *Poaceae undifferentiated) Andropogon, Aristida, Axonopus, Crytopodium, Elionurus, Eragrostis, Mayaca, Panicum, Paspalum

Lower to upper montane broad-leaved forest (yungas, selva nublada; mostly eastern slopes of the Andes, <1000–3400 m, MAT 14°C, MAP 900–2500 mm)—Alsophila, *Blechnum, *Dicksonia, *Equisetum, *Hemitelia, *Lycopodium, *Pteris, *Selaginella, *Araucaria, *Podocarpus, *Acacia (edaphically dry habitats), Acrocomia, Aechmea, *Allophylus, Alnus, Amburana, *Amicia, Anadenanthera, Apuleia, Aspidosperma, *Astronium, Baccharis, Balfourodendron, Bastardiopsis, Begonia, Bidens, Blepharocalyx, Bocconia, Boehmeria, Bombax, Bomarea, Buddleja, Cabralea, Calceolaria, Calycophyllum, *Casearia, *Cassia, *Cecropia, *Cedrela, *Celtis, Cestrum, Chlorophora, Chorisia, *Chrysophyllum, Chusquea, *Citharexylum, Cnicothamnus, *Cordia, Cordyline, Crinodendron, *Cupania, *Diatenopteryx, Doxantha, Duranta, Elephantopus, *Enterolobium, Epidendrum, Erythrina, Eugenia (as Eugenia/Myrcia), Euterpe, *Fagara, *Ficus, Gentianella, Gleditsia, Guadua, Holocalyx, Hydrocotyle, *Ilex, *Inga, Iresine, Lippia, *Lonchocarpus, Malaxis, Mandevillea, Manettia, Merostachys, *Miconia, *Mimosa, Muntingia, Myrocarpus, Myroxylon, *Nectandra, *Ocotea, Olyra, Oncidium, Oplismenus, Parapiptadenia, *Passiflora, Patagonula, Pavonia, Pennisetum, Pharus, Philodendron, *Phoebe, Phyllostylon, Piper, Pisonia, *Pithecellobium, Poaceae (numerous), Pogonopus, Polygala, *Polylepis, Polymnia, Pouteria, Pseudocaryophyllus, *Pterogyne, Rhipsalis, *Ruprechtia, Salvia, Sambucus, Sapium, *Schinus, Seemania, Senecio, Sibthorpia, Solanum, Stellaria, Stevia, Syagrus, *Tabebuia, Tagetes, *Terminalia, Tessaria, Tibouchina, Tillandsia, *Tipuana, *Trichilia, Urera, Vernonia, Vriesea

Tundra (Dominio Antarctum; sea level –4400 m, páramo above ca. 3400 m, MAT to −13.4 °C, MAP 100–500 mm)—mosses, lichens; *Colobanthus, Descampsia*

Of the ca. 300 representative genera listed for the above vegetation types, ca. 75 or 25% are known as fossils from the Cono del Sur and, thus, were utilizing at least the northern part of the Magellan Land Bridge sometime during the Cenozoic. This is less than for the boreal land bridges because after ca. 35 Ma, ice cover began reducing the land surface available for plants in the far southern part.

Indigenous People

Upon crossing into South America, the first arrivals from the narrowing landscape of Central America encountered an expansive land of almost unparalleled physical, climatic, biological, and resource diversity. They radiated into the geographically isolated and near-isolated habitats and now have differentiated into approximately 366 cultural and linguistic groups that include about 4,606,475 remaining indigenous people (People-Groups, accessed 2016; see also International Work Group for Indigenous Affairs [IWGIA], accessed 2016). For the Cono del Sur the largest population, at the time of Magellan's landing in 1520, was the Tehueche (Patagóns, now assimilated into the European culture). The indigenous people and their language groups are shown in Fig. 6.4.

Peopling of South America (from the North)

The currently accepted pattern of human migration into South America is that people came from the north and radiated outward, including (1) movement back north into the Lesser Antilles and then to the Greater Antilles, where they met people coming eastward from Yucatán and northern Central America (Belize, Honduras; see chapter 4, above); and (2) movement to the south along the lowlands primarily as hunters and gatherers and then into the highlands, initially establishing seasonal encampments for gathering food and materials (e.g., obsidian; Méndez et al., 2012) and later permanent settlements for religious, agricultural, and defensive purposes. Although there is consensus about the general direction and pattern of these movements, the chronology is uncertain. As in almost every other

6.4. Indigenous peoples of South America and their language groups. From Wikipedia, public domain.

part of the world, the timeline is being pushed back. The controversy for North America is whether the first Americans crossed Beringia soon after 18,000 or around 14,500 BP, and whether they followed coastal, inland, or both routes to the south. This debate about early inhabitants pales in comparison to South America. The claims and responses about the earliest human occupation generate news coverage in which it is often difficult to distinguish between science and entertainment. There are suggestions that humans were hunting sloths 30,000 years ago and that some charcoal is the result of human fires set 48,000 years ago. In the Serra da Capivara

National Park in northeastern Brazil, the age of rock paintings and supposed stone tools is estimated from 9000 to 22,000 to 100,000 BP. The
responses to these claims range from suggestions that the stones may have
been chipped by monkeys to "saying that monkeys produced the tools is
stupid." This recalls reaction to a theory advanced some years ago that because early humans had a proclivity for coastal habitats, and made sea voyages, they were descended from a line of aquatic apes (Morgan, 1997). The
suggestion was mostly ignored, primarily because there was no evidence,
but a symposium was held in 1987 to debate the issue and the proceedings published by the Souvenir Press (Roede et al., 1991). Responses to the
South American claims are often accommodating: "I'm pretty sure we have
some significant surprises waiting for us. Maybe people killing sloths at
30,000 years ago is one of them, maybe it's not—but it certainly isn't going
to hurt to have it on our collective radar screen as we continue to contem-
plate the peopling of the New World." (Interview quotes are from the National Geographic, 2013; New York Times, 2014; and ScienceNews, 2013).

Less controversial evidence and better dates are gradually accumulating
for the early presence of humans in South America. The oldest is stone tools
and remains of edible plants at an ephemeral summer site at Monte Verde
in Chile dating to 14,500–18,500 BP (Dillehay et al., 2015). Rademaker
(2014; Rademaker et al., 2013) has found evidence of human occupation
of the Cuncaicha Rock Shelter in the high Andes of Peru at 12,400 BP. In
the shelter there are animal bones (deer, vicuñas), tools with plant starch
on the edges, and a fragment of a human skull (without DNA and its age is
uncertain). DNA was recovered from skeletal remains at Lauricocha, Peru,
giving a date of ca. 9000 BP (Fehren-Schmitz et al., 2015), and there are
fishtail or Fell-type projectile points in Río Negro and Salta provinces, Argentina, in Pleistocene-Holocene transition deposits estimated at 11,000–
10,000 BP (Miotti et al., 2010; Aráoz and Nami, 2014). Fell projectile
points have convex blade surfaces and distinctive fluted channels (Loponte
et al., 2015). Middens of the Yaghan people are known from the seasonally
occupied Bahia Wulaia Dome site, Isla Navarino along the Beagle Channel,
Patagonian Chile (Hogan, 2008). The estimated date is ca. 10,000 BP.

The expanse of landscape and diversity of habitats in South America
are vast compared to the relatively sparse archaeological evidence, making it difficult to estimate when humans began having an impact on the
vegetation—that is, measurably altering the modern analogs used for interpreting the paleocommunities. Unless there are archaeological sites in
the immediate vicinity of the fossil plant localities, the estimates must be
a balance between the agricultural traditions early people brought with

them from the north and the novel opportunities for maintaining hunting and gathering activities prolonged by the rich biota encountered in South America. As cultures formed sustained settlements in the mesic highlands, their populations began to noticeably expand around 5000 BP, and for the Cono del Sur region this probably corresponds approximately to the time of significant impact on the vegetation: "The unique extent of humanity's ability to modify its environment to markedly increase carrying capacity in South America is therefore an unexpectedly recent phenomenon" (Goldberg et al., 2016). This correlates well with the previous estimates of 6000–7000 BP for modification of the Mexican vegetation (chapter 5, above), 5000–6000 BP for the Antilles (chapter 4, above), and 5000 BP for South America. After that, trade, transport, exchange, dispersal by humans, domestication, cultivation, and extinction of native species in South America increased markedly.

Antarctica

Geographic Setting and Climate

When in 1578 Francis Drake saw the passage of open water south of the Magellan Strait, he was observing part of what makes Antarctica such a frigid land. Nearly 98% is covered by glaciers, the ice averages 1.6 km in thickness, and the weight has depressed the underlying bedrock to 2.5 km below sea level. Drake was navigating the wind-driven ACC that flows clockwise around the continent carrying 130 Sv ($10^6 m^3 S^{-1}$), or 600 times the flow of the Amazon River, and isolates it from warm waters to the north. Antarctica is 14 million km^2 in area, the fifth largest landmass after Asia, Africa, North America, and South America, and is divided into East and West Antarctica by the Transantarctic Mountains, extending between the Ross Sea and the Weddell Sea (Fig. 6.2). In West Antarctica the Vinson Massif at 4897 m in the Ellsworth Mountains is the highest point, and a depression called the Bentley Subglacial Trench is the lowest at −2.5 km. The MAT of the four coldest months is −63°C with an extreme of −89.2°C recorded at an altitude of 3900 m near the Vostok Station on 21 July 1983. For comparison, recall that at Yakutsk in Siberia the coldest temperature recorded is −43°C with an extreme of −89.9°C. The warmest temperature at the South Pole was −12°C on 25 December 2011. MAP is ca. 166 mm (6.5 in) along the coast and essentially 0 mm inland. There is no permanent population, and visitors to the research stations and seasonal field camps maintained by several nations (e.g., Argentina, Brazil, Chile, China,

Ecuador, Poland, South Korea, Peru, Russia, and Uruguay) for various reasons (Walker, 2013, xiv) number between 1000 and 4000 per year, plus about 30,000 tourists. Governance is through the Antarctic Treaty System, signed by 53 nations, barring mining, military activities, nuclear testing, and waste disposal, and including other regulations aimed at protecting the environment (compare with the Arctic; see the section on indigenous people in chapter 2, above).

It is impossible to read about the geography of Antarctica without being caught up in the discoveries, astonishing fortitude, and ultimate fate of the early explorers. Their diaries and logs reflect conditions at the southern terminus of the Magellan Land Bridge and are captured in several accounts: the Swedish South Polar Expedition (1901–5); the Scott Expeditions (1901–4 and 1911–13), in the latter of which the party of 3 died just 11 miles short of their camp at Cape Evans; the Australasian Antarctic Expedition under Douglas Mawson (1912), where 2 of the 3 explorers died within 20 miles of their camp at Cape Denison (Roberts, 2013); and Ernest Shackleton's voyage on the *Endurance* through the Drake Passage (1914–17; Lansing, 1959, with an introduction by Nathaniel Philbrick). The highlights of Shackleton's voyage are as follows:

Sir Ernest Henry Shackleton (1874–1922; Fig. 6.5) had participated in two previous trips to Antarctica, and on 8 August 1914 he left London with 27 men on his own Imperial Trans-Atlantic Expedition with the goal of being the first to walk across the Antarctic Continent. Their ship was trapped and destroyed in severe pack-ice in the Weddell Sea on 23 October 1915, forcing them to spend the winter on an ice floe in tents and makeshift lean-tos built from wreckage of the ship. There was no means of communication, the nearest settlement being 1200 miles away, and it was assumed by the outside world that they had perished.

When the ice started to break apart in the spring, Shackleton faced the dangerous task of leading the group across the Drake Passage in open boats intended for short excursions from the *Endurance*. At the outset, Frank Worsley wrote optimistically about leaving the ice floe: "We feel as pleased as Balboa when, having burst through the forest of the Isthmus of Darien, he beheld the Pacific." Their situation was quite different, however, and after barely surviving the crossing, they landed on the deserted Elephant Island. Facing starvation, severe frostbite, amputations, hallucinations resulting from the cold and lack of food, and debilitating illnesses, a party of 3 left in an attempt to reach South Georgia Island, from where they had started out 522 days before. While on Elephant Island temperatures reached −36.6°C (−34°F), and on occasion there were hurricane winds

6.5. Sir Ernest Henry Shackleton, 1916. From Wikipedia, public domain.

estimated at over 320 km/hr (200 mph). After landing on South Georgia Island, to reach the whaling station at Stromness on the other side, they had to walk another 29 miles across previously impenetrable glaciers: "Not one man had ever crossed the island—for the simple reason it could not be done. When they came to a massive vertical cliff there was no need to explain the situation: if they stayed where they were they would freeze. So Shackleton decided they would slide—three tobogganers clinging together without a toboggan" (Lansing, 1959).

When they walked into Stromness, their hair was down to their shoulders, their faces black from months of dirt and cooking oil left as protection from the unrelenting wind and cold; they were filthy, and dressed in rags of what once had been clothes. Two small boys ran away when the strangers approached and asked them for the foreman, Mathias Andersen. He took the men to the station's manager, Thoralf Sørlle. Andersen knocked on the door and after a moment Sørlle opened it. He was startled at the sight of the three men and moved back, "Who the hell are you?" The man in the center stepped forward.

"My name is Shackleton," he said. It is recorded that the burly whalers and sea captains at a reception held later to honor him had tears in their eyes as they stepped forward to greet him and hear of his exploits. After a brief recovery they started the return trip to Elephant Island to rescue the rest of the party. The voyage was nearly as bad as the earlier crossing of the Drake Passage, and by the time the island came into view, they had gone 80 hours without sleep. As the boat neared the shore, Shackleton simply called out:

"Are you all right?"

"All is well," was the reply.

The 28 men were safely reunited, and Shackleton returned to England on 29 May 1917, where he was knighted by King Edward VII. Soon afterward he began to organize another expedition to Antarctica with the intent of circumnavigating the continent. Many of the same men signed on for the voyage. Shackleton died on that trip, on 5 January 1922, on St. Georgia Island, where he is buried.

Fernández-Armesto's (2006) assessment of this and other voyages is that "Shackleton's attempted transantarctic expedition . . . was pointless from a scientific point of view. It was justifiable only on grounds of unalloyed adventure. It became another heroic failure" (395). That seems harsh, like his characterizing of Bering's efforts as excelling fiction for human interest and farce for human foibles (see chapter 1, above). At the very least people like Bering, Magellan, Shackleton, and others made new geographic discoveries and documented the need for vastly improved maps, provisions sufficient for years rather than weeks of travel, and preparations and personalities adequate for leading a highly diverse group through some of the harshest conditions on Earth.

Geology

The breakup of Gondwana in the Middle Jurassic ca. 180 Ma had an immense effect on the climatic and biotic history of the Earth. The plate movements are shown in Lawver and Gahagan (2003) and in Fig. 6.6 from Fitzgerald (2002). The north-south drift from the Late Cretaceous onward was not great (a few degrees, northern edge of the continent; compare maps 1 and 20 in Smith et al., 1994; Plates Project, University of Texas Institute for Geophysics/Animations/Antarctica), but more than the northward drift of the boreal lands (1° to 2°; see chapter 2, above). Throughout

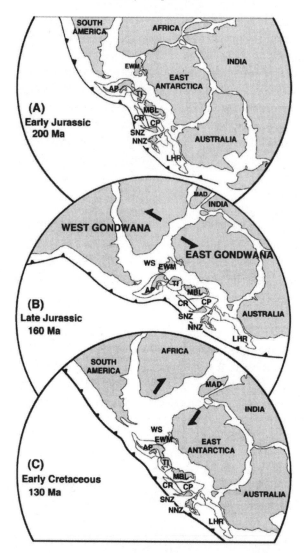

6.6. Gondwana reconstruction for the Early Jurassic (A),
Late Jurassic (B), and Early Cretaceous (C). From Fitzgerald,
2002, Royal Society of New Zealand. Used with permission.

its early history five subplates making up West Antarctica and the craton
of East Antarctica were fusing and drifting southwest to the polar position,
which the continent essentially attained by the end of the Cretaceous. The
subsequent isolation of Antarctica from other continents of the Southern
Hemisphere and its separation from South America was another signature

geologic event. In the Late Jurassic–Early Cretaceous, 162–130 Ma, there is evidence of stretching of ocean strata between Antarctica and South America, resulting in the opening of the Rochas Verdes Basin in southern Chile. Early in the Cenozoic spreading from the West Antarctic Rift System compressed the continental crust, uplifting the Transantarctic Mountains between 55 and 45 Ma and establishing the West Antarctic and East Antarctic provinces (Fitzgerald, 2002). The accompanying volcanism in the Cretaceous (Manfroi et al., 2015), Eocene (Kim et al., 2006; with recent eruptions in 1842, 1967, 1969, and 1970 in the South Shetland Islands; Berrocoso et al., 2006), and in the Neogene (15–2 Ma) in West Antarctica (Haywood et al., 2009, table 10.2) created fires (Cantrill and Poole, 2012, 286–88), open areas, and various edaphic habitats, as mentioned for Patagonia. Substrate and soil types included weathered ash from volcanism, gravel and sand from outwash in warm times of glacial melting, and gelicols (or gleysols) in the cold intervals (Brockheim, 2013). During the Early/Late Cretaceous (ca. 120 Ma) to the latest Eocene (ca. 49 Ma), Australia was separating from Antarctica, followed in the Oligocene at 35–30.5 Ma by subsidence to create the Powell Basin between the Antarctic Peninsula and South America. Further sea-floor spreading at 34–29 Ma formed the Drake Passage (ca. 31 +/− 2 Ma, fide Lawver and Gahagan, 2003), and a full-flowing ACC consisting still of relatively warm waters. According to Barker (2001), "the reconstruction to 40 Ma creates a compact, cuspate continental connection between South America and the Antarctic Peninsula at the subducting Pacific margin." Part of this tectonic activity resulted in the separation and uplift of Shackleton's Elephant Island in the Pliocene (Galindo-Zaldívar et al., 2006b; López-Martínez et al., 2006).

One of the critical paradigm shifts in global climate occurred ca. 55 Ma toward the end of the EECO. Global temperatures entered a cooling phase primarily through decreasing CO_2 concentration as mega-movements from the initial breakup of Pangaea slowed. The earliest ice and initial ice caps (as opposed to permanent glaciers and extensive ice caps) date from ca. 40–35 Ma. Haywood et al. (2009, 414) mention "the onset of Antarctic glaciation during the Oligocene," and Passchier et al. (2013) cite a cooling of ca. 8°C between the early Eocene (54–52 Ma) and the middle Miocene (15–13 Ma). The polar position of Antarctica and formation of the ACC in the Oligocene was augmented by colder climates after the MMCO at ca. 15–14 Ma. Atmospheric cooling in the mountains is estimated at ca. 20°C (Haywood et al., 2009). The ice cap expanded, the exchange of organisms with other continents was disrupted, and the effects were felt well beyond the polar region. Ice discharged from the margins cooled the oceans, and the Humboldt Current, which had formed at ca. 30 Ma, began transporting progressively colder water northward along the west coast of South America between ca. 22–17 Ma. This decreased MAP and, along with topographic effects (orographic deflection of moisture-laden winds), contributed to the development of arid vegetation landward as in the Atacama Desert. Studies of these older deposits, along with more recent marine, lake, and ice core data, are published in symposia volumes of the Scientific Committee on Antarctic Research (SCAR; Fütterer et al., 2006). The beginning of the earliest glaciations on the Antarctic Peninsula ('the last refugium') is reviewed by Anderson et al. (2011):

> Mountain glaciation began in the latest Eocene (approximately 37–34 Ma), contemporaneous with glaciation elsewhere on the continent and a reduction in atmospheric CO_2 concentrations. This climate cooling was accompanied by a decrease in diversity of the angiosperm-dominated vegetation that inhabited the northern peninsula during the Eocene. A mosaic of southern beech and conifer-dominated woodlands and tundra continued to occupy the region during the Oligocene (approximately 34–23 Ma). By the middle Miocene (approximately 16–11 Ma), localized pockets of limited tundra still existed at least until 12.8 Ma. The northernmost Peninsula was overridden by an ice sheet in the early Pliocene (approximately 5.3–3.6 Ma).

At the Eocene-Oligocene transition at 34 Ma there is evidence of ice sheet variability in the Ross Sea responding to atmospheric CO_2 levels falling below 750 ppmv (Galeotti et al., 2016). There were also orbitally influenced changes at 34–31 Ma, but stable continental ice did not form until 32.8 Ma when CO_2 fell below 600 ppmv, showing that at this level it became a principal forcing mechanism for glacial conditions at the far-southern pole.

During the Pliocene there were several advances and retreats of the East Antarctic ice sheet, reflecting cool (early Pliocene) to warmer (mid-Pliocene) to cold (late Pliocene) intervals (Reinardy et al., 2015). Estimates of the extent of the changes vary widely. Haywood et al. (2009, 404) state that "the stability of the Antarctic Ice Sheets during the Late Miocene and Pliocene has been the subject of almost continuous debate for more than 20 years," dividing the participants into "stabilists" and "dynamicists" (reminiscent of the "fixist" and "mobilist" models for movement of the Antilles through the Central American seaway; II, 72). The debate was initiated by the discovery of *Nothofagus* fossils in glacial deposits of Pliocene age in West Antarctica (Carlquist, 1987; Hill et al., 1996), suggesting that temperatures were warmer than previously supposed. Changes in the physiognomy of *Nothofagus* (tall trees to gnarled understory shrubs) also appear consistent with fluctuations described by Galeotti et al. (2016) for the Eocene/Oligocene and by Reinardy et al. (2015) for the early Pliocene. By the late Pliocene widespread terrestrial vegetation had disappeared from Antarctica.

Although by the beginning of the Quaternary at ca. 2.5 Ma migration of vascular plants across Antarctica had ceased, mention is made of the ice core data because that information, along with conditions and vegetation responses in the Tertiary, is widely used for predicting things to come in response to global warming. An increase in the amount of dust along the core means colder/drier conditions; heavier isotopes of oxygen (O^{18} versus O^{16}) mean greater evaporation and warmer temperatures; and increases in atmospheric CO_2 and other heat-trapping greenhouse gases like methane are closely associated with higher MATs. The principal research stations for these studies are at Vostok (Russia), with cores tracking the climate record back 400,000 years, and at Concordia (France and Italy as part of the European Project for Ice Coring in Antarctica, or EPICA) with a core extending to 800,000 years ago.

Among the findings from the Vostok core are that glaciations follow a periodicity of about 100,000 years (the Milankovitch variations). The current interpretation is that these trigger temperature changes, which are then amplified by trends in CO_2 concentration. The cold intervals from buildup to termination last about 100,000 yr with warm intervals of 10,000–20,000 yr. The latest warm interval (that is, the present interglacial or Holocene) began about 11,500 BP with the next glacial interval probably to be delayed by current high CO_2 emissions. The concern of climatologists is that when the cold intervals begin, they are exceptionally turbulent. The highest peak of CO_2 along the cores was ca. 290 ppmv compared to the present value of ca. 400.

The warming at low elevations on the Bellingshausen Sea (westward) coast of the Antarctic Peninsula is as large as any increase observed on Earth over the last 50 years, and at Faraday (now Vernadsky) station the annual mean surface temperature has risen by about 2.5° C since the 1950s. (Turner et al., 2005)

The East Antarctic Ice Sheet is presently stable, while the West Antarctic Ice Sheet is sliding over the wet underlying rock into the ocean. As it continues to melt, net sea level will rise by an estimated 1 m by 2100 and 15 m by 2500 (DeConto and Pollard, 2016), with over 2 m in a fully retreated configuration of the Totten Glacier alone (Aitken et al., 2016). A new study

(Zeebe et al., 2016) suggests the increase in carbon at the PETM, the highest release in the last 66 Ma and elevating temperatures by ca. 5°C, occurred over an interval of 4000 years. The present rate is 10 times faster and legitimizes provocative warnings like: "Antarctic model raises prospect of unstoppable ice collapse" (Tollefson, 2016). As a reminder, long-term climate records are conditioned by Milankovitch variations and CO_2 concentration, and are locally to regionally modified by topography (flat extensive plains, mountainous landscapes, island configuration), warm or cold offshore currents, drift into different climate zones, and the slant of incident light (greater at the poles with more absorption of solar heat than at the equator). Consequently, the records are comparable but not identical temporally or in intensity in the boreal, equatorial, and austral regions.

Land Bridges and Island Biogeography

In addition to the human impact of sea-level changes, there are indications that such changes may also influence biodiversity (Crawford and Archibald, 2017; Ricklefs and Bermingham, 2008). "Island biogeography" is an equilibrium theory holding that richness is a function of an island's isolation (distance from source areas) and size (accommodation of migrants). It applies well to land-bridge islands currently separated from the mainland by high sea levels and less so to volcanic islands that were never part of a larger continent (Fernández-Palacios, 2016). Factors other than isolation and size are now being recognized, and these include trends and fluctuations in climate, ocean currents, wind patterns, configuration of the islands, elevations, and past fusions and separation. When these additional factors are incorporated into revised models (Weigelt et al., 2016), one of the findings is that the number of angiosperm endemics is higher on islands that were larger during the LGM with lesser effect of isolation and climate change during that interval. The role of climate may be significant, however, even though size emerges as more important in the models, because periods of climate change also reduce diversity (and presumably endemics), especially on small islands (Karnauskas et al., 2016). In a review of Karnauskas et al. (2016), Dasgupta (2016) says that "climate change may dry out 73 percent of the world's small islands by 2050" and that "some of the island groups to be affected the most . . . include the Juan Fernandez Islands, Chile and the Lesser Antilles."

These observations would presumably apply to times beyond the LGM, so information on vegetation in the deeper past may also be relevant to analyses about the cause(s) of current diversity, and vice versa—that is, current patterns of diversity reflect past events and may provide insight into reconstructing past vegetation (e.g., the tropical rain forest) in areas where paleobotanical information is meager (the Amazon Basin) or absent (the Lesser Antilles). This is particularly interesting, given that all the New World land bridges either (1) are now and always have been islands (ALB); or (2) were once islands or peninsulas that have become contiguous (CALB); or (3) were contiguous land masses that separated in the Jurassic/Early Cretaceous (MLB) or in the Eocene (NALB) into land parcels of various sizes, or that individually fluctuated in size with sea-level changes in the late Pliocene and Quaternary (BLB). Whether residues of these older histories are imprinted on modern island biotas like those of the Quaternary (Weigelt et al., 2016) as a kind of environmental memory, analogous to the genetic memory of experiences past, with regard to island biogeography would be an interesting question to investigate. Meiri (2017) believes that continental land bridges require more investigation to balance the greater emphasis on oceanic islands (Hawaii, Galapagos Islands) in the original theory and in subsequent studies. In this connection it is noted that all land bridges mentioned here are or were continental islands.

The topics of modern vegetation, indigenous people, and peopling of Antarctica (and South America from the south) can be addressed succinctly. The modern vegetation consists of only 2 vascular plants, *Colobanthus quitensis* and *Deschampsia antarctica*. At present there are no indigenous people or permanent residents (only long-term support staff associated with the research stations), and there never have been, because long before humans evolved, the climates were frigid in the extreme and the continent was covered by ice. That makes all the more remarkable the vegetation that existed before Antarctica drifted slightly to its current polar position (warm-temperate to subtropical at 100 Ma), during the formation of the ACC (mesic at 34–29 Ma), and during climate change at the end of the MMCO at ca. 15 Ma (cool to cold-temperate). There was disappearance of vegetation episodically during the Pliocene, with decline (until the late Anthropocene) of $p\mathrm{CO_2}$ concentrations.

Utilization of the Magellan Land Bridge

Cono del Sur

The previous discussion provides a context for considering movement of organisms through the austral region during the Late Cretaceous and Cenozoic. Reconstructing the biota and paleoenvironmental history generally (Rabassa, 2008; Ruzzante and Rabassa, 2011; Woodburne et al., 2014, especially pp. 5–22 for fossil floras), and for the Middle Cenozoic of the Cono del Sur specifically, is facilitated by information available from the Sarmiento Formation in the Gran Barranca of Chubut Province (Madden et al., 2010) and the Santa Cruz Formation in Santa Cruz Province (Vizcaíno et al., 2012) of Patagonian Argentina. The Sarmiento Formation contains a sequence of fossil faunas and floras existing between 42 and 18 Ma (middle Eocene to early Miocene). During this interval there was only slight movement of South America, so changes observed in the biota are attributable to alterations in the climate and landscape. These alterations include progression from a more flat topography in the Eocene to increasing relief in the Oligocene through rise of the Andes. The climatic trends evident from various faunal assemblages and sedimentology are as follows:

- Gastropods reflecting warm and humid conditions in the late middle Eocene, followed later in the Oligocene and Miocene by cooler and drier seasonal climates consistent with global trends
- Marsupials that show a more dramatic change attributed to a drop in temperature between the latest Eocene and the early Oligocene
- Notoungulate mammals showing a change from brachydont and mesodont to hypsodont dentation with a thick coating of cement. This is interpreted to mean an Eocene-to-Miocene trend toward more open habitats and grassland with a diet richer in silicates (phytoliths; but see Palazzesi and Barreda, 2012)
- Ichnofossils (indirect/trace evidence; burrows, tracks, etc.) interpreted to mean drier/colder climates with a sparser vegetation cover
- Isotopes from fossil tooth enamel indicating a decrease in MAT of $2°-3°C$ at the Eocene–Oligocene transition
- Sedimentation and edaphic evidence suggesting a trend toward drier to semi-arid conditions in the early Miocene, although this is before the MMCO, after which climates globally became more seasonal and colder before onset of extensive glaciations in the Pliocene and Quaternary

The Santa Cruz Formation is 18.5–16 Ma (Perkins et al., 2012, 39), or near the beginning of the MMCO interval between 21[17]–15 Ma. Sedimentological and paleontological evidence (ichnofssils, amphibians, reptiles, birds, and mammals) is generally consistent with a mosaic of vegetation and habitats, interrupting the trend toward cool and dry conditions, with a brief return of warmer and moister forested environments

(MMCO), then resumption of cooler, drier climates and more open, shrubby vegetation.

Fossil floras are well represented in the Cono del Sur and have been studied extensively in Argentina by Sergio Archangelsky (1963 et seq.) and colleagues (e.g., Romero, 1977), including a regional summary by Barreda and Palazzesi (2010; Figs. 6.7, 6.8), and for Patagonia by Iglesias et al. (2011). The angiosperms appear in the latter part of the Early Cretaceous (Aptian ca. 110 Ma; Archangelsky et al., 2009). There are aquatic-marsh plants from the Late Cretaceous La Colonia Formation (Cúneo et al., 2014), including *Azolla*-type, *Dicksonia*-type, *Mirasolia*, *Nelumbo*, *Regnellidium*, and *Typha*-type. Throughout the Tertiary there is the same shift in vegetation

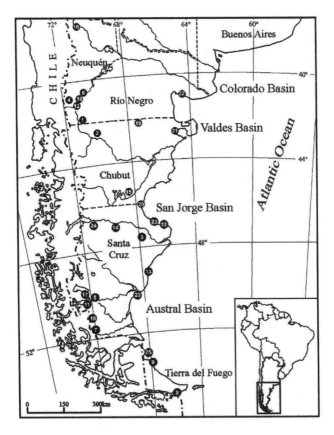

6.7. Location map for paleobotanical/paleopalynological studies in southern South America. From Barreda and Palazzesi, 2010, in Madden et al. (eds), The Paleontology of Gran Barranca, Cambridge University Press, Cambridge. Used with permission.

		Coeval Mammal Faunas (SALMAs)	Botanical information	Inferred vegetation	% entire-margin leaves	Inferred paleoclimate
Miocene	M / Bu. / E / Aq.	"Pinturan," Colhuehuapian	20, 24, 25	New increase of megatherm elements – swamp communities and humid forests	No record	Warm–temperate Humid to subhumid
			20, 21, 22, 23	Arid-adapted vegetation expanded	No record	Temperate – Subhumid to semi-arid
			16, 17, 18, 19, 20	Patches of seasonally dry forests	No record	Temperate to warm–temperate–subhumid
Oligocene	L	Deseadan	11, 12, 13, 14, 15	Meso-megatherm forests –first occurrences of modern shrubby and herbaceous lineages – first Asteracean record (sunflower family)	No record	Warm–temperate Humid
	E	"La Cantera" Tinguirirican	8, 9, 10	Spreading of microtherm Nothofagus gymnosperm dominated forests	~27%	Temperate to cold–temperate Humid
Eocene	L / M	Mustersan Barrancan	5, 6, 7	First irruption of micro-mesotherm Nothofagus forests – some megatherm lineages still prevail – first poaceae records (grass family)	~40%	Temperate to Warm–temperate Humid
	E		1, 2, 3, 4	Megatherm communities	~69%	Warm–temperate Humid MAT 14-18°C MAP 1050-1250 mm

6.8. Inferred plant communities, climates, and South American Land Mammal Ages (SALMAs), Eocene through Miocene, southern South America. From Barreda and Palazzesi, 2010, in Madden et al. (eds.), The Paleontology of Gran Barranca, Cambridge University Press, Cambridge. Used with permission.

from warm and humid to colder and drier with a decrease in leaf size as is generally evident elsewhere. The early Eocene Laguna del Hunco and Río Pichileufú floras are dated at 52.2 Ma and 47.7 Ma (Knight and Wilf, 2013) and show high diversity (Wilf et al., 2003) with a predominance of megathermal plants like palms, *Dacrycarpus* (presently southeast Asian and Australasian rainforests; Wilf, 2012), *Atherospermophyllum guinazui* related to the extant genera *Daphnandra* and *Doryphora* of Australia, *Cupania, Coprosma, Gymnostoma, Monimiophyllum callidentatum* related to *Wilkiea* of Australia, *Myrcia*, and *Schmidelia*. Mixed in are plants suggestive of drier habitats (*Cassia, Schinopsis, Celtis*) attributed to open habitats and looser soils resulting from fires and volcanic activity. This was the maximum southern extent of warm-temperate to subtropical vegetation, comparable to that observed for the boreal regions in the early Eocene. The MAT is estimated at 14°–18°C and the MAP at 1050–1250 mm (Wilf et al., 2005). Later in the Eocene mesothermal to microthermal plants like Araucariaceae, Podocarpaceae, Caryophyllaceae, Cunoniaceae, Gunneraceae, Myrtaceae, Proteaceae, and *Nothofagus* appear. By the early Miocene the collective vegetation is indicative of colder and drier climates with *Ephedra, Acacia, Anadenanthera, Caesalpinia, Casuarina, Cressa/Wilsonia, Prosopis*, Asteraceae, Calyceraceae, and Chenopodiaceae present, along with some megathermal species persisting possibly in gallery forests (Barrreda and Palazzesi, 2010).

In the Santa Cruz Formation (Brea et al., 2012) there is wood of *Araucaria marensii*, *Doroteoxylon* (cf. Caesalpinioideae), *Eucryphiaceoxylon eucryphioides* (*Eucryphia*), *Laurinoxylum atlanticum* (cf. *Persea*), *Myrceugenia chubutense*, *Nothofagoxylon triseriatum* (*Nothofagus*), and aff. *Xilotype* 3 (*Sophora*). After a brief reprieve in the middle Miocene the vegetation becomes shrubbier, scattered, and eventually disappears after the MPCO.

Antarctica

I had never before had the experience of beholding scenic beauty so dazzling.

—J. Franzen (2016)

Antarctica is the coldest, highest, driest, most isolated, and most vegetation-sparse of all the continents, but the long view is that "for much of its existence Antarctica has been ice-free and supporting a wealth of plant and animal life" and that "East Antarctica, a craton, and West Antarctica, comprising a number of crustal blocks, are separated by the West Antarctica Rift System and have evolved within very different geological and climatic frameworks" (Cantrill and Poole, 2012, 17). In the Late Cretaceous ca. 100 Ma the continent was a unified land mass and essentially in its present position. The climatic and biotic changes it later experienced were due primarily to global (pCO$_2$ concentration) and extra-global (Milankovitch) forcing, modified by formation of the ACC and by alterations in regional topography. The Early to Middle Cretaceous pCO$_2$ concentrations have been estimated at 2000 ppmv rising to 4000 ppmv (Berner and Kothavala, 2001), although these estimates may be high. Breecker et al. (2010) believe the values are closer to 1000 ppmv in the Middle Cretaceous and 800 ppmv by the end of the period (but see McElwain et al., 2016). Still, at the minimum, that was double present values and there was no ACC preventing warm waters from circulating into the region. MATs were from 10°C to 16°C, and MAP was ca. 3000 mm.

The history of the vegetation is extensively reviewed by Cantrill and Poole (2012). Compared to other New World connections, severance was more complete and began early (from the Oligocene onward) because of the climatic hurdle created by the ACC, frigid conditions on the continent itself, and the expanding ice. Data is limited to exposures along the margins of the continent, such as on Alexander, James Ross, and Seymour islands of the Antarctic Peninsula, and to cores through the ice and from offshore. In the Early and Middle Cretaceous, vegetation included fern and gymnosperm holdovers from the earlier Mesozoic (Bennettitales, Cycadales). At the beginning of the Late Cretaceous angiosperms were added, along with gymnosperms like *Ginkgo*, Araucariaceae (Fig. 6.9), Cupressaceae, and Podacarpaceae (Fig. 6.10). The plant-eating Cretaceous dinosaur *Antarctopelta oliveroi* was the first discovered on the continent (70 Ma, James Ross Island; Olivero et al., 1991), and later the carnivorous *Crylophosaurus ellioti* was found with the remains of a plant-eating prosauropod inside. The first evidence of flowering plants in Antarctica is from the early Albian (ca. 112 Ma) in the form of a few monocolpate pollen grains of *Clavatipollenites* of chloranthaceous affinity on James Ross Island (Dettmann and Thomson, 1987; see Archangelsky and Taylor, 1993, for Patagonia). The presence of a greater number and higher diversity of angiosperms near the same time in the adjacent Cono del Sur make it improbable that an extensive interchange of flowering plants was taking place across the Magellan Land Bridge before ca. 100 Ma:

> It is unlikely that the Antarctic Peninsula acted as a biotic gateway for dispersal across Gondwana, at least in the initial stages of radiation. The steep climatic gradient must have acted as a barrier to any potential (pre-) Aptian migration of angiosperms through South America and into Antarctic via this route. (Cantrill and Poole, 2002, 152; also Cantrill and Poole, 2012)

But see Poropat et al. (2016), regarding dinosaur movement across the region:

> Interestingly, floral evidence suggests that a sharp climatic barrier existed between Antarctica and South America during the Aptian and early Albian. Thus, the climatic conditions of the land routes across both Patagonia-West Antarctica and East Antarctica-Australia would not have been conducive to sauropod dispersal during this interval. As a corollary of the above scenario, we hypothesize that the appearance of somphospondylan sauropods in Australia in the late Albian and Cenomanian-early Turonian reflects climatic shifts that removed these barriers to dispersal via this relatively high latitude route. Global warming during the late Albian-Turonian flattened the latitudinal thermal gradient, which in turn would have enabled sauropods to disperse from South America, across Antarctica, to Australia via a set of suitable habitats. (22–23)

By floral evidence temperatures had fallen and rainfall decreased by the Late Cretaceous and cooler temperate taxa diversified along with holdovers

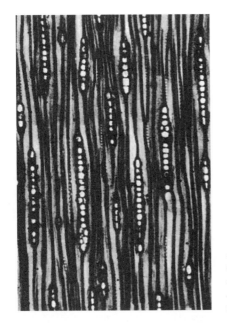

6.9. Fossil wood of *Araucarioxylon* from the Cretaceous of Antarctica. From Falcon-Lang and Cantrill, 2001, Cretaceous Research, Academic Press (Elsevier). Used with permission.

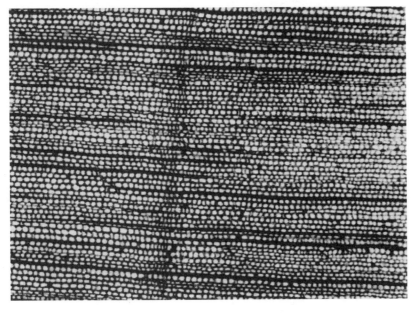

6.10. Fossil wood of *Podocarpoxylon* with faint growth ring from the Cretaceous of Antarctica. From Falcon-Lang and Cantrill, 2001, Cretaceous Research, Academic Press (Elsevier). Used with permission.

6.11. Fossil leaf of *Nothofagus* from the Paleo-gene of Antarctica. From Cantrill and Poole, 2012, The Vegetation of Antarctica through Geological Time, Cambridge University Press, Cambridge. Used with permission.

from warmer mid-Cretaceous times. Present were tree ferns of the Cyathea-ceae, Dicksoniaceae, and Gleicheniaceae; the podocarps *Dacrycarpus* and *Dacrydium*; and angiosperms of the Aquifoliaceae, Bombacaceae, Dille-niaceae, Gunneraceae, Loranthaceae, Myrtaceae, Olacaceae, Palmae, Peda-liaceae, Proteaceae, and Sapindaceae (e.g., *Sapindus*). *Nothofagus* (Fig. 6.11) first appears in Antarctica in the early Campanian (88–80 Ma) and then in South America, indicating the land bridge was functioning for those with at least moderate means of dispersal and that some migration was taking place from south to north. (The age estimated for the crown group of Nothofagaceae is 66(–58) to ca. 43(–35(?21)) Ma, and this is "decidedly after the first fossil records" as compiled in Stevens [2001 et seq.].) At the Cretaceous/Tertiary boundary there is evidence of the asteroid impact from high iridium levels in deep-sea sediments, but the effect on vegetation compared to the Northern Hemisphere was less. Podcarpaceae-Proteaceae-*Nothofagus* woodlands continued to dominate over wide areas. Diversifica-tion of *Nothofagus* peaked in the Oligocene before the full separation of Antarctica from South America and formation of the ACC.

The Tertiary opened with a brief drop in global MATs in the early Paleocene, sea ice formed, and sea levels fell by 50–125 m. This was soon followed by the PETM (55 Ma), which peaked at the EECO at ca. 55 Ma with MATs increasing by $5°–8°C$ from $10°C$ to$18°–22°C$ due to a CO_2 rise from ca. 1000 to possibly 1700 ppmv. Prominent plants in this warm interlude included *Lygodium*, *Nypa* and other palms, Atherospermataceae (*Laurelia*), Bixaceae, Chrysobalanaceae, Cochlospermaceae (*Cochlospermum*), Lauraceae (*Nectandra, Ocotea*), Menispermaceae (as *Menispermites*), *Nothofagus*, Olacaceae, Pandanaceae, Sapindaceae (*Cupania*), Sterculiaceae (*Sterculia*), and Symplocaceae (*Symplocos*). The several modern genera ascribed to the late Paleocene of Seymour Island by Dušen (1908) have been mostly reassigned by Tosolini et al. (2013) to other genera (*Fagus* to *Nothofagus*) or to higher taxonomic categories. Ice volume was reduced, then came and went beginning at about 45 Ma before becoming sustained near the Eocene/Oligocene boundary ca. 32.8–40 Ma when CO_2 concentration fell to between 200 and 500 ppmv. Griener et al. (2013) suggest a decline in MAP at 35.9 Ma with a shift from dense *Nothofagus* woodland to sparser woodlands and tundra, consistent with other evidence for cooling. These Paleogene assemblages are reminiscent in their mixed-ecotype compositon of the boreotropical flora (in other words, an "austrotropical" flora), although neither this concept nor Antarctic geofloras have been applied to the vegetation.

During the transition interval of the latest Eocene through the Oligocene to the earliest Miocene, many of these plant groups persisted, varying in abundance and distribution with the episodic fluctuations in climate but trending toward colder conditions. For example, "the contrast between the early and late Oligocene flora on the edge of Victoria Land was not as great as previously thought, and in general this distinctive flora remained largely unchanged throughout the Oligocene" (Cantrill and Poole, 2012, 420). Plant microfossils from the early to middle Miocene section of the ANDRILL 2A (AND-2A) core from the Ross Sea show an increase in abundance and diversity corresponding to a period of ice minima: "We found that palynomorph assemblages reflecting generally warmer conditions are largely associated with 400-kyr eccentricity maxima, while assemblages indicative of colder conditions coincide with 400-kyr eccentricity minima. These data are consistent with other findings that indicate the early to middle Miocene climate was eccentricity-paced" (Griener et al., 2015). If climates were as sensitive to the Milankovitch cycles as they clearly were to fluctuations in atmospheric CO_2 concentration, this is one more indication that the ice and biota were more dynamic throughout the Late Cretaceous and Cenozoic than often depicted. In the middle Miocene global climates warmed and MAP increased to the MMCO, and this is shown in the Antarctic vegetation (Haywood et al., 2009). There was a significant increase in podocarps and *Nothofagus*, including a trend from shrubby forms to taller trees, and *Isöetes, Drosera, Sparganium*, and *Typha* were present, probably reflecting increased MAP associated with the rising temperature. Then, temperatures dropped by ca. 20°C, diversity decreased, *Nothfagus* species were gnarled and prostrate, and tundra became well established. There was the brief warm interval of the MPCO, after which climates became colder. In the late Pliocene terrestrial vegetation disappeared from Antarctica.

In comparing the northern and southern portions of the Magellan Land Bridge, appendix 6 in the online supplementary materials shows some of the taxa found as fossils in sediments in the Cono del Sur. Recall that the list is for specimens referred to modern genera, or to artificial taxa where a living generic equivalent or near-equivalent is specified rather than cited generally for morphological comparison. A more complete listing is provided by Cantrill and Poole (2012) and in references in the text and legends to online appendices 6 and 7.

Approximately 311 taxa are listed for the Late Cretaceous and Tertiary of the Cono del Sur. There is some overlap from taxa being recorded in more than one geologic interval. Of this number, ca. 90 occur in the Cretaceous and Paleocene (there was a warm interval in the Middle Cretaceous), 134 in the Eocene (there was a warm interval at the PETM), and 1 in the Oligocene, for a total of 224 or 72% in the Cretaceous and Paleogene through the Oligocene with a sharp drop in the Oligocene. There are 75 in the Miocene and 9 in the Pliocene for a total of 84 or ca. 28% in the Neogene with the big break occurring in the Oligocene. This is generally consistent with atmospheric CO_2 concentrations and terrestrial glacial and marine geochemical records from the proximity of the Magellan Land Bridge (intensification of cooling), acknowledging the uneven availability of the paleobotanical data geographically from the Cono del Sur (more studied from southern Santa Cruz and Chubut provinces than from northern Argentina, Chile, Paraguay, and Uruguay) and stratigraphically (more studied from the Cretaceous through Eocene than from the Miocene and Pliocene). There was an overall reduction in diversity and a simplification in structure and composition of the vegetation from the Paleogene through the Neogene.

Appendix 7 in the online supplementary materials lists representative fossils from the Cretaceous through the Pliocene for Antarctica within the time, space, and localities available. As expected, the same global trends as for the Cono del Sur, previously discussed, are evident but more intense.

Approximately 90 genera are listed as representative of the Cretaceous and Tertiary vegetation of Antarctica, and most are from the Antarctic Peninsula. Of this number, ca. 76 or 84% are from the Cretaceous and Paleogene, ca. 14 or 16% from the Miocene and Mio-Pliocene, and none from the middle and late Pliocene. The figures are general, and a more complete inventory for Antarctica and the Cono del Sur might change the numbers to some extent, but the relative percentages, general trends, and consistency with independent evidence are likely to remain the same.

The online appendices may be examined for fossil taxa north, south, or on the MLB in the past, and some modern distributions examined where their current biogeography suggests a place of origin and subsequent directional movement (for additional references see I, 191; II; and Cantrill and Poole, 2012). *Araucaria* is a holdover from an older and widely distributed lineage of gymnosperms, and its present distribution involves more than one factor. It occurs at present from New Guinea, eastern Australia, and New Zealand to southern Brazil and Chile. By far the greatest number of the ca. 19 species occurs in New Zealand with 13 endemics. This suggests movement from that region across Antarctica into South America, and the modern affinities are consistent with the fossil record. The family is known from the late part of the Early Cretaceous (Albian) of Australia, where it was already established when the angiosperms arrived, introduced first in the Early Cretaceous (Albian) into Antarctica, then appearing in the Eocene of Argentina. *Podocarpus* occurs in the latest Cretaceous of Antarctica and in the Eocene of Argentina. *Nothofagus* is found in the early Campanian (Late Cretaceous, 88–80 Ma) in Antarctica and in the Eocene of Argentina. The brackish-water mangrove palm *Nypa* is presently represented by one species in Ceylon, Malaysia, and tropical Australia. Among the many fossil records it is found in the Late Cretaceous of the Borneo region, early Paleocene of Patagonia, and in the Eocene at the Mexico/Texas boundary before becoming extinct in the New World. The fossil *Antherospermophyllum*, related to *Daphnandra* and *Doryphora* (Atherospermataceae) of Australia, and the modern *Dacrycarpus*(?) (Podocarpaceae) of southeastern Asia and Australasian rainforests occur in the Late Cretaceous of Antarctica and in the early Eocene Laguna del Hunco flora of Argentina (see also Knight and Wilf, 2013). These records document extensive movement across the Magellan Land Bridge in the Late Cretaceous and Paleocene when conditions and connections were suitable. It is likely that the migration from Australia and Antarctica to the Cono del Sur also reflects the prevalent westerly winds, an extensive source area to the south and west, only a narrow barrier across the Drake Passage, and a land tapering but soon expanding into the large target land of central and northern South America. There undoubtedly was some movement to the west, although the fossils are often not attributable to extant genera (Kooyman et al., 2014). *Beauprea* (at present in New Zealand) may be an example, with *B.*-type pollen found in the Cretaceous of Antarctica. Even so, the diversity and number of organisms moving south were restricted, beginning from the equator by diminishing land and some degree of a barrier formed by declining temperatures. In the Oligocene the genera having moved northward into the Cono del Sur were isolated to go their own way, which more often involved evolutionary stasis or extinction rather than the extensive diversification and expansion that often accompanies movement into new areas (Uribe-Convers and Tank, 2015), perhaps due in part to the limited fragment of the gene pool brought from the Australasian source area.

The unique character of the MLB was imparted by the early disconnection of Antarctica from South America and Australia in the Oligocene ca. 35–31 Ma; essentially the disappearance of vegetation on Antarctica early (after the MMCO) and completely (after the MPCO), its remaining so even during periods of low sea level in the Quaternary, and no evidence the MLB was ever occupied or used by humans as a pathway between the austral lands of the New World.

References

Aitken, A. R. A., et al. (+ 7 authors). 2016. Repeated large-scale retreat and advance of Totten Glacier indicated by inland bed erosion. Nature 533: 385–89 (plus additional supplementary information).

Anderson, J .B., et al. (+15 authors). 2011. Progressive Cenozoic cooling and the demise of Antarctica's last refugium. Proceedings of the National Academy of Sciences USA 108: 11356–60.

Aragón, E., F. J. Goin, Y. E. Aguilera, M. O. Woodburne, A. A. Carlini, and M. F. Roggiero. 2011. Palaeogeography and palaeoenvironments of northern Patagonia from the Late Cretaceous to the Miocene: the Palaeogene Andean gap and the rise of the North Patagonian High Plaetau. In D. E. Ruzzante and J. Rabassa (eds.), Palaeogeography and Palaeoclimatology of Patagonia: Implications for Biodiversity. Biological Journal of the Linnean Society 103: 305–15.

Aráoz, C. J. P., and H. G. Nami. 2014. The first Paleoindian fishtail point find in Salta Province, northwestern Argentina. Archaeological Discovery 2: 26–30.

Archangelsky, S. 1963. Notas sobre la flora fóssil de la zona de Ticó, Provincia de Santa Cruz. Ameghiniana 3: 57–62.

Archangelsky, S., and T. N. Taylor. 1993. The ultrastructure of in situ *Clavatipollenites* pollen from the Early Cretaceous of Patagonia. American Journal of Botany 80: 879–85.

Archangelsky, S., et al. (+ 9 authors). 2009. Early angiosperm diversification: evidence from southern South America. Cretaceous Research 30: 1073–82.

Askin, R. A. 1989. Endemism and heterochroneity in the Late Cretaceous (Campanian) to Paleocene palynofloras of Seymour Island, Antarctica: implications for origins, dispersal and palaeoclimates of southern floras. In J. A. Crame (ed.), Origins and Evolution of the Antarctic Biota. Geological Society of London, Special Publications 47: 107–19.

Barker, P. F. 2001. Scotia Sea regional tectonic evolution: implications for mantle flow and palaeocirculation. Earth-Science Reviews 55: 1–39.

Barreda, V., and L. Palazzesi. 2007. Patagonian vegetation turnovers during the Paleogene–early Neogene: origin of arid-adapted floras. Botanical Review 73: 31–50.

———. 2010. Vegetation during the Eocene-Miocene interval in central Patagonia: a context of mammal evolution. In R. H. Madden, A. A. Carlini, M. G. Vucetich, and R. F. Kay (eds.), The Paleontology of Gran Barranca. Cambridge University Press, Cambridge. Pp. 375–82.

Bergreen, L. 2003. Over the Edge of the World. HarperCollins, New York.

Berner, R. A., and Z. Kothavala. 2001. GEOCARB III: a revised model of atmospheric CO_2 over Phanerozoic time. American Journal of Science 301: 182–204.

Berrocoso, M., et al. (+ 7 authors). 2006. Geodynamical studies on Deception Island: DECVOL and GEODEC projects. In D. K. Fütterer, D. Damaske, G. Kleinschmidt, H. Miller, and F. Tessensohn (eds.), Antarctica. Springer, Berlin. Pp. 283–88.

Boorstin, D. J. 1983. The Discoverers. Random House, New York.

Bowman, V. C., J. E. Francis, R. A. Askin, J. R. Riding, and G. T. Swindles. 2014. Latest Cretaceous–earliest Paleocene vegetation and climate change at the high southern latitudes: palynologial evidence from Seymour Island, Antarctic Peninsula. Palaeogeography, Palaeoclimatology, Palaeoecology 408: 26–47.

Brea, M., A. F. Zucol, and A. Iglesias. 2012. Fossil plant studies from the late early Miocene of the Santa Cruz Formation: paleoecology and paleoclimatology at the passive margin of Patagonia, Argentina. In S. F. Vizcaíno, R. F. Kay, and M. Susana Bargo (eds.), Early Miocene Paleobiology in Patagonia. Cambridge University Press, Cambridge. Pp. 104–28.

Breecker, D. O., Z. D. Sharp, and L. D. McFadden. 2010. Atmospheric CO_2 concentrations during ancient greenhouse climates were similar to those predicted for A.D. 2100. Proceedings of the National Academy of Sciences USA 107: 576–80.

Brockheim, J. G. 2013. Soil formation in the Transantarctic Mountains from the Middle Paleozoic to the Anthropocene. Palaeogeography, Palaeoclimatology, Palaeoecology 381–82: 98–109.

Cabrera, A. L. 1971. Fitogeografía de la República Argentina. Boletin Argentina Botanica 14: 1–42.

———. 1976. Regiones fitogeograficas Argentinas. Enciclopedia Argentina Agricultura y Jardineria. 2d ed. Vol. 2. Editorial Acme, Buenos Aires.

Cantrill, D. J., and I. Poole. 2002. Cretaceous patterns of floristic change in the Antarctic Peninsula. In J. A. Crame and A. W. Owen (eds.), Palaeobiogeography and Biodiversity Change: The Ordovician and Mesozoic-Cenozoic Radiations. Geological Society of London Special Publication 194, London. Pp. 141–52.

———. 2012. The Vegetation of Antarctica through Geologic Time. Cambridge University Press, Cambridge.

Carlquist, S. 1987. Pliocene *Nothofagus* wood from the Transantarctic Mountains. Aliso 11: 571–83.

Clarke, J. D. A. 2006. Antiquity of aridity in the Chilean Atacama Desert. Geomorphology 73: 101–14. ["The San Pedro Group contains gypsum, andydrite and halite, and is perhaps the most spectacular occurrence of Neogene evaporates in the Atacama Desert. These have been interpreted as having been deposited in an alluvial-fan to playa-lake environment" (109).]

Compagnucci, R. H. 2011. Atmospheric circulation over Patagonia from the Jurassic to present: a review through proxy data and climatic modeling scenarios. In D. E. Ruzzante and J. Rabassa (eds.), Palaeogeography and Palaeoclimatology of Patagonia: Implications for Biodiversity. Biological Journal of the Linnean Society, London. Pp. 229–49.

Crawford, D. J., and J. K. Archibald. 2017. Island floras as model systems for studies of plant speciation: prospects and challenges. Journal of Systematics and Evolution 55: 1–15.

Cúneo, N. R., M. A. Gandolfo, M. A. Zamaloa, and E. Hermsen. 2014. Late Cretaceous aquatic plant world in Patagonia, Argentina. PLoS ONE 9 (8): e104749. doi: 10.1371/journal.pone.0104749.

Dasgupta, S. 2016. Climate change is drying up small islands, study says. https://news/mongaby.com/2016/04/climate-change-drying-small-islands-study-says.

DeConto, R. M., and D. Pollard. 2016. Contribution of Antarctic to past and future sea-level rise. Nature 531: 591–97 (plus additional supplementary information).

Dettmann, M. E., and M. R. A. Thomson. 1987. Cretaceous palynomorphs from the James Ross Island area, Antarctica—a pilot study. British Antarctic Survey Bulletin 77: 13–59.

Dillehay, T. D., et al. (+ 13 authors). 2015. New archaeological evidence for an early human presence at Monte Verde, Chile. PLoS ONE 10 (11): e0923.0141923/journal.pone.0141923.

Dusén, P. 1908. Über die Tertiare Flora der Seymour Insel. In O. Nordenskjöld (ed.), Wissenchaftliche Ergebnisse der Schwedischen Südpolar-Expedition 1901–1903. Geologie und Paläontologie. Norstedt & Söner, Stockholm. Pp. 1–27.

Falcon-Lang, H. J., and D. J. Cantrill. 2001. Gymnosperm woods from the Cretaceous (mid-Aptian) Cerro Negro Formation, Byers Peninsula, Livingstone Island, Antarctica: the arborescent vegetation of a volcanic arc. Cretaceous Research 22: 277–93.

Fehren-Schmitz, L., et al. (+ 15 authors). 2015. A re-appraisal of the early Andean human remains from Lauricocha in Peru. PLoS One. doi: 10.1371/journal.pone.0127141.

Fernández-Armesto, F. 2006. Pathfinders: A Global History of Exploration. W. W. Norton & Company, New York.

Fernández-Palacios, J. M. 2016. Shaped by sea-level shifts. Nature. doi: 10.1038/nature 17880.

Fitzgerald, P. 2002. Tectonics and landscape evolution of the Antarctic plate since the breakup of Gondwana, with an emphasis on the West Antarctic Rift System and the Transantarctic Mountains. Royal Society of New Zealand Bulletin 35: 453–69.

Folguera, A., et al. (+ 7 authors). 2011. A review of Late Cretaceous to Quaternary palaeogeography of the Southern Andes. In D. E. Ruzzante and J. Rabassa (eds.), Palaeogeography and Palaeoclimatology of Patagonia: Implications for Biodiversity. Biological Journal of the Linnean Society 103: 250–68.

Franzen, J. 2016. The end of the end of the world. New Yorker, 23 May: 44–55.

Fütterer, D. K., D. Damaske, G. Kleinschmidt, H. Miller, and F. Tessensohn (eds.). 2006. Antarctica. Springer, Berlin.

Gajardo, R. 1994. La Vegetación Natural de Chile, Classificación y Distribución Geográfica. Editorial Universitaria, Santiago.

Galeotti, S., et al. (+ 12 authors). 2016. Antarctic Ice Sheet variability across the Eocene-

Oligocene Boundary climate transition. Science, First Release, 10 March. www.science mag.org.

Galindo-Zaldívar, J., et al. (+ 7 authors). 2006a. Crustal thinning and the development of deep depressions at the Scotia-Antarctic plate boundary (southern margin of Discovery Bank, Antarctica). In D. K. Fütterer, D. Damaske, G. Kleinschmidt, H. Miller, and F. Tessensohn (eds.), Antarctica. Springer, Berlin. Pp. 237–42.

Galindo-Zaldívar, J., A. Maestro, J. López-Martínez, and C. Sanz de Galdeano. 2006b. Elephant Island recent tectonics in the framework of the Scotia–Antarctic–South Shetland Block triple junction (NE Antarctic Peninsula). In D. K. Fütterer, D. Damaske, G. Kleinschmidt, H. Miller, and F. Tessensohn (eds.), Antarctica. Springer, Berlin. Pp. 271–76.

Goldberg, A., A. M. Mychajliw, and E. A. Hadly. 2016. Post-invasion demography of prehistoric humans in South America. Nature. doi: 10.1038/Nature 17176.

Graham, A. 2009. The Andes: a geological overview from a biological perspective. Annals of the Missouri Botanical Garden 96: 371–85.

Graham, A., K. M. Gregory-Wodzicki, and K. L. Wright. 2001. Studies in neotropical paleobotany. XV. A Mio-Pliocene palynoflora from the Eastern Cordillera, Bolivia: implications for the uplift history of the Central Andes. American Journal of Botany 88: 1545–57.

Gregory-Wodzicki, K. M. 2000a. Relationship between leaf morphology and climate, Bolivia: implications for estimating paleoclimate from fossil floras. Paleobiology 26: 668–88.

———. 2000b. Uplift history of the Central and Northern Andes: a review. Geological Society of America Bulletin 112: 1091–105.

Gregory-Wodzicki, K. M., W. C. McIntosh, and K. Velásquez. 1998. Climate and tectonic implications of the late Miocene Jakokkota flora, Bolivian Altiplano. Journal of South American Earth Sciences 11: 533–60.

Griener, K. W., D. M. Nelson, and S. Warny. 2013. Declining moisture availability on the Antarctic Peninsula during the late Eocene. Palaeogeography, Palaeoclimatology, Palaeoecology 383–84: 72–78.

Griener, K. W., S. Warny, R. Askin, and G. Acton. 2015. Early to middle Miocene vegetation history of Antarctica supports eccentricity-paced warming intervals during the Antarctic icehouse phase. Global and Planetary Change 127: 67–78.

Haywood, A, M., et al. (+ 12 authors). 2009. Middle Miocene to Pliocene history of Antarctica and the Southern Ocean. In F. Florindo and M. Siegert (eds.), Antarctic Climate Evolution. Developments in Earth and Environmental Sciences 8: 401–63.

Hill, R. S., D. M. Harwood, and P. N. Webb. 1996. *Nothofagus beardmorensis* (Nothofagaceae), a new species based on leaves from the Pliocene Sirius Group, Transantarctic Mountains, Antarctica. Review of Palaeobotany and Palynology 94: 11–14.

Hogan, C. M. 2008. Bahia Wulais Dome middens. Megalithic Portal, www.megalithic.co .uk. (Retrieved 21 March 2016.)

Iglesias, A., A. E. Artabe, and E. M. Morel. 2011. The evolution of Patagonian climate and vegetation from the Mesozoic to the present. In D. E. Ruzzante and J. Rabassa (eds.), Palaeogeography and Palaeoclimatology of Patagonia: Implications for Biodiversity. Biological Journal of the Linnean Society 103: 409–22.

International Work Group for Indigenous Affairs (IWGIA). http://www/iwgia.org/ regions/latin-america/indigenous-peoples-in-latin-america. (Accessed 2016.)

Karnauskas, K. B., J. P. Donnelly, and K. J. Anchukaitis. 2016. Future freshwater stress for island populations. Nature Climate Change. doi: 10.1038.nclimate2987.

Kim, S. B., Y. K. Sohn, and M. Y. Choe. 2006. The Eocene volcaniclastic Sejong Formation, Barton Peninsula, King George Island, Antarctica: evolving arc volcanism from
precursory fire fountaining to vulcanian eruptions. In D. K. Fütterer, D. Damaske,
G. Kleinschmidt, H. Miller, and F. Tessensohn (eds.), Antarctica. Springer, Berlin.
Pp. 261–70.

Knight, C. L., and P. Wilf. 2013. Rare leaf fossils of Monimiaceae and Atherospermataceae
(Laurales) from Eocene Patagonian rainforests and their biogeographic significance.
Palaeontologia Electronica 16, article number 26A, 39 pp.

Kooyman, R. M., et al. (+ 13 authors). 2014. Paleo-Antarctic rainforest into the modern
Old World tropics: the rich past and threatened future of the "southern wet forest
survivors." American Journal of Botany 101: 1–15.

Lansing, A. 1959. (With an introduction by Nathaniel Philbrick.) Endurance. Basic
Books, New York.

Lawver, L. A., and L. M. Gahagan. 2003. Evolution of Cenozoic seaways in the circum-
Antarctic region. Palaeogeography, Palaeoclimatology, Palaeoecology 198: 11–37.

López-Martínez, J., et al. (+ 6 authors). 2006. Tectonics and geomorphology of Elephant
Island, South Shetland Islands. In D. K. Fütterer, D. Damaske, G. Kleinschmidt,
H. Miller, and F. Tessensohn (eds.), Antarctica. Springer, Berlin. Pp. 277–82.

Loponte, D., M. Carbonera, and R. Silvestre. 2015. Fishtail projectile points from South
America: the Brazilian record. Anthropological Discovery 3: 85–103.

Madden, R. H., A. A. Carlini, M. G. Vucetich, and R. F. Kay (eds.). 2010. The Paleontology
of Gran Barranca: Evolution and Environmental Change through the Middle Cenozoic of Patagonia. Cambridge University Press, Cambridge.

Malumián, N., and C. Náñez. 2011. The Late Cretaceous–Cenozoic transgressions in Patagonia and the Fuegian Andes: foraminifera, palaeoecology, and palaeogeography.
In D. E. Ruzzante and J. Rabassa (eds.), Palaeogeography and Palaeoclimatology of
Patagonia: Implications for Biodiversity. Biological Journal of the Linnean Society
103: 269–88.

Manfroi, J., T. L. Dutra, S. Gnaedinger, D. Uhl, and A. Jasper. 2015. The first report of a
Campanian palaeo-wildfire in the West Antarctic Peninsula. Palaeogeography, Palaeoclimatology, Palaeoecology 418: 12–18.

Marticorena, C., and M. Quezada. 1985. Catálogo de la plantas vascular de Chile. Gayana
Botánica 42: 1–157.

Marticorena, C., and R. Rodríguez (eds.). 1995 et seq. Flora de Chile. Universidad de
Concepción.

Martínez, O. A., and A. Kutschker. 2011. The "Rodados Patagónicos" (Patagonian shingle
formation) of eastern Patagonia: environmental conditions of gravel sedimentation.
In D. E. Ruzzante and J. Rabassa (eds.), Palaeogeography and Palaeoclimatology of
Patagonia: Implications for Biodiversity. Biological Journal of the Linnean Society
103: 336–45.

Mautino, L. R., and L. M. Anzotegui. 2014. Novedades palinológicas de las formacio-
nes San José y Chiquimil (Miocene medio y tardío), noroeste de Argentina. Revista
Museo Argentino Ciencias Naturales, n.s., 16: 143–64.

McElwain, J. C., I. Montañez, J. D. White, J. P. Wilson, and C. Yiotis. 2016. Was atmospheric CO_2 capped at 1000 ppm over the past 300 million years? Palaeogeography,
Palaeoclimatology, Palaeoecology 441: 653–58.

McLoughlin, S. 2001. The breakup history of Gondwana and its impact on pre-Cenozoic
floristic provincialism. Australian Journal of Botany 49: 271–300.

Meiri, S. 2017. Oceanic island biogeography: nomothetic science of the anecdotal. Frontiers of biogeography 9.1: e32081. doi: 10.21425/F59132081.

Méndez M., C. A., C. R. Stern, O. R. Reyes B., and F. Mena L. 2012. Early Holocene long-distance obsidian transport in central-south Patagonia. Chungara, Revista de Antropología Chilena 44: 363–75.

Miotti, L., D. Hermo, and E. Terranova. 2010. Fishtail points, first evidence of Late Pleistocene hunter-gatherers in Somuncurá Plateau (Río Negro Province, Argentina). Current Research in the Pleistocene 27: 22–24.

Morgan, E. 1997. The Aquatic Ape Hypothesis. Penguin Press, London.

National Geographic. 2013. Were people killing giant sloths in South America 30,000 years ago? http://news.nationalgeographic.com/news/2013/11/13120-giant-sloths-people-americas-ancient-archaeology-science.html.

New York Times. 2014. Discoveries challenge beliefs on humans' arrival in the Americas. https://www.nytimes.com/2014/03/28/world/americas/discoveries-challenge-beliefs-on-humans-arrival-in-the-americas.html.

Nullo, F., and A. Combina. 2011. Patagonian continental deposits (Cretaceous-Tertiary). In D. E. Ruzzante and J. Rabassa (eds.), Palaeogeography and Palaeoclimatology of Patagonia: Implications for Biodiversity. Biological Journal of the Linnean Society 103: 289–304.

Olivero, E., Z. Gasparini, C. Rinaldi, and R. Scasso. 1991. First record of dinosaurs in Antarctica (Upper Cretaceous, James Ross Island): paleogeographical implications. In M. R. Thomson, J. A. Crame, and J. W. Thomson (eds.), Geological Evolution of Antarctica. Cambridge University Press, Cambridge. Pp. 617–22.

Palazzesi, L., and V. Barreda. 2012. Fossil pollen records reveal a late rise of open-habitat ecosystems in Patagonia. Nature Communications 3. doi: 10.1038/ncomms2299.

Passchier, S., et al. (+7 authors). 2013. Early Eocene to middle Miocene cooling and aridification of East Antarctica. Geochemistry, Geophysics, Geosystems 14: 1399–410.

PeopleGroups. http://www.peoplegroups.org/explore/ClusterDetails.aspx?rop2=C0201. (Accessed 2016.)

Perkins, M. E., et al. (+ 6 authors). 2012. Tephrochronology of the Miocene Santa Cruz and Pinturas Formations, Argentina. In S. F. Vizcaíno, R. F. Kay, and M. Susana Bargo (eds.), Early Miocene Paleobiology in Patagonia: High-Latitude Paleocommunities of the Santa Cruz Formation. Cambridge University Press, Cambridge. Pp. 23–40.

Plates Project. s.d. http://www-udc.ig.utexas.edu/external/plates/recons.htm. [Scroll to Animations, Antarctica: Keystone of Gondwana (Powerpoint Animation, 1999-06-25).]

Poole, I., R. J. Hunt, and D. J. Cantrill. 2001. A fossil wood flora from King George Island: ecological implications for an Antarctic Eocene vegetation. Annals of Botany 88: 33–54.

Poropat, S. F., et al. (+ 10 authors). 2016. New Australian sauropods shed light on Cretaceous dinosaur palaeobiogeography. Nature, Scientific Reports 6. doi: 10.1038/srep34467.

Rabassa, J. (ed.). 2008. The Late Cenozoic of Patagonia and Tierra del Fuego. Elsevier, Amsterdam. [See papers by A. M. J. Coronato et al., V. Barreda et al., M. C. Salemme, and L. L. Miotti.]

Rabassa, J., A. Coronato, and O. Martínez. 2011. Late Cenozoic glaciations in Patagonia and Tierra del Fuego: an updated review. In D. E. Ruzzante and J. Rabassa (eds.), Palaeogeography and Palaeoclimatology of Pagtagonia: Implications for Biodiversity. Biological Journal of the Linnean Society 103: 316–35.

Rademaker, K. 2014. Late Ice-Age human settlement of the high-altitude Peruvian Andes. Mitteilungen der Gesellschaft für Urgeschichte 23: 13–35.

Rademaker, K., M. D. Glascock, B. Kaiser, D. Gibson, D. R. Lux, and M. G. Yates. 2013. Multi-technique geochemical characterization of the Alca obsidian source, Peruvian Andes. Geology 41: 779–82.

Ricklefs, R., and E. Bermingham. 2008. The West Indies as a laboratory of biogeography and evolution. Philosophical Transactions Royal Society of London B, Biological Sciences, 363: 2393–413.

Reinardy, B. T. I., et al. (+ 6 authors). 2015. Repeated advance and retreat of the East Antarctic Ice Sheet on the continental shelf during the early Pliocene warm period. Palaeogeography, Paleoclimatology, Palaeoecology 422: 65–84.

Roberts, D. 2013. Alone on the Ice. W. W. Norton & Company, New York.

Roede, M., J. Wind, J. Patrick, and V. Reynolds (eds.). 1991. The Aquatic Ape: Fact or Fiction. Proceedings from the Valkenburg Conference. Souvenir Press, London.

Romero, E. J. 1977. Polen de Gimnospermas y Fagáceas de la Formación Río Turbio (Eoceno), Santa Cruz, Argentina. Centro de Investigaciones en Recursos Geológicos (CIRGEO), Buenos Aires, Argentina.

Ruzzante, D. E., and J. Rabassa (eds.). 2011. Palaeogeography and Palaeoclimatology of Patagonia: Implications for Biodiversity. Biological Journal of the Linnean Society 103: 221–529.

ScienceNews. 2013. Disputed finds put humans in South America 22,000 years ago. http://www.sciencenews.org/article/disputed-finds-put-humans-south-america -22000-years-ago.

Smith, A. G., D. G. Smith, and B. M. Funnell. 1994. Atlas of Mesozoic and Cenozoic Coastlines. Cambridge University Press, Cambridge.

Stevens, P. 2001 et seq. Angiosperm Phylogeny Website. http://www.mobot.org/MOBOT/ research/APweb.

Suárez, M., R. de la Cruz, and C. M. Bell. 2000. Timing and origin of deformation along the Patagonian fold and thrust belt. Geological Magazine 137: 345–53.

Tollefson, J. 2016. Antarctic model raises prospect of unstoppable ice collapse. Nature 531: 562.

Tosolini, A.-M. P., D. J. Cantrill, and J. E. Francis. 2013. Paleocene flora from Seymour Island, Antarctica: revision of Dusén's (1908) angiosperm taxa. Alcheringa (Association of Australasian Palaeontologists) 37: 1–26.

Turner, J., et al. (+ 8 authors). 2005. Antarctic climate change during the last 50 years. International Journal of Climatology 25: 279–94.

Uribe-Convers, S., and D. C. Tank. 2015. Shifts in diversification rates linked to biogeographic movement into new areas: an example of a recent radiation in the Andes. American Journal of Botany 102: 1854–69.

Vizcaíno, S. F., R. F. Kay, and M. Susana Bargo (eds.). 2012. Early Miocene Paleobiology in Patagonia, High-latitude Paleocommunities of the Santa Cruz Formation. Cambridge University Press, Cambridge.

Walker, G. 2013. Antarctica: An Intimate Portrait of a Mysterious Continent. Bloomsbury Publishers, London.

Weigelt, P., M. J. Steinbauer, J. S. Cabral, and H. Kreft. 2016. Late Quaternary climate change shapes island biodiversity. Nature 532: 99–102 (plus additional supplementary information).

Wilf, P. 2012. Rainforest conifers of Eocene Patagonia: attached cones and foliage of the extant southeast Asian and Australasian genus *Dacrycarpus* (Podocarpaceae). American Journal of Botany 99: 1–23.

Wilf, P., N. R. Cúneo, K. R. Johnson, J. F. Hicks, S. L. Wing, and J. D. Obradovich. 2003. High plant diversity in Eocene South America: evidence from Patagonia. Science 300: 122–25.

Wilf, P., K. R. Johnson, N. R. Cúneo, M. E. Smith, B. S. Singer, and M. A. Gandolfo. 2005. Eocene plant diversity at Laguna del Hunco and Río Pichileufú, Argentina. American Naturalist 165: 634–50.

Woodburne, M. O., et al. (+ 7 authors). 2014. Paleogene land mammal faunas of South America: a response to global climatic changes and indigenous floral diversity. Journal of Mammal Evolution 21: 1–73.

Zeebe, R. E., A. Ridgwell, and J. C. Zachos. 2016. Anthropogenic carbon release rate unprecedented during the past 66 million years. Nature Geoscience. doi: 10.1038/ngeo2681.html.

Zuloaga, F. O., O. Morrone, and M. J. Belgrano (eds.). 2008. Catálogo de las Plantas Vasculares del Cono Sur. Monographs in Systematic Botany from the Missouri Botanical Garden 107. St. Louis.

Additional References

Agostini, C. A., P. H. Brown, and A. Roman. 2008. Estimating Poverty for Indigenous Groups in Chile by Matching Census and Survey Data. William Davidson Institute, University of Michigan, Ann Arbor.

Amazon Conservation Association (ACA) and Conservation Amazónica (ACCA). 2016. Number 29. Construction of new road between Manu National Park and Amarakaeri Communal Reserve (Madre de Dios). amazonconservation.org.

Andrews, E., T. White, and C. del Papa. 2017. Paleosol-based paleoclimate reconstruction of the Paleocene-Eocene Thermal Maximum, northern Argentina. Palaeogeography, Palaeoclimatology, Palaeoecology 471: 181–95.

Antonelli, A., J. A. A. Nylander, C. Persson, and J. Sanmartín. 2009. Tracing the impact of the Andean uplift on neotropical plant evolution. Proceedings of the National Academy of Sciences USA 106: 9749–54.

Artabe, A. E., A. B. Zamuner, and D. W. Stevenson. 2004. Two new petrified cycad stems, *Brunoa* gen. nov. and *Worsdellia* gen nov., from the Cretaceous of Patagonia (Bajo de Santa Rosa, Río Negro Province), Argentina. Botanical Review 70: 121–33.

Askin, R. A., and R. A. Spicer. 1995. The Late Cretaceous and Cenozoic history of vegetation and climate at northern and southern high latitudes. In National Research Council (U.S.) Panel on Effects of Past Global Change on Life, Effects of Past Global Change on Life. Chap. 9. National Academies Press, Washington, DC.

Astorga, G., and M. Pino. 2011. Fossil leaves from the last interglacial in central-southern Chile: inferences regarding the vegetation and paleoclimate. Geologia Acta 9: 45–54.

Atchison, G. W., et al. (+ 7 authors). 2016. Lost crops of the Incas: origins of domestication of the Andean pulse crop tarwi, *Lupinus mutabilis*. American Journal of Botany 103: 1592–606.

Bakker, P., P. U. Clark, N. R. Golledge, and A. Schmittner. 2017. Centennial-scale Holocene climate variations amplified by Antarctic Ice Sheet discharge. Nature 541: 72–76.

Barker, P. F., and E. Thomas. 2006. Potential of the Scotia Sea region for determining the onset and development of the Antarctic Circumpolar Current. In D. K. Fütterer, D. Damaske, G. Kleinschmidt, H. Miller, and F. Tessensohn (eds.), Antarctica. Springer, Berlin. Pp. 433–40.

Beck, C. 2016. Why is a blue cloud appearing over Antarctica? Christian Science Monitor, 3 December.

Callaway, E. 2015. South America settled in one go. Nature 520: 598–99.

Cantrill, D. J., and M. A. Hunter. 2005. Macrofossil floras of the Latady Basin, Antarctic Peninsula. New Zealand Journal of Geology and Geophysics 48: 537–53. [Jurassic.]

Cenizo, M. M., et al. (+ 5 authors). 2015. A new Pleistocene bird assemblage from the Southern pampas (Buenos Aires, Argentina). Palaeogeography, Palaeoclimatology, Palaeoecology 420: 65–81.

Clark, J. A., et al. (+ 9 authors). 2016. Fossil evidence of the avian vocal organ from the Mesozoic. Nature. doi: 10.1038/nature19852. [Antarctica.]

Conniff, R. 2011 (updated 3/24/2016). The wall of the dead: a memorial to fallen naturalists. https://strangebehaviors.wordpress.com/2011/01/14/the-wall-of-the-dead.

Cooper, A., and 10th ISAES Editorial Team (eds.). 2008. Antarctica: A Keystone in a Changing World. National Academies Press, Washington, DC.

Cracraft, J. 1973. Continental drift, paleoclimatology, and the evolution and biogeography of birds. Journal of Zoology 169: 455–543.

———. 1985. Historical biogeography and patterns of differentiation within the South American avifauna: areas of endemism. Ornithological Monographs 36: 49–84.

Cuéneo, R., and M. A. Gandolfo. 2005. Angiosperm leaves from the Kachaike Formation, lower Cretaceous of Patagonia, Argentina. Review of Palaeobotany and Palynology 136: 29–47.

Dana, R. H. 1840. Two Years before the Mast. Harper, New York. [A classic book about a sea voyage from Boston to California by Richard Hendry Dana around Cape Horn starting in 1834, and 80 years before Ernest Shackleton's adventures in crossing the Drake Passage. A movie version was released in 1946.]

Davies, B. (last updated 03/03/2014). Antarctic Peninsula ice sheet evolution. Antarctic Glaciers.org. http://www/antarcticglaciers.org/glacial-geology/icesheet_evolution.

Day, D. 2012. Antarctica: A Biography. Oxford University Press, Oxford.

Del Fueyo, G. M., et al. (+ 9 authors). 2006. Biodiversidad de las paleofloras de Patagonia austral durante el Cretácico Inferior. Ameghiniana, Publicación Especial 11: 101–22.

de Wet, C. B., L. Godfrey, and A. P. de Wet. 2015. Sedimentology and stable isotopes from a lacustrine-to-palustrine limestone deposited in an arid setting, climatic and tectonic factors: Miocene-Pliocene Opache Formation, Atacama Desert, Chile. Palaeogeography, Palaeoclimatology, Palaeoecology 426: 46–67.

Donovan, M. P., A. Iglesian, P. Wilf, C. C. Labandeira, and N. Rubén Cúneo. 2016. Rapid recovery of Patagonian plant-insect associations after the end-Cretaceous extinction. Nature Ecology and Evolution 1. doi: 10.1038/s41559-016-0012.

Fajardo, A. 2016. Are trait-scaling relationships invariant across contrasting elevations in the widely distributed treeline species Nothofagus pumilio? American Journal of Botany 103: 821–29.

Fehren-Schmitz, L., and W. Haak. 2014. Climate change underlies global demographic, genetic, and cultural transitions in pre-Columbian southern Peru. Proceedings of the National Academy of Sciences USA. http://www/pnas.org/content/111/26/9443.

Florindo, F., and M. Siegert (eds.). 2009. Antarctic Climate Evolution: Developments in Earth and Environmental Sciences. Elsevier, Amsterdam.

Francis, J. E., et al. (+10 authors). 2009. From greenhouse to icehouse—the Eocene/Oligocene in Antarctica. In F. Florindo and M. Siegert (eds.), Antarctic Climate Evolution. Developments in Earth and Environmental Sciences 8. Elsevier Science. Pp. 309–67.

Fraser, B. 2014. The first South Americans: extreme living. http://www/nature.com/news/the-first-south-americans-extreme-living-1.16038.

Gayó, E., L. F. Hinojosa, and C. Villagrán. 2005. On the persistence of tropical paleofloras in central Chile during the early Eocene. Review of Palaeobotany and Palynolgy 137: 41–50.

Gonzalez, C. C., M. A. Gandolfo, M. C. Zamaloa, N. R. Cúneo, P. Wilf, and K. R. Johnson. 2007. Revision of the Proteaceae macrofossil record from Patagonia, Argentina. Botanical Review 73: 235–66.

Hervé, F., H. Miller, and C. Pimpirev. 2006. Patagonia-Antarctica connections before Gondwana break-up. In D. K. Fütterer, D. Damaske, G. Kleinschmidt, H. Miller, and F. Tessensohn (eds.), Antarctica. Springer, Berlin. Pp. 217–28.

Herzfeld, U. C. 2012. Atlas of Antarctica: Topographic Maps from Geostatistical Analysis of Satellite Radar Altimeter Data. Springer, Berlin.

Hinojosa, L. F., J. J. Armesto, and C. Villagrán. 2006. Are Chilean coastal forests pre-Pleistocene relicts? Evidence from foliar physiognomy, palaeoclimate, and phytogeography. Journal of Biogeography 33: 331–41.

Hinojosa, L. F., and C. Villagrán. 2005. Did South American mixed paleofloras evolve under thermal equability or in the absence of an effective Andean barrier during the Cenozoic? Palaeogeography, Palaeoclimatology, Palaeoecology 217: 1–23.

Holgate, G. R., I. R. K. Sluiter, and J. Taglieri. 2017. Eocene-Oligocene coals of the Gippsland and Australo-Antarctic basins—paleoclimatic and paleogeographic context and implications for the earliest Cenozoic glaciations. Palaeogeography, Palaeoclimatology, Palaeoecology 472: 236–55.

Hughes, C. E., and R. Eastwood. 2006. Island radiation on a continental scale: exceptional rates of plant diversification after uplift of the Andes. Proceedings of the National Academy of Sciences USA 103: 10334–39.

IndigenousNews.org. 2011. Indigenous peoples in Chile. http://indigenousnew.org/indigenous-peoples/chile.

Jomelli, V., et al. (+ 18 authors). 2014. A major advance of tropical Andean glaciers during the Antarctic cold reversal. Nature 513: 224–28 (plus additional supplementary information).

Keough, P., and R. Keough. 2002. Antarctica. Limited Edition, Antarctica Explorer Series 1. Nahanni Productions, Salt Spring Island, BC.

Kutzinski, V. M., and O. Ette (eds. and translators). 2012. Views of the Cordilleras and Monuments of the Indigenous Peoples of the Americas. University of Chicago Press, Chicago.

Leane, E. 2016. South Pole, Nature and Culture. Distributed for Reaktion Press by the University of Chicago Press, Chicago.

Lear, C. H., and D. J. Lunt. 2016. How Antarctica got its ice. Science Online. First release, 11 March 2016. www.sciencemag.org.

Levy, R., et al. (+ 24 authors and SMS Science Team). 2016. Antarctic ice sheet sensitivity to atmospheric CO_2 variations in the early to mid-Miocene. Proceedings of the National Academy of Sciences USA. www.pnas.org/cgi/doi/10.1073/pnas.1516030113.

Li, X., D. M. Holland, E. P. Gerber, and C. Yoo. 2014. Impacts of the north and tropical Atlantic Ocean on the Antarctic Peninsula and sea ice. Nature 505: 538–42 (plus additional supplementary information).

Limarino, C. O., M. G. Passalia, M. Llorens, E. I. Vera, V. S. Perez Loinaze, and S. N. Césari. 2012. Depositional environments and vegetation of Aptian sequences affected by volcanism in Patagonia. Palaeogeography, Palaeoclimatology, Palaeoecology 323–25: 22–41.

Mackey, B. G., S. L. Berry, and T. Brown. 2008. Reconciling approaches to biogeographical regionalization: a systematic and generic framework examined with a case study of the Australian continent. Journal of Biogeography 35: 213–29.

Mangaravite, E., et al. (+ 6 authors). 2016. Contemporary patterns of genetic diversity of *Cedrela fissilis* offer insight into the shaping of seasonal forests in eastern South America. American Journal of Botany 103: 307–16.

McGrath, M. 2017. Hugh Antarctic iceberg poised to break away. BBC News, Science and Environment, 6 January.

Milne, R. I. 2006. Northern Hemispheric plant disjunctions: a window on Tertiary land bridges and climate change? Annals of Botany 98: 465–72.

Mooney, C. 2016a. Scientists race to study massive Antarctic glacier. Washington Post, 20 October. [Thwaites Glacier, West Antarctic Ice Sheet.]

———. 2016b. This stunning Antarctic lake is buried in ice: And that could be bad news. Washington Post, 12 December.

Morueta-Holme, N., K. Engelmann, P. Sandoval-Acuña, J. D. Jones, R. M. Segnitz, and J.-C. Svenning. 2015. Strong upslope shifts in Chimborazo's vegetation over two centuries since Humboldt. Proceedings of the National Academy of Sciences USA 112: 12741–45.

Nature. 2017a. Antarctic pull-out. Nature 541: 13–19. ["The British Antarctic Survey (BAS) will pull its winter staff out of the Halley VI Research Station . . . between March and November . . . and relocate 23 kilometres away owing to a growing crack in the ice shelf it rests on."]

———. 2017b. Sub-ice-shelf sediments record history of twentieth-century retreat of Pine Island Glacier. Nature 541: 77–80.

Nature Research. 2016. Glaciology: Antarctica warmed up fast in the past. Nature Research 540: 173.

NatureServe and EcoDecisión. 2015. Ecosystem Profile, Tropical Andes Biodiversity Hotspot. For Submission to the Donor Council, Critical Ecosytem Partnership Fund. www.cepf.net.

New York Times Editorial Board. 2016. The danger of a runaway Antarctica: editorial. New York Times, 31 March. http://www/nytimes.com/2016/04/01/opinion/the-danger-of-a-runaway-antarctica.html.

Palazzesi, L., V. Barreda, J. I. Cutiño, M. V. Guier, M. C. Telleria, and R. Ventura Santos. 2014. Fossil pollen records indicate that Patagonian desertification was not solely a consequence of Andean uplift. Nature Communications. doi: 10.1038/ncomms4558.

Pancel, L., and M. Köhl. 2016. Tropical Forestry Handbook. Springer, Berlin.

Pigafetta, A. 1969. The Voyage of Magellan, the Journal of Antonio Pigafetta. Translated by Paula Spurlin Paige from the edition in the William L. Clements Library, University of Michigan, Ann Arbor. Englewood Cliffs, NJ, Prentice-Hall.

Pivel, M. A. G., A. C. A. Santarosa, F. A. L. Toledo, and K. B. Costa. 2013. The Holocene onset in the southwestern South Atlantic. Palaeogeography, Palaeoclimatology, Palaeoecology 374: 164–72.

Plumstead, E. P. 1962. Fossil floras of Antarctica (with an appendix on Antarctic fossil wood by R. Kräusel). Trans-Antarctic Expedition 1955–1958, Scientific Reports No. 9, Geology. Transantarctic Expedition Committee, London. 154 pp.

Poole, I., and D. J. Cantrill. 2006. Cretaceous and Cenozoic vegetation of Antarctica integrating the fossil wood record. In J. E. Francis, D. Pirrie, and J. A. Crame (eds.), Cretaceous-Tertiary High-Latitude Palaeoenvironments, James Ross Basin, Antarctica. Geological Society of London, Special Publication 258. Pp. 63–81.

Poole, I., D. J. Cantrill, and T. Utescher. 2005. A multi-proxy approach to determine Antarctic terrestrial palaeoclimate during the Late Cretaceous and early Tertiary. Palaeogeography, Palaeoclimatology, Palaeoecology 222: 95–121.

Quattrocchio, M. E., W. Volkheimer, A. M. Borromei, and M. A. Martínez. 2011. Changes of the palynobiotas in the Mesozoic and Cenozoic of Patagonia: a review. In D. E. Ruzzante and J. Rabassa (eds.), Palaeogeography and Palaeoclimatology of Patagonia: Implications for Biodiversity. Biological Journal of the Linnean Society 103: 380–96.

Raven, P. H., and D. I. Axelrod. 1975. History of the flora and fauna of Latin America. American Scientist 63: 420–29.

Reguero, M. A., and S. A. Marenssi. 2010. Paleogene climatic and biotic events in the terrestrial record of the Antarctic Peninsula: an overview. In R. H. Madden, A. A. Carlini, M. G. Vucetich, and R. F. Kay (eds.), The Paleontology of Gran Barranca. Cambridge University Press, Cambridge. Pp. 383–97.

Roberts, E. M., et al. (+ 8 authors). 2014. Stratigraphy and vertebrate paleoecology of Upper Cretaceous–?lowest Paleogene strata on Vega Island, Antarctica. Palaeogeography, Palaeoclimatology, Palaeoecology 402: 55–72.

Schiermeier, Q. 2016a. Giant ocean reserve is a go (International agreement to create world's largest marine protected area near Antarctica is hailed as a diplomatic breakthrough). News in Focus. Nature 539.

———. 2016b. Speedy Antarctic drills start hunt for Earth's oldest ice. Nature 540: 18–19.

Schoepfer, S. D., T. Tobin, and J. D. Witts. 2017. Intermittent euxinia in the high-latitude James Ross Basin during the latest Cretaceous and earliest Paleocene. Palaeogeography, Palaeoclimatology, Palaeoecology 477: 40–54.

Sérsic, A. N., et al. (+7 authors). 2011. Emerging phylogeographical patterns of plants and vertebrates from Patagonia. In D. E. Ruzzante and J. Rabassa (eds.), Palaeogeography and Palaeoclimatology of Patagonia: Implications for Biodiversity. Biological Journal of the Linnean Society 103: 475–94.

Steig, E. J. 2016. Cooling in the Antarctic. Nature 535: 358.

Stonehouse, B. (ed.). 2002. Encyclopedia of Antarctica and the Southern Oceans. John Wiley & Sons, New York.

Strelin, J. A., T. Sone, J. Mori, C. A. Torielli, and T. Nakamura. 2006. New data related to Holocene landform development and climatic change from James Ross Island, Antarctic Peninsula. In D. K. Fütterer, D. Damaske, G. Kleinschmidt, H. Miller, and F. Tessensohn (eds.). Antarctica. Springer, Berlin. Pp. 455–60.

Strömberg, C. A. E., R. E. Dunn, R. H. Madden, M. J. Kohn, and A. A. Carlini. 2013. Decoupling the spread of grasslands from the evolution of grazer-type herbivores in South America. Nature Communications 4: 1478. doi: 10.1038/ncomms2508. www/nature.com/naturecommunications.

Strugnell, J. 2016. Profiles: Kudos for female Antarctic researchers. Nature 536: 148.

Stump, E. 2011. The Roof at the Bottom of the World: Discovering the Transantarctic Mountains. Yale University Press, New Haven.

Tambussi, C. P. 2011. Paleoenvironmental and faunal inferences based on the avian fossil record of Patagonia and Pampa: what works and what does not. In D. E. Ruzzante and J. Rabassa (eds.), Palaeogeography and Palaeoclimatology of Patagonia: Implications for Biodiversity. Biological Journal of the Linnean Society 103: 458–74.

Taylor, T. N., and E. L. Taylor (eds.). 1990. Antarctic Paleobiology, Its Role in the Reconstruction of Gondwana. Springer, Berlin. [The studies deal mostly with the Paleozoic
and Early Mesozoic; for paleovegetation, see papers by R. A. Spicer, G. T. Creber, E. M.
Truswell, A. N. Drinnan and P. R. Crane, and E. L. Taylor and T. N. Taylor.]

Tollefson, J. 2016. Dry amazon could see record fire season. Nature 535: 18–19.

———. 2017. Giant crack in Antarctic ice shelf spotlights advances in glaciology. Nature
542: 402–3.

Truswell, E. M. 1991. Antarctica: a history of terrestrial vegetation. In R. J. Tingey (ed.),
The Geology of Antarctica. Oxford Monographs on Geology and Geophysics 17. Oxford University Press, Oxford. Pp. 499–537.

Truswell, E. M., P. G. Quilty, M. McMinn, M. K. MacPhail, and G. E. Wheeler. 2005. Late
Miocene vegetation and palaeoenvironments of the Drygalski Formation, Heard Island, Indian Ocean: evidence from palynology. Antarctic Science 17: 427–42.

Turner, J., et al. (+ 9 authors). 2016. Absence of 21st century warming on Antarctic Peninsula consistent with natural variability. Nature 535: 411–15 (plus additional supplementary information).

Wade, L. 2016. Humans spread through South America like an invasive species. Science.
doi: 10.1126/Science.aaf4091.

Warny, S., C. M. Kymes, R. A. Askin, K. P. Krajewski, and P. J. Bart. 2016. Remnants of Antarctic vegetation on King George Island during the early Miocene Melville Glaciation.
Palynology 40: 66–82. [Palynomorphs sparse, majority reworked.]

Wilson, D. S., and B. P. Luyendyk. 2006. Bedrock plateaus within the Ross Embayment
and beneath the West Antarctic ice sheet, formed by marine erosion in Late Tertiary
time. In D. K. Fütterer, D. Damaske, G. Kleinschmidt, H. Miller, and F. Tessensohn
(eds.), Antarctica. Springer, Berlin. Pp. 123–28.

Witze, A. 2016. Iconic Antarctic lab gets the boot. Nature 534: 448.

Zimmermann, C., G. Jouve, R. Pienitz, P. Francus, and N. I. Midana. 2015. Late Glacial
and Early Holocene changes in paleowind conditions and lake levels inferred from
diatom assemblage shifts in Laguna Potrok Aike sediments (southern Patagonia, Argentina). Palaeogeography, Palaeoclimatology, Palaeoecology 427: 20–31.

SEVEN

Case Studies

The history of New World land bridges and representative plants utilizing them at different points in time is preserved in the various archives of the past. The fossil record is only one way to assess these movements, and land bridges are not the only way plants get around. The fossils document some of the taxa that used the connections, while molecular phylogenies and time calibrations of extant organisms, alone or in combination with fossils, suggest that others followed alternative routes employing various modes of dispersal. The following case studies illustrate the many different ways patterns of distribution may be achieved.

De Queiroz (2005; Fig. 7.1) cites instances where molecular dating and gene flow indicate recent long-distance exchange over substantial stretches of ocean. The estimated time for some events is so recent that global geography was essentially in its modern configuration, arguing against distributions being primarily the residue of ancient continental positions, and for others it is so old as to preclude initial introduction by humans. That leaves gradual movement across continuous land surfaces or, in the absence of fossils, the possibility of long-distance transport over, through, or around the bridges by one of the different means available (see legends to Fig. 7.1 and to De Queiros, 2005, fig. 1).

Acridocarpus (Malpighiaceae) has a distribution and a molecular-dated phylogeny suggesting it originated in Africa and dispersed over water to Madagascar and New Caledonia. In one of the many challenges posed by the fossil record, fruits referred to as cf. *Acridocarpus* sp. are reported from the Plio-Pleistocene of El Salvador (Lötschert and Mädler, 1975; II, app. 2.2, 595; II, table 6.1, 324): "The genus *Acridocarpus* is today African, but with the close relationship between the floras, it is not inconceivable it was represented in the pre-Glacial in Central America" (Lötschert

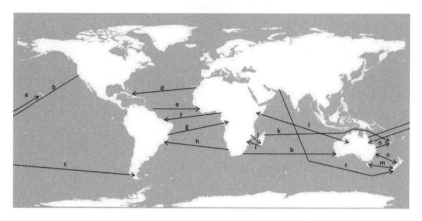

7.1. Oceanic dispersals (see text for discussion). a = *Scaveola*, b = *Lepidium*,
c = *Myosotis*, d = *Tarentola* (geckos), e = *Maschalocephalus*, f = monkeys, g = melastomes.
h = *Gossypium*, I = chameleons, j = several frogs, k = *Acridocarpus*, l = *Adansonia*,
m = various plant species (Tasmania–New Zealand), n = plant species (Australia–
New Zealand), o = *Nemuaron*. From A. De Queiroz, 2005, Trends in Ecology and
Evolution, Cell Press (Elsevier), Cambridge, MA. Used with permission.

and Mädler, 1975, 129, translation). However, according to Schatz (pers. comm., July 2016), the fruits are similar to several African and Central American Malpighiaceae. On balance, recent overwater dispersal within the Old World bypassing land bridges seems probable, and a re-ssessment of the El Salvador fossils by specialists and entry of their status into a centralized database would be useful.

Adansonia (Bombaceae) is native to Africa, and according to Wickens (2008, 313), "no fossil *Adansonia* pollen has ever been found," so the history of its distribution must be inferred from extant taxa. Baum et al. (1998, 181) believe that "a maximum likelihood analysis of branching times shows that the dispersal between Africa and Australia occurred well after the fragmentation of Gondwana and therefore involved overwater dispersal."

Gossypium (Malvaceae, Tribe *Gossypieae*) is a genus of about 50 species with much taxonomic attention devoted to the cultivated cotton. It presumably originated in Africa possibly 20 Ma (molecular calculations), but there is no corroborating information from the fossil record. Long-distance, oceanic dispersal is considered likely: "In this respect, seeds of many species of *Gossypium* are tolerant of prolonged periods of immersion in salt water. Remarkably, seeds of the Hawaiian endemic cotton G. *tomentosum* are capable of germination after three years of immersion in artificial sea water" (Wendel et al., 2010, 11, and papers citied therein). After introduction into the New World (possibly to western Mexico where currently there is great diversity), it moved presumably by drift and possibly over continuous land surfaces to South America. In the New World fossil pollen is listed for the Quaternary (recent) of Mexico (coast of Tabasco) and El Salvador (II, app. 2.2, table 4.3, 273, and table 6.1, 326; see also below on *Thespesia*, Gossypieae). Fossils of Malvaceae reported since the compilation in II (app. 2.2) include fossil wood (*Guazumaoxylon miocenica, Periplanetoxylon panamense*) from the lower Miocene Cucaracha Formation of Panama (Rodríguez-Reys et al., 2014); wood of *Wataria* from the middle Miocene of southwest China (Li et al., 2015); leaves of *Malvaciphyllum macondicus* (Malvoideae) from the Paleocene Cerrejón Formation of Colombia (Carvalho et al., 2011); and fruits and leaves of *Burretiodendron* from the upper Miocene of Yunnan, China; see also Hinsley (2005 et seq.) In the distribution of this group ocean currents appear primary and land bridges useful but not essential for dispersal.

Lepidium (Brassicaceae/Cruciferae) is thought to have moved from North America to Australia to Africa (Fig. 7.1; De Queiroz, 2005, fig. 1). The seeds of *L. campestre* are ballistically ejected and wind dispersed (Thiede and Augspurger, 1996). In the New World it is reported from the Quaternary (recent) of Mexico, Argentina, and Colombia (II, app. 2.2, 591; table 4.1, 253; table 7.6, 438). If that is accurate, then substantial overwater transport was involved.

Maschalocephalus (Rapateaceae; Givnish et al., 2004, S35): "Among rapateads, the divergence of west African *Maschalocephalus dinklagei* from its closest South American relatives implies that *Maschalocephalus* resulted via long-distance dispersal 7 Ma, not ancient continental drift."

Myosotis (Boraginaceae) ancestors are estimated to be of Northern Hemisphere origin arriving in the Southern Hemisphere ca. 1.2–4.9 Ma (Winkworth et al., 2002, 184). The seeds are enclosed in hooked fruits and presumably can be dispersed over long distances by a variety of vectors.

Scaevola (Goodeniaceae) is calculated to have moved 3 times from Australia to Hawaii. The major Hawaiian Islands are relatively recent in age (28 to 0.4 Ma) and have been isolated since their origin. The seeds of species within the center of diversity in Australia and Polynesia are typically dry, while those elsewhere are enclosed in fleshy fruits dispersed by frugivores.

An accurate taxonomy and phylogeny can suggest a great deal about the time and place of origin of a taxon and how it likely moved about. In turn, paleobotanical, geological, and climatological histories set parameters for proposing the means, feasible conditions, patterns of diversification, radiation, disjunctions, and the origin of geographic affinities for taxa. A com-

262 / Chapter Seven

bination of these approaches is important because new finds, such as the oldest *Pinus* reported in 2016, the first *Todea* in Patagonia (2013), and Asteraceae in Antarctica (2015); the wide range in divergence estimates based on molecular calculations; and selective suitability of different techniques for different purposes (e.g., plastid versus nuclear DNA) can quickly turn the avant-garde into the old guard. Estimates of origin, diversification, and dispersal directions based on indirect (nonfossil) calibrations have been called "rough" (Filipowicz and Renner, 2012, 10, re *Brunfelsia*) and at best "only useful ballpark estimates" (Wendel et al., 2010, 3, re *Gossypium*). The oldest angiosperm fossils are 130–136 Ma (Friis et al., 2006; Soltis et al., 2008), while some interpretation of divergence DNA sequence data indicates the *crown* group of extant angiosperms "to be Early to Middle Jurassic (179–158 Ma), and the origin of eudicots is resolved as Late Jurassic to mid-Cretaceous (147–131 Ma)" (Wikström et al., 2001, 2211). Divergence and diversification times between and within various clades are equally approximate, and among these some range from very old (Early Cretaceous) to quite young (early Tertiary or later), depending on the molecular and numerical analyses used. Also important in establishing the subsequent means and migration route(s) followed are the dispersal vector; the direction, distance, and seasonality of vector's migration(s); and seed viability after deposition. This information is known for only a relatively few modern plants, and the vector is rarely known for their fossil progenitors. Estimates of when, how, which, and where are rough, indeed.

In the examples below, principal references are the lead citation and others mentioned therein (e.g., Jud et al., 2008, for *Todea*; Terry et al., 2016, for *Callitropsis/Hesperocyparis*; etc.). Additional reports of fossils for the New World are in I (tables and index), II (app. 2.2 and pages cited), and III (illustrations and supplementary discussions). Although most records from the older literature are unconfirmed, and many are unlikely, they are currently being used. In other instances they have been missed. Their existence needs to be acknowledged and their validity assessed as part of any proposed new reconstructions.

Ferns and Allied Groups

Todea (Fig. 7.2; Osmundaceae; Jud et al., 2008; II, app. 2.2, 585, as *Todiosporites*, Cretaceous, Oligocene). This is an Old World tropical rainforest genus of 2 species at present found in Africa, Australia, New Zealand, and New Guinea. Its fossil record consists of *Todea tidwellii* from the Late Cre-

7.2. *Todea amissa pinnule* from the early Eocene, Laguna del Hunco, Patagonia, Argentina. From M. Caravalho et al., 2013, *American Journal of Botany*, Botanical Society of America. Used with permission.

taceous of British Columbia (Jud et al., 2008) and *T. amissa* from the early Eocene Laguna del Hunco flora of Patagonian Argentina (Carvalho et al., 2013). The Osmundaceae occupy a basal position among ferns, and the presence of *Todea* in the Cretaceous of Laurasia and the Eocene of Gondwana suggests that the past and present patterns of occurrence are residues of early continental configuration and breakup. Its disappearance at the end of the Cretaceous in the far Northern Hemisphere, and at the end of the early Eocene in far southern South America, suggests a response to cooling and drying climates (Fig. 2.4).

Gymnosperms

Callitropsis and *Hesperocyparis* (Cupressaceae; Terry et al., 2016; I, table 7.6, 253, table 7.8, 259; II, app. 2.2, 586). These are New World cypresses formerly placed in *Cupressus*. The origin of the ancestral complex is estimated as in the Late Cretaceous (74 Ma) with DNA sequence data and divergence time using BEAST suggesting colonization from Asia through

Beringia in the Late Cretaceous or early Cenozoic. The presence of Cupressaceae in northwestern Canada (Vancouver Island, Alberta, Saskatchewan; McIver, 1994; McIver and Aulenbeck, 1994; McIver and Bassinger, 1987) in the Late Cretaceous and early Tertiary support the interpretation. Once established in North America, the plants migrated as climate cooled after the EECO from the northwest in the middle Eocene (ca. 45.3 Ma) to the southeast. Dispersal is primarily by wind and rain/flooding into rivers and coastal waters. The principal time of diversification was in the late Miocene (6 Ma).

Pinus (Pinaceae; Falcon-Lang et al., 2016; I, index; II, app. 2.2, 586) is the most widespread tree in the Northern Hemisphere with 126 species accepted in *The Plant List* of the Missouri Botanical Garden/Kew, 4 of which extend as far south as Nicaragua (*P. caribaea, P. maximinoi, P. oocarpa,* and *P. tecunumanii*). The oldest fossil is *P. mundayi* from the Early Cretaceous Chaswood Formation of Canada (Valanginian, ca. 133–140 Ma). "Their preservation as charcoal and the occurrence of resin ducts, which produce flammable terpenes in modern pines, show that *Pinus* has co-occcurred with fire since its Mesozoic origin" (Falcon-Lang et al., 2016, 303). In an ecological study Ledig et al. (2015) found that *Pinus rigida* in the northern latitudes was more precocious, highly fecund, and had more seeds than populations to the south. These data support the view that *Pinus* originated in the Northern Hemisphere, adapted well to the cooling and drying trends of post-PETM times, migrated extensively across the boreal and subboreal latitudes using the BLB and NALB, and moved southward in the mid- and late Tertiary.

Angiosperms

Monocotyledons

Ceroxylon (Arecaceae/Palmae; Sanín et al., 2016; Trénel et al., 2007) includes 12 species growing in moist cloud forests of the Northern Andes at ca. 2000 m average elevation. Seed/fruit dispersal (in *C. klopstockia*) is by birds (*Aulocornychus sulcatus,* toucanet) and primates (Zona and Henderson, 1989; Braun, 1976). It has no fossil record, and its origin, migration, and diversification have been inferred from analyses of the modern species by niche evolution and phylogenetic niche conservation methods. The tribe Ceroxyleae (Madagascar and the Comoro Islands, Juan Fernández Islands, Australia, Andean South America) is thought to have dispersed through Antarctica into South America during the Eocene (crown group constrained at 29–[17]–7 Ma). Isolation of the preadapted Andean endemic *Ceroxylon* in the mesic to cool environments of the Andes is considered the result of subsequent northward migration (stem/crown nodes inferred at 16.8–[12.3]–8.3 Ma and 11.6–[8.3]–5.3 Ma, respectively), rather than being a relict of ancient Gondwana separation. Although estimates for the principal divergence and radiation of many angiosperm groups vary from Late Cretaceous (Wikström et al., 2001) to Early Tertiary (Magallón and Sanderson, 2001), as noted, "many angiosperm lineages are still too young to have been influenced by break-up of larger landmasses, and it is therefore likely that continental history may have had less of an effect on the modern distributions of [some] flowering plants than had previously been assumed" (Richardson et al., 2004, 1495). There is a complex of reasons for the occurrence of each group of angiosperms, and at the subfamily and generic level within the Arecaceae, *Cocos/Nypa* may be examples where past continental positions were a factor and for *Ceroxylon* less so. Colonization of uplands by *Ceroxylon* in the late Miocene/Pliocene is consistent with the uplift history of the Andes.

cf. *Cocos/Nypa* (Arecaceae/Palmae; Harley, 2006; Gomez-Nararro et al., 2009; II, app. 2.2, 597–98). The family is possibly Australasian in origin, and the *crown* group estimated at ca. 110 Ma, so past and present occurrences of older progenitors likely involve previous regional continental connections. There was extensive radiation of the family in the Paleocene and Eocene corresponding to the EECO. Both cf. *Cocos* and *Nypa* are found in lowland/swamp/marsh environments of the Paleocene Cerrejón Formation of Colombia at 60 Ma and *Nypa* in the Eocene at the Texas/Mexico border (Westgate and Gee, 1990). Although the fruits of both are distributed by ocean currents, the range of *Nypa* contracted after the PECO and the one species (*N. fruticans*) is currently found in Southeast Asia, while *Cocos nucifera* has become widespread.

Bromeliaceae (Givnish et al., 2011; II, app. 2.2, 589): "Bromeliads arose in the Guyana Shield ca. 100 million years ago (Ma), spread centrifugally in the New World beginning ca. 16–13 Ma, and dispersed to West Africa ca. 9.3 Ma. Modern lineages began to diverge from each other roughly 19 Ma. Nearly two-thirds of extant bromeliads belong to two large radiations: the core tillandsioids, originating in the Andes ca. 14.2 Ma, and the Brazilian Shield bromelioids, originating in the Serro do Mar and adjacent regions ca. 9.1 Ma" (Givnish et al., 2011, 872–73). Dispersal units include plumose seeds and fleshy fruits, and the recent (ca. 9.3 Ma) introduction to Africa was presumably by long-distance transoceanic transport. The genus *Puya* likely originated in central Chile with speciation driven by Pleistocene glaciation cycles (Jabaily and Sytsma, 2013).

Costus (Costaceae) has been studied from the standpoint of molecular phylogeny of the neotropical species by ITS, ETS, *rps 16*, *trnL-F*, *trnK*. and *CaM* (Salzman et al., 2015). One clade is native to tropical Africa and a derived one to the New World, with the ancestral location inferred as along the Pacific coast of Mexico and Central America. Seed dispersal is not well known, but the red bracts enclosing the fruits spread, revealing a contrasting white capsule that probably attracts birds (Kubitzki, 1998; Surget-Groba and Kay, 2013), and the seeds are possibly consumed by fish (Gottsberger, 1978). There was diversification at ca. 40 Ma, and the origin of New World *Costus* is suggested to have been by a single (very) long-distance dispersal event from Africa ca. 34 Ma (Specht, 2006) with radiation at ca. 22 Ma. Kay et al. (2005) believe introduction into Central America of some species was ca. 1.1–5.4 Ma, with divergence between clades at 0.3–1.6 Ma. As Salzman et al. (2015) note, outgroup fossil dates are needed to better explain the biogeography of *Costus*.

Cyperaceae (Jiménez-Mejias et al., 2016; II, app. 2.2, p. 591). The authors have significantly clarified the pre-Pleistocene fossil record of *Carex*. Among their finds are that (1) ca. 83 names belonging to different *Carex* groups are applied to remains that are reliable (from 550 sites) and 23 names to fossils that are doubtful; (2) the oldest fossil reported (*C. tsagajanica*, Cretaceous-Paleocene, Siberia, Krassilov, 1976) is likely incorrect, the oldest in Siberia being Miocene and Pliocene; (3) the very oldest are from the late Eocene of England, becoming abundant globally in the middle Miocene and (4) throughout its geographic and stratigraphic range, the genus is consistently associated with freshwater habitats as at present. Viljoen et al. (2013) suggest that the ancestral area of the Schoeneae clade of the Cyperaceae is Australia at ca. 50 Ma (middle Eocene) and that it migrated across the austral oceans through 29 long-distance dispersal events.

Orchidaceae (Ramírez et al., 2007; II, app. 2.2, 597) are represented as fossils by a pollinarium (*Meliorchis caribea*) attached to the extinct stingless bee *Proplebeia dominicana* in the Miocene Dominican amber (15–20 Ma). Affinities are with the subfamily Orchidoideae and possibly with extant *Kreodanthus* and *Microchilus*. Cladistics analysis of a morphological character matrix and a calibrated molecular tree suggest that the most recent ancestor of modern Orchidaceae is Late Cretaceous in age (84–76 Ma) with radiation shortly after the K/T boundary event (see also Gustafsson et al., 2010, for the South American genus *Hoffmannseggella*, subfamily Epidendroideae). The minute seeds of orchids are predominantly wind distributed, but those of the Asian/Australasian genus *Cyrtosia* are dispersed by birds. Many orchids have fleshy fruits, and Suetsugu et al. (2015) suggest that avian seed dispersal may be more common than previously noticed. The presence of orchid fossils in the Dominican Republic documents the ALB as a pathway of migration. Geological reconstructions indicate that 20 Ma is near the time western Hispaniola was still connected to Cuba (see chapter 4, above), climates were suitable, and given the short distances involved, along with the wind and possible avian dispersal of the seeds, movement across the bridge is feasible. Although rarely cited, there is a report of an orchid *Orchidamasulitis schlechteri* from the Oligocene of Venezuela (Di Giacomo, 1985).

Poaceae (Gramineae; Christin et al., 2014; II, app. 2.2, 592–93) were used as a model for testing the efficacy of various techniques in revealing age and diversification among groups with a meager fossil record. Remains of grasses are common in sediments, but generally they cannot be identified to genus. If calibration is based on macrofossils, the core group dates to 55–51 Ma, but if microfossils (phytoliths) are included, it extends to 82–74 Ma. Molecular dating estimates range from 86–52 to 39–23 Ma with BEAST-based estimates of the crown group at 54.9 Ma compatible with the macrofossil but not the microfossil record. Specifically, if phytoliths assigned to the Oryzaea are included (67 Ma), an older age of divergence of the grass clade results, with variation of between 82.4 and 79.1 Ma depending on whether BEAST or MULTIDIVTIME is used. Fifty-five Ma versus 82–79 Ma are critical intervals for interpreting the biogeography of the group. As Christin et al. (2014, 162) note: "If splits occurred at or after 55 Ma, then the grass lineages must have spread from their Gondwana center(s) of origin long after the breakup of this southern supercontinent, pointing to long-distance dispersal as an important mechanism by which grass lineages achieved their world-wide distribution. In contrast, under the phytolith-based age hypothesis, these divergences would have occurred during a time when there were still land connections between the southern continents; hence, vicariance may have played a larger role in early grass diversification." And "based on analyses that did not include the fossil phytoliths . . . it has been suggested that core Pooideae evolved cold tolerance in response to climatic cooling following the Eocene-Oligocene boundary (33.9 Ma), which is compatible with our analyses without phytolith fossils. If the phytolith-based ages are used, core Pooideae are significantly older than 33.9 Ma, and would have evolved in the warm, middle Eocene." Finally, "unfortunately, the estimates of evolutionary rate variation (linked to the model assumptions) and divergence times of key nodes (linked to the placement of fossils) are tightly connected and one can be confidently estimated only with an accurate knowledge of the other" (Christin et al., 2014, and references cited therein).

Dicotyledons

Acanthaceae (Tripp and McDade, 2014; Daniel and McDade, 2014; II, app. 2.2, 587). Study of the taxonomy, phylogeny, and fossil-calibrated estimates of divergence were based on 51 fossil records. As expected, the results varied widely with the age of the fossils and 8 reports were considered sufficiently reliable to include in the analysis. The oldest was from the middle Bartonian (middle Eocene, ca. 42 Ma). Long-distance dispersal better explained the present distributions than ancient vicariance (consistent with the earlier separation of the major continents), or migration over either Gondwanan or Laurasian land bridges. Movement was predominantly from the Old World to the New World (13 dispersal events) and only one in the opposite direction (*Staurogyne*, 145 species, from South America to Africa, Europe, Australia, India, China, Taiwan).

Annonaceae/Rhamnaceae (I, 214, 221; II, app. 2.2, 588). Richardson et al. (2004) conclude that bridges (*sensu lato*) between continents and suitable climates (tropical for Anonaceae, xeric for Rhamnaceae), and long-distance dispersal best explain the present patterns of distribution and diversity because the families are considered too young for past continental movements to be definitive. The Annonaceae (stem 91–82 Ma; connections between South America and Africa severed at ca. 105–90 Ma but with intervening islands) are sparsely represented in Australia. The fossil record for the Annonaceae is diffuse across Gondwana and Laurasia (Africa–South America; Paleocene of Egypt and Colombia; Eocene of England, Argentina, and Vancouver Island with modern forms similar to those in southeast Asia; Richardson et al., 2004, 1497–500). West Gondwana is considered a plausible place of origin for the family. Vectors for long-distance dispersal between these impressively distant regions are unspecified,but modern on-land dispersal is by primates. Expansion of the Annonaceae during the warm and moist periods of the Paleocene with interchange across the BLB and the NALB with range contraction during the cooler and drier times of the Neogene seems evident. However, there is considerable latitude in the age estimates for this lineage based on existing molecular techniques, and the fossil record only establishes the minimum but not the maximum age; that is, the very least, but not the very first. Given the size of the family (ca. 2500 species) and its widespread distribution in the past, a role for drift in the latter stages of major continental breakup, especially between Africa and northern South America, in the Early to Middle Cretaceous (ca. 120–105 Ma; II, 97) is possible.

Guatteria (Annonaceae; Erkens et al., 2007, 2009; II, app. 2.2, 588) is the third-largest genus of neotropical trees (after Inga and Ocotea), and the history of its migration and diversification is important for understanding species richness in the lowland tropical rainforest, as well as use of the New World land bridges. Seed dispersal (see Erkens et al., 2007, fig. 1B) was studied for G. atabapensis, and the time escape method was found to be important (Guzmán and Stevenson, 2011; Stevenson, 2000). This method means a long fruiting time so that after predators (mostly invertebrates) consume the first masting and move on, enough fruits are still produced for later dispersal (mostly by vertebrates). Lagothrix lagothricha (wooly monkeys) were important, consuming seeds of 112 different plant species, including Guatteria, with deposition of more than 25,000 seeds/km²/ day. The average dispersal distance by Lagothrix was only 100–500 m with a maximum of up to 1.5 km from the parent tree. Predation by tortoises was also mentioned so dispersal across great distances must have been a slow process, but ca. 15–20 million years were available (see later discussion). Erkens et al. (2007) state there is no fossil record and did not mention G. culebrensis (Berry, 1918) described from the Miocene of Panama (II, 588, 323). They conclude from Bayes-DIVA analyses that the genus had an African origin, dispersed across the NALB in the early to middle Eocene, then into North and Central America (and ultimately South America) in the Miocene, before final and complete closure of the CALB. Temporally this is consistent with the geology and the Miocene plant record. There was major diversification in South America and then reradiation back to Central America. In seeking calibration dates, need sometimes confronts data and, as the authors note (2007, 404), "although objections exist against the use of the age of strata *on* which endemic taxa occur for calibrating phylogenetic trees, it is the only calibration point possible within *Guatteria* since there are no known fossils." The cautionary "although" is well placed, and an alternative position is that calibration in such situations cannot be done.

Regarding the Rhamnaceae, the summary by Richardson et al. (2004) suggests that although various subgroups occur in South America, Madagascar, India–southeast Asia–Africa, and North America (Gondwana to Laurasia, tropical to temperate), age estimates based on molecular techniques are not sufficiently constrained by fossil evidence to offer a clear explanation of the distribution and place of origin. Laurasia is plausible for the Rhamnaceae, and molecular-based estimates for the ziziphoid group yield a maximum age of 40.9 Ma. Current dispersal is by multiple vectors, including wind and birds, and modern patterns of distribution and diversification are thought to involve drying after the MMCO. The capacity for long-distance dispersal reduces the need for invoking past continental movements and land bridges, but connections of the GAARlandia and Madrean-Tethyan type are suggested (Richardson et al., 2004). These possibilities have been discussed in chapter 4, above.

Atherospermataceae are commonly referred to as the southern sassafrases and are known from 2 species in Chile and 12 in Australasia. "Calibration of *rbcl* substitution rates with fossils suggests an initial diversification of the family 100–140 million years ago (MYA), probably in West Gondwana, early entry into Antarctica, and long-distance dispersal to New Zealand and New Caledonia at 50–30 MYA" (Renner et al., 2000).

Asteraceae (Compositae; Katinas et al., 2007; II, app. 2.2, 588, 590–91; Funk et al., 2005) include the southern South American subfamily Barnadesioideae regarded as basal and Patagonia as the ancestral area for the family. Barreda et al. (2015) report fossil pollen with similarities to the Barnadesioideae from the Cretaceous of Antarctica at 76–66 Ma and note that diversification was after the K/T boundary coinciding with the PECO. The discoveries indicate an origin of the family in southern South America and document use of the MLB as a pathway of interchange between Antarctica and Patagonia at a time when both suitable climates and physical connections existed.

Bignoniaceae/Solonaceae/Verbenaceae have calculated stem lineage ages suggesting that the three families originated in the Late Cretaceous in South America (Olmstead, 2013; II, app. 2.2, 600, 601). Long-distance dispersal to North America, Africa, and elsewhere was presumably by wind with subsequent dispersal there of the fleshy- and spiny-fruited species by animals. Minimum divergence times for the stem clade of Bignoniaceae are 47 Ma (68 Ma fide Bremer et al., 2004, whose estimates are typically older but still within the accepted range for the angiosperms), and for the crown clade 40–45 (to 48) Ma. The stem clade figure for the Verbenaceae is 48–62 Ma and 40 Ma for the crown. South America was narrowly separated from North America across Panama until 3.5 Ma, with little evidence of dry habitats favorable especially to the Verbenaceae, spiny Solonaceae, and some savanna Bignoniaceae, so dispersal of these was probably over the CALB by wind and birds. South America was connected physically to Antarctica until ca. 35–29 Ma, with cold climates developing with onset of

the ACC especially after the MMCO at 21[17]–15 Ma. Antarctica was separating from Australia between the latest Cretaceous and the late Eocene
(49 Ma). Some overland movement of Bignoniaceae, combined with wind
dissemination for certain Solonaceae and Verbenaceae, seems plausible in
the history of these lineages.

Solonaceae (Olmstead, 2013; I, II) have a stem clade date of 65–85 Ma
(Late Cretaceous), based on divergence time estimates, and a crown clade
of ca. 35 Ma. The family originated and diversified in South America from
a presumed Gondwana ancestral complex. The fruits are either fleshy and
dispersed by animals or dry dehiscent capsules and distributed by wind.
Fossils of Solonaceae are reported from the Oligo-Miocene Pie de Vaca
flora of Puebla, Mexico, and from several sites of Quaternary age (II, app.
2.2, 600), but most are not diagnostic and evidence for use of New World
land bridges by any Solonaceae is equivocal. An exception (for the family) is the broadly identified *Solanum/Physalis* type from the late to early
Miocene Mary Sachs Gravel on southern Banks Island of the NWT (I, 226–
27, table 6.16). *Brunfelsia* (Filipowicz and Renner, 2012) has ca. 50 species
concentrated in the Antilles, Andean South America, and in the Amazon
Basin and Guiana Shield of South America. Plastid and nuclear DNA sequences suggest an age of 21–16 Ma, with the genus entering the Antilles
from South America. The seeds are fleshy capsules and probably dispersed
by birds. Some species are hallucinogenic, toxic, or horticulturally important, so their distribution after ca. 5–6 ka has probably been altered by
humans.

Boraginales (primary woody Boraginales, Ehretiaceae, Cordiaceae, Heliotropiaceae of ca. 1000 species; Gottschling et al., 2004; II, app. 2.2, 589)
have centers of diversity in South America, Africa, and from India to Australasia. A Cretaceous origin is proposed (Upper Cretaceous 81–77 Ma;
Euasterid I clade 110 Ma, Wikström et al., 2001). Leaf fossils of the Cordiaceae are known from western North America and Asia (vicinity of a
broadly defined BLB) with the oldest being Eocene in age; pollen of Heliotropiaceae (*Tournefortia*) occurs in Oligocene deposits from Puerto Rico
(ALB); and endocarps of Ehretiaceae (*Ehretia*) in the Eocene London Clay
flora (NALB). Initial diversification of the woody Boraginales was in South
America. With specific reference to the equatorial land bridges, distribution
is regarded as mostly the result of long-distance dispersal of the drupaceous
fruits over, rather than primarily across, the ALB (*Tournefortia* excepted;
Graham and Jarzen, 1969) and the CALB. Given the Cretaceous estimated
age of origin (110 Ma) and the modern distributions noted above, former
continental positions likely played a role.

Campanula/Campanulaceae (Crowl et al., 2016; II, app. 2.2, 589) have molecular and chromosomal features suggesting Southern Hemisphere out-of-Africa dispersal and diversification. Fossils include *Campanula* seeds from the Miocene of Poland at 17–16 Ma, and using this as a calibration point, the origin of the family (split between the Campanulaceae and Rousseaceae) is estimated at 76 Ma (86–67 Ma), followed by numerous dispersal events into the Nearctic and neotropics. This is well after the breakup of Pangaea, so present patterns of distribution are not primarily residues of past continental connections. Nine dispersal events out of the neotropics are inferred from the data (vectors unspecified). The authors note that the time of these movements (40–50 Ma) was when islands existed between North and South America, so the ALB, CALB, and later the NALB were presumably involved. Not all distributions are fully explained, for example, those between Africa and the Nearctic, which is often the case when a major component of the explanatory network is absent or meager (in this case, the fossil record).

Chrysobalanaceae (Bardon et al., 2013; II, app. 2.2, 590). "We found that Chrysobalanaceae most probably originated in the Palaeotropics about 80 Mya. The family dispersed into the Neotropics at least four times beginning 40–60 Mya [late Paleocene/Eocene], with at least one back-dispersal to the Palaeotropics; migration into South America after the breakup of Gondwana is highly likely" (19, 33). Long-distance dispersal is considered likely "since six genera have species that can disperse via flotation; many of the more recent dispersal events do not seem to be explicable by dispersal via continents or land bridges" (28).

Inga (Fabaceae/Leguminosae; Richardson et al., 2001; I, index; II, app. 2.2, 594–95) is a genus of ca. 300 tree species of neotropical rainforests. The seed viability is 1–2 weeks, and they are distributed by primates, suggesting lengthy across-land distribution using continuous or near-continuous equatorial land bridges. The genus is said to have no fossil record, which influences age and diversification estimates based on molecular evidence (14–10 Ma), but there are several unverified reports dating it back to the Eocene (e.g., II, table 4.1, 252, table 5.1, 305–6, table 6.1, 323–24, table 7.1, 363, 369, table 7.2, 391–94, etc.). These, along with records for many other plant groups mentioned, speak again to the need for a systematic reexamination of fossil plants, especially in the published literature, to establish a database of authenticated (or at least reevaluated) records.

Gesneriaceae (subfamily Gesnerioideae; Perret et al., 2013; II, app. 2.2, 592) are represented in the New World by ca. 1000 species. Molecular

dating suggests that ancestors of the family originated in the Late Cretaceous in South America (crown age ca. 71 Ma), with occurrences in the Old
World tropics and Australasia being the result of long-distance dispersal in
the Eocene and Oligocene, respectively. Distribution within the New World
beginning ca. 34 Ma extended to the tropical Andes, Brazilian Atlantic forest and cerrado, Central America, and the West Indies, indicating use of the
equatorial land bridges. The study by Perret et al. (2013) suggests early colonization of Central America and the Caribbean by at least 26 Ma. At that
time the ALB provided a stepping-stone pathway of discontinuous land,
and Central America was narrowly separated from South America (see
chapters 4 and 5, above). The seeds are either dry and dehiscent (wind dispersed) or fleshy (bird dispersed). It is said there is no fossil record for the
family, but there are reports from the Oligocene/Oligo-Miocene of Mexico
(II, table 4.1, 251, family only) and the Miocene of Chile (II, table 7.2,
390, *Mitraria*) that need to be taken into account. A stem age older than
71 Ma, if accurate, makes the ancestors younger than the time of continental separation (South America from Africa in the south at 135–105 Ma, in
the north at 119–105 Ma; II, fig. 2.47).

Hedycarya (Fig. 7.3; Monimiaceae; Conran et al., 2016; II, app. 2.2, 596)
is a basal member of the Laurales. Fossils resembling leaves, flowers, fruits,
and pollen tetrads (*Planarpollenites fragilis*) were recovered from the Miocene (23 Ma) of southern New Zealand. Leaves (*Moniophyllum*) possibly
representing the family are reported from the Late Cretaceous–early Paleogene of West Antarctica and in the early Eocene of Patagonia (Birkenmajer
and Zastawniak, 1989; Knight and Wilf, 2013), there is wood (*Hedycaryoxy-
lon*) from the Cretaceous of West Antarctica (Poole and Gottwald, 2001),
and numerous other mostly unverified occurrences (Conran et al., 2016,
951–52; II, app. 2.2, 596). An estimated divergence time for the family
from the Lauraceae is ca. 90 Ma (*Hedycarya* 23.7 Ma). The molecular evidence is said to suggest a residual signal of continental breakup in East but
not West (Africa/South America) Gondwana (Renner et al., 2010), and the
fossils indicate use of the MLB as a route of dispersal. The fruits of *Hedycarya* are fleshy yellow drupes that in some species turn red upon ripening
and are distributed over short distances by birds.

Hedyosmum (Chloranthaceae; Antonelli and Sanmartín, 2011; possible
modern analog of *Asteropollis* and *Clavatipollenites* fossil pollen; and *Cla-
vainaperturites*, Martínez et al., 2013; II, app. 2.2, 590). There are only ca.
75 species in the family, and *Hedyosmum* has an extensive fossil record. The
distribution of modern *Hedyosmum* is intriguing, with 40–45 species in the
neotropics (foothills of the Andes, Central Cordillera) and 1 in the paleo-

7.3. *Hedycarya* flower from the early Miocene of New Zealand. From J. Conran et al., 2016, American Journal of Botany, Botanical Society of America. Used with permission.

tropics (southeastern Asia). There is also a potential gap of nearly 80 Ma between the Chloranthaceae stem group (Early Cretaceous, 112 Ma) and crown group (*Hedyosmum,* 29–60 Ma), raising the question of whether this was due to actual low diversification and recent radiation, or was the result of extensive extinction (Antonelli and San Martín, 2011, 596). The varied analyses (paleontological, relaxed-clock molecular dating, diversification, parametric ancestral area reconstruction) indicate origin of the Chloranthaceae ancestors and *Hedyosmum* stem lineages in the Holarctic of Laurasia during the Early Cretaceous (*Chloranthus* in the Late Cretaceous), and high extinction rate as a result of Cenozoic climate change. The crown group of *Hedyosmum* is thought to have originated 36–43 Ma and moved into South America by the early-middle Miocene ca. 20 Ma (it is reported from the middle Pliocene Paraje Solo Formation of Veracruz, Mexico; Graham, 1976; see also Jaramillo et al., 2014, tables 1 and 4; Jaramillo and Rueda, 2013; Hoorn, 1994). There it diversified with uplift of the Northern

Andes in the mid-Miocene especially after ca. 10 Ma, which is the same as for neotropical birds (Weir, 2006). There were three reradiations back to Central America "possibly before the final closure (uplift) of the Panama Isthmus ~3.5 Ma (as indicated by Bayes-DIVA)," two after closure, and dispersal to Brazil and the Guiana Shield area in the late Miocene-Pliocene (Weir, 2006, 605). All this is consistent with climates and equatorial land bridge histories and Andean orogeny. The BLB and the NALB were likely important in the early radiation of the Chloranthaceae and *Hedyosmum* across Laurasia, and the ALB and CALB useful but not critical to their migration back and forth to South America. The seed is a drupe with a fleshy mesocarp dispersed by birds (Kubitzki et al., 1993) and monkeys (Grow et al., 2014). The presence of 1 species in Africa (*H. orientale*) is interpreted as a relict from former widespread distribution, but its phylogenetic position and time of appearance in Africa is not clear.

Humiriaceae (Herrera et al., 2010, 2014; II, app. 2.2, 593) grow in the neotropics, mostly in the lowlands up to ca. 1400 m (*Humiria* to 2300 m where it interfaces with forested savanna). The family is represented in the fossil record by endocarps and pollen of the extant genera *Vantanea* (Eocene and Miocene), *Sacoglottis* (early Oligocene, early Miocene, late Pliocene), *Humiria* (Miocene and Pliocene), *Humiriastrum* (Miocene and early Pliocene), and *Ducksia* (Oligocene), and by wood of *Humiriaceoxylon* and fruits of *Lacunofructus* (late Eocene, Panama). The fossil pollen described in the stratigraphic literature as *Psilabrevitricolporites devriesii* (Columbia, Brazil, Venezuela, Panama) from the early Miocene onward is similar to *Humiria*, and fruits of *Sacoglottis* are known from the Oligocene of Puerto Rico. The fruits are a drupe consumed by rodents, tapirs, primates, birds, and bats. Herrera et al. (2010) further note that the endocarps of extant species may be transported downstream by rain and rivers, and distributed by ocean currents. Molecular divergence evidence suggests an Early to Middle Cretaceous origin; the oldest known fossils are from the late Eocene of Panama (Herrera et al., 2014). The Eocene age of Berry's (1929) Peru specimens is now revised to Oligocene. The collective reports from South America, Central America (Panama, Costa Rica), and the Antilles (Puerto Rico) document the ALB and the CALB as migration routes.

Ilex (Aquifoliaceae; II, app. 2.2, 588) is a cosmopolitan genus of ca. 400 species, and Manen et al. (2002) found its history "very complex" with several "distribution peculiarities." The fossil record (based on the Global Plotter Program for *Ilex* and *Ilexpollenites*) and different DNA sequences (plastid phylogeny based on loci, *atp*B-*rbc*L spacer *trn*L-*trn*F, versus nuclear phylogeny based on the ribosomal ITS and the 5S RNA spacer) did not agree (see below). The current primary centers of distribution are eastern Asia and North America, with extensive diversification in South America (only 1 species is found each in Africa and Australia and 4 in Europe). Hybridization and interspecific introgression is widespread. The plastid phylogeny, but not the nuclear, indicates eastern Asia as the ancestral home and shows no direct relationship between taxa in South America and Asia. Conclusions pertaining to place of origin, subsequent migration, and disjunct floristic affinities are as follows:

- *Ilex* . . . is a perfect example of an Arcto-Tertiary lineage (see chapter 3, above, discussion of Arcto-Tertiary Geoflora).
- Early identifications of the leaves in the Americas and used in reviewing Aquifoliaceae are in great doubt; there is no work from cuticle studies and SEM that the leaf form of *Ilex* is very similar to presumed fossilized taxa; also the wood structure is not unique.
- Reliable published records of *Ilex* macrofossils are lacking from southern hemisphere pre-Pliocene sediments.
- Molecular data do not sustain a Cretaceous origin of *Ilex* but rather a more recent [Eocene] one.
- Several evidences are in favor of southerly migration to South America late in the Tertiary. (Manen et al., 2002; Burnham and Graham, 1999)

If the origin of fthe amily was in Asia, then migration of *Ilex* into North America in the Eocene and afterward, and its subsequent movement into South America in the Neogene, implies utilization of the BLB, ALB, and CALB. The fossil record is consistent with this scenario to the extent that fossils of *Ilex* occur extensively but never abundantly on the land bridges themselves (e.g., the Antilles, Mexico, Costa Rica, Panama; Graham 1976, 1985, 1987, 1988a,b, 1989, 1991a,b; Graham and Jarzen, 1969; Manen et al., 2002, table 4). However, the red fleshy fruits are consumed by birds, so its expansion is probably the combined result of land bridge use across and avian transport over the connections (Fig. 5.2).

Lythraceae (Fig. 7.4) is a moderate-size family of 28 genera and ca. 600 species in which the morphological taxonomy, phylogeny, and divergence are (becoming) well understood (S. Graham, 2013). The pollen is exceptionally diverse (A. Graham et al., 1985, 1987, 1990), and the family has a good fossil record consisting of seeds/fruits, wood, and pollen that has recently been augmented (Estrado-Ruiz et al., 2009; Grimsson et al., 2011) and summarized (S. Graham, 2013). Forty-four genera are described as lythracean and 24 are presently accepted (14 extant and 10 extinct). Thus,

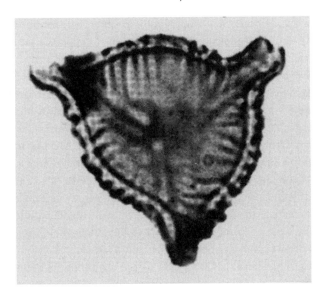

7.4. *Cuphea* pollen from the late Miocene of Alabama.
From S. Graham, 2013, Botanical Review: The New York
Botanical Garden, Springer, Berlin. Used with permission.

it might be expected that the time, place of origin, and ancestral group of the family is well established. That is not the case, however, and numerous questions remain unanswered. This brings into focus the considerable latitude involved in trying to reconstruct the history of other plant groups when relationships are not as clearly established and the fossil record is less well known.

Although the terminal clades generally have strong support in all analyses to date, the earliest branching events have little or no support and remain to be firmly established (S. Graham, pers. comm., 2016). The oldest fossils are pollen of *Lythrum/Peplis* from the Campanian (Cretaceous, 82–81 Ma) of Wyoming and seeds of *Decodon* from the Campanian (73.5 Ma) of northern Mexico. Phylogenetic analyses place *Lythrum/Peplis* and *Decodon* either on first emerging branches in the family (S. Graham et al., 2005; maximum likelihood tree of 20 lythracean genera) or, in more recent analyses including 26 of the 28 genera of the Lythraceae, in one of two major clades of the family, primarily among genera of Asian and African distribution (Bayesian consensus tree; Morris, 2007; Berger et al., 2016). In the Paleocene *Sonneratia, Lagerstroemia,* and the extinct *Sahnianthus* are known from India, and the extinct *Hemitrapa* from the Paleocene of northwestern

North America. The record is too fragmentary and the distributions too diffuse to confidently reconstruct the early history of the family.

With regard to land bridges, however, 2 genera have a modern distribution and habitat occurrence, dispersal potential, and fossil record suggestive of movements to and from and within the continents of the New
World. *Decodon* is a monospecific genus of eastern North America occurring in swamps and along lake margins. Fossils are known from the Late
Cretaceous Cerro del Pueblo Formation of Coahuila, Mexico; middle Eocene Princeton Chert of British Columbia; the middle Eocene Clarno Beds
of Oregon; and the early Oligocene Bridge Creek flora of Oregon. In the
Miocene it is found in the Brandon lignite of Vermont and in Alaska and
the Northwest Territories; in the Pliocene at even higher latitudes in the
Arctic Islands and Iceland; in Europe from the Paleogene of England, Isle
of Wight, Russia, and Germany, and in the Neogene of Denmark, Italy,
and Poland (S. Graham, 2013, 59–63). This distribution includes fossil occurrences on either side of and directly on the NALB and the BLB, when
warmer climates and land continuity prevailed through the middle Eocene;
the cold tolerance of the genus allows for its presence in later times; and
the seeds (and those of *Lythrum*) have the potential for wide dispersal—
spongy float tissue that composes a large part of the seed in *Decodon* and
emergent mucilaginous hairs in *Lythrum* that enlarge the float surface and
may also adhere to other objects.

Cuphea is a large and advanced New World genus of 240–50 species
with centers of diversity in Brazil and Mexico. It appears in the fossil record
relatively recently, being known from the Miocene of Alabama, Mexico,
and Trinidad, and in the Pleistocene at several sites in Mexico, Cuba, and
South America (Brazil, Colombia; S. Graham, 2013, 57–59; II, app. 2.2,
595). The genus likely originated in western South America, moved rapidly
and diversified in the cerrado of Brazil, migrated on and over the CALB to
a secondary center of diversification in western Mexico (fossils in Trinidad
and Mexico), and along the ALB into the southeastern United States (fossils
in Trinidad, Cuba, Alabama; S. Graham, 2003). Recent, rapid, and moderate long-distance transport is feasible in *Cuphea*, given the sticky hairs in
the seed coat that extrude when the seeds are wet. There are other modern
occurrences in the southern and eastern United States that likely include
transport along storm tracks and in ship ballast. *Cuphea* is a good example
of an upland (not lowland bog, swamp, marshy) plant that moved from
south to north on or over the CALB and the ALB in the late Tertiary.

Malpighiaceae (Davis et al., 2002, fig. 1; I, table 7.8, 259; II, app. 2.2,
595) are widely distributed between the New World (85% of the species)

and the Old World (15%; Africa, Australasia). The assumption has been that vicariance was important in explaining its past history, but molecular evidence indicates an age of 63.6 +/−5.8 Ma, and if accurate, this is well after the ca. 105 Ma separation of Africa from South America. The suggestion is that migration took place from South America northward across Laurasia (North Atlantic), and although there are no confirmed Malpighiaceae fossils from the NALB, which would be helpful in evaluating the possibility, there is a report of *Hiraea* from the Miocene of southeastern Oregon (Graham, 1963, 1965). *Hiraea* is a genus of ca. 65 species found from Mexico to Argentina and in the Lesser Antilles. The seed is a schizocarp of 3 samaras each with 3 lateral wings (one reduced, giving the appearance of a bilateral seed), so dispersal by wind over moderate distances is probable. There are numerous reports of several genera of the Malpighiaceae in Latin America from the Eocene onward (II, app. 2.2, 595), so dispersal over and/ or on the CALB and the ALB is documented.

With reference to the Clusioid clade (Bonnetiaceae, Calophyllaceae, Clusiaceae, Hypericaceae, Podostemonaceae; fossils *Paleoclusia, Pachydermites*), Ruhfel et al. (2016) believe that the current distribution "reflects extensive recent dispersal during the Cenozoic (<65 Ma), most of which occurred after the beginning of the Eocene (~56 Ma)."

Pelliceria/Rhizophora (New World mangroves) and *Nypa* (see above) have been considered earlier with reference to their history (Graham, 1977, 1995, 2006; II, app. 2.2, 598–99; III, 222–23; see also Castillo-Cárdenas et al., 2016; Kennedy et al., 2016). *Pelliceria* is at present found along the western coasts of Central America and northern Colombia, but earlier it extended from northern South America (early Eocene), to Jamaica (middle Eocene), Panama (late Eocene), Puerto Rico (Oligocene), and Mexico (Oligo-Miocene of Chiapas). The present widespread *Rhizophora* appears first and slightly (?) earlier in the Eocene of Australasia, then soon afterward in the Eocene of South America rapidly spreading via coastal currents to Central America, Mexico, and the Antilles. The CALB and the ALB were involved in their migration, because fossils are found on fragments of those connections. For *Nypa* and *Rhizophora* the margins of the MLB were likely used as pathways into the New World, because Paleogene climates were suitable and currents were flowing eastward along the connection.

Phytocreneae (Fig. 7.5; Icacinaceae; Stull et al., 2012; I, 174; II, app. 2.2, 593) at present include plants in the lowland tropical forests of Africa, Madagascar, and Indo-Malaysia but are more widespread as fossils in the Paleogene. The tribe is known in the late Paleocene of western North America (*Palaeophytocrene piggae*; ca. 58 Ma) and the middle to late Paleocene of

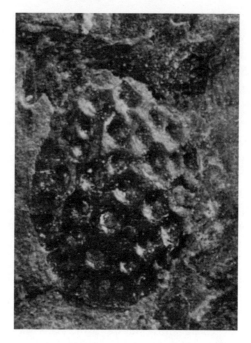

7.5. cf. *Phytocrene* sp., from Colombia. From Stull
et al., 2012, Systematic Botany, American Society
of Plant Taxonomists. Used with permission.

Colombia (*P. hammenii*; ca. 60–58 Ma), which are the oldest records, as
well as the extant *Pyrenacanthus austroamericana* from the late early Oligocene of Peru and the Eocene of North America (early Eocene of Alaska;
I, 174) and Europe. Among the fruit types are large aggregate drupes with
seeds presumably dispersed by primates. Considering the group is an early
branching lineage of Lamiids, together with its present and past distribution, this suggests earlier continental configurations (might have been)
and most of the New World land bridges were (probably; no fossils on
the ALB) involved in its migrations. Post-EECO cooling and drying more
clearly determined its present distribution.

Thespesia (Malvaceae) is a genus of 16 small tropical trees and shrubs, including the widespread coastal *T. populnea*. Areces-Berazain and Ackerman
(2016) believe the genus originated in Southeast Asia–Oceania ca. 30 Ma,
differentiating principally in the Miocene, and dispersing into Africa ca.
11 Ma and the Antilles ca. 9 Ma. Their analyses further indicate "a much
earlier origin than previously reported for the eumalvoid clade and its

tribes Gossypieae, Malveae and Hibisceae suggesting that vicariance might have had an important role early in the history of these groups" (171).

Ulmaceae includes the extinct genus *Cedrelospermum* widespread as fossils in Europe and North America and now known from the late Miocene of Yunnan, China (Jia et al., 2015). It was part of the temperate forest of the Northern Hemisphere, utilizing the BLB and the NALB.

Voyria (Gentianaceae; Merckx et al., 2013; II, app. 2.2, 592) is a genus of 19 species with 18 in tropical America (Mexico, the Antilles, Bolivia, Brazil, Paraguay) and 1 (*V. primuloides*) in West and Central Africa. They are mycoheterotrophic (photosynthates derived from surrounding plants via connecting mycorrhizal fungi), the fruits are capsules, and the numerous "dust" or winged seeds are probably dispersed by wind. Estimates are that it originated in the neotropics in the early Eocene and reached its transoceanic distribution using "the Miocene North Atlantic Land Bridge [cold and disrupted; see chapter 3, above], or via a long-distance dispersal event" (Merckx et al., 2013, 716).

Weigela (Caprifoliaceae; Liang et al., 2013; II, app. 2.2, 590). The seeds of *Weigela* are winged, and the oldest are "from the Oligocene of West Siberia, Asia. The genus later expanded into East Asia and Europe as well as into Arctic/Subarctic areas of North America via the North Atlantic Land Bridge and/or Bering Land Bridge in the Miocene. *Weigela* disappeared from Europe and North America probably during the Pleistocene glaciations, but survived in Asia until now" (1009).

These case studies document some of the plants and vegetation types on the New World and other land bridges during the Late Cretaceous and Cenozoic, and their probable direction and mode of dispersal. Several include the kind of recent paleomonographic studies needed to confirm the identity and age of the numerous unrevised plant fossils in the literature. In combination with taxonomic and phylogenetic studies on the modern representatives, and emerging geologic histories documenting suitable physical and climatic environments at the relevant time for a particular clade, they establish plausible pathways for migration. The survey also reveals the fragmented nature of the fossil record, and together with the partial inventory of the extant record through extinction, it highlights for most plants the tentative nature of estimates of time, place, and movement when based on individual lines of inquiry.

References

Antonelli, A., and I. Sanmartín. 2011. Mass extinction, gradual cooling, or rapid radiation? Reconstructing the spatiotemporal evolution of the ancient angiosperm genus *Hedyosmum* (Chloranthaceae) using empirical and simulated approaches. Systematic Biology 60: 596–615.

Areces-Berazain, F., and J. D. Ackerman. 2016. Phylogenetics, delimitation and historical biogeography of the pantropical tree genus *Thespesia* (Malvaceae, Gossypieae). Botanical Journal of the Linnean Society 181: 171–98.

Bardon, L., et al. (+ 5 authors). 2013. Origin and evolution of Chrysobalanaceae: insights into the evolution of plants in the Neotropics. Botanical Journal of the Linnean Society 171: 19–37.

Barreda, V. D., L. Palazzesi, M. C. Telleria, E. B. Olivero, J. I. Raine, and F. Forest. 2015. Early evolution of the angiosperm clade Asteraceae in the Cretaceous of Antarctica. Proceedings of the National Academy of Sciences USA 112: 10989–94.

Baum, D. A., R. L. Small, and J. F. Wendel. 1998. Biogeography and floral evolution of baobabs (*Adansonia*, Bombaceae) as inferred from multiple data sets. Systematic Biology 47: 181–207.

Berger, B. A., R. Kriebel, D. Spalink, and K. J. Sytsma. 2016. Divergence times, historical biogeography, and shifts in speciation rates of Myrtales. Molecular Phylogenetics and Evolution 95: 116–36.

Berry, E. W. 1918. The fossil higher plants from the Canal Zone. In Contributions to the Geology and Paleontology of the Canal Zone, Panama, and Geologically Related Areas in Central America and the West Indies. Bulletin of the U.S. National Museum 103: 15–44.

———. 1929. Early Tertiary fruits and seeds from Belén, Peru. Johns Hopkins University Studies in Geology 10: 137–72.

Birkenmajer, K., and E. Zastawniak. 1989. Late Cretaceous–early Tertiary floras of King George Island, West Antartica: their stratigraphic distribution and palaeoclimatic significance. In J. A. Crame (ed.), Origins and Evolution of the Antarctic Biota. Geological Society of London, Special Publications, Vol. 47. Geological Society, London. Pp. 227–40.

Braun, A. 1976. Various observations on *Ceroxylon klopstockia*. Principes 20: 158–66.

Bremer, K., E. M. Friis, and B. Bremer. 2004. Molecular phylogenetic dating of asterid flowering plants shows early Cretaceous diversification. Systematic Biology 53: 496–505.

Burnham, R. J., and A. Graham. 1999. The history of neotropical vegetation: new developments and status. Annals of the Missouri Botanical Garden 86: 546–89.

Carvalho, M. R., F .A. Herrera, C. A. Jaramillo, S. L. Wing, and R. Callejas. 2011. Paleocene Malvaceae from northern South America and their biogeographical implications. American Journal of Botany 98: 1337–55.

Carvalho, M. R., P. Wilf, E. J. Hermsen, M A. Gandolfo, N. Rubén Cúneo, and K. R. Johnson. 2013. First record of *Todea* (Osmundaceae) in South America, from the early Eocene paleorainforest of Laguna del Hunco (Patagonia, Argentina). American Journal of Botany 100: 1831–48.

Castillo-Cárdenas, M., O. Sanjur, and N. Toro-Perea. 2016. Differences in sculpture and size of pollen grains: new morphological evidence of diversification in *Pelliciera rhizophorae*, an ancient neotropical mangrove species. Palynology 40: 302–7.

Christin, P.-A., E. Spriggs, C. P. Osborne, C. E. Strömberg, N. Salamin, and E. J. Edwards.

2014. Molecular dating, evolutionary rates, and the age of the grasses. Systematic Biology 63: 153–65.

Conran, J. G., J. M. Bannister, D. C. Mindenhall, and D. E. Lee. 2016. *Hedycarya* macrofossils and associated *Planarpollenites* pollen from the early Miocene of New Zealand. American Journal of Botany 103: 938–56.

Crowl, A. A., et al. (+ 6 authors). 2016. A global perspective on Campanulaceae: biogeographic, genomic, and floral evolution. American Journal of Botany 103: 233–45.

Daniel, T. F., and L. A. McDade. 2014. Nelsonioideae (Lamiales: Acanthaceae): revision of genera and catalog of species. Aliso 32: 1–45.

Davis, C. C., C. D. Bell, S. Matthews, and M. J. Donoghue. 2002. Laurasian migration explains Gondwana disjunctions: evidence from Malpighiaceae. Proceedings of the National Academy of Sciences USA 99: 6833–37.

De Queiroz, A. 2005. The resurrection of oceanic dispersal in historical biogeography. Trends in Ecology and Evolution 20: 68–73.

Di Giacomo, E. de. 1985. *Orchidamasulitis schlechteri* nov. gen. et sp.—Fósil de orchidea (masula) de edad Oligoceno encontrada en Venezuela. Proceedings of the VI Congreso Geológico Venezolano 1: 530–35.

Erkens, R. H. J., L. W. Chatrou, J. W. Maas, T. van der Niet, and V. Savolainen. 2007. A rapid diversification of rainforest trees (*Guatteria*; Annonaceae) following dispersal from Central into South America. Molecular Phylogenetics and Evolution 44: 399–411.

Erkens, R. H. J., J. W. Maas, and T. L. P. Couvreur. 2009. From Africa via Europe to South America: migrational route of a species-rich genus of neotropical lowland rain forest trees (*Guatteria*, Annonaceae). Journal of Biogeography 36: 2338–52.

Estrada-Ruiz, E., L. Calvillo-Canadell, and S. R. S. Cevallos-Ferriz. 2009. Upper Cretaceous aquatic plants from northern Mexico. Aquatic Botany 90: 282–88.

Falcon-Lang, H. J., V. Mages, and M. Collinson. 2016. The oldest *Pinus* and its preservation by fire. Geology 44: 303–6.

Filipowicz, N., and S. S. Renner. 2012. *Brunfelsia* (Solonaceae): a genus evenly divided between South America and radiations on Cuba and other Antillean islands. Molecular Phylogenetics and Evolution 64: 1–11.

Friis, E. M., K. R. Pedersen, and P. R. Crane. 2006. Cretaceous angiosperm flowers: innovation and evolution in plant reproduction. Palaeogeography, Palaeoclimatology, Palaeoecology 232: 251–93.

Funk, V. A., et al. (+ 11 authors). 2005. Everywhere but Antarctica: using a super tree to understand the diversity and distribution of the Compositae. Biologiske Skrifter 55: 343–73.

Givnish, T. J., et al. (+ 6 authors). 2004. Ancient vicariance or recent long-distance dispersal? Inferences about phylogeny and South American–African disjunctions in Rapateaceae and Bromeliaceae based on *nhh*F sequence data. International Journal of Plant Sciences 165 (supplement): S35–S54.

Givnish, T. J., et al. (+ 14 authors). 2011. Phylogeny, adaptive radiation, and historical biogeography in Bromeliaceae: insights from an eight-locus plastid phylogeny. American Journal of Botany 98: 872–95.

Gomez-Navarro, C., C. Jaramillo, F. Herrera, S. L. Wing, and R. Callejas. 2009. Palms (Arecaceae) from a Paleocene rainforest of northern Colombia. American Journal of Botany 96: 1300–1312.

Gottsberger, G. 1978. Seed dispersal by fish in the inundated regions of Humaltá, Amazonia. Biotropica 10: 170–83.

Gottschling, M., N. Diane, H. Hilger, and M. Weigend. 2004. Testing hypotheses on disjunctions present in the primarily woody Boraginales: Ehretiaceae, Cordiaceae, and Heliotropiaceae, inferred from ITS1 sequence data. International Journal of Plant Sciences 165 (supplement): S123–35.

Graham, A. 1963. Systematic revision of the Sucker Creek and Trout Creek Miocene floras of southeastern Oregon. American Journal of Botany 50: 921–36.

————. 1965. The Sucker Creek and Trout Creek Miocene Floras of Southeastern Oregon. Kent State University Press, Kent, OH.

————. 1976. Studies in neotropical paleobotany. II. The Miocene communities of Veracruz, Mexico. Annals of the Missouri Botanical Garden 63: 787–842.

————. 1977. New records of *Pelliceria* (Theaceae/Pelliceriaceae) in the Tertiary of the Caribbean. Biotropica 9: 48–52.

————. 1985. Studies in neotropical paleobotany. IV. The Eocene communities of Panama. Annals of the Missouri Botanical Garden 72: 504–34.

————. 1987. Miocene communities and paleoenvironments of southern Costa Rica. American Journal of Botany 74: 1501–18.

————. 1988a. Studies in neotropical paleobotany. V. The lower Miocene communities of Panama—the Culebra Formation. Annals of the Missouri Botanical Garden 75: 1440–66.

————. 1988b. Studies in neotropical paleobotany. VI. The lower Miocene communities of Panama—the Cucaracha Formation. Annals of the Missouri Botanical Garden 75: 1467–79.

————. 1989. Studies in neotropical paleobotany. VII. The lower Miocene communities of Panama—the La Boca Formation. Annals of the Missouri Botanical Garden 76: 50–66.

————. 1991a. Studies in neotropical paleobotany. IX. The Pliocene communities of Panama—angiosperms (dicots). Annals of the Missouri Botanical Garden 78: 201–23.

————. 1991b. Studies in neotropical paleobotany. X. The Pliocene communities of Panama—composition, numerical representations, and paleocommunity paleoenvironmental reconstructions. Annals of the Missouri Botanical Garden 78: 465–75.

————. 1995. Diversification of Gulf/Caribbean mangrove communities through Cenozoic time. Biotropica 27: 20–27.

————. 2006. Paleobotanical evidence and molecular data in reconstructing the historical phytogeography of Rhizophoraceae. Annals of the Missouri Botanical Garden 93: 325–34.

Graham, A., S. Graham, J. W. Nowicke, V. Patel, and S. Lee. 1990. Palynology and systematics of the Lythraceae. III. Genera *Physocalymma* through *Woodwardia*, addenda and conclusions. American Journal of Botany 77: 159–77.

Graham, A., K. M. Gregory-Wodzicki, and K. L. Wright. 2001. Studies in neotropical paleobotany. XV. A Mio-Pliocene palynoflora from the Eastern cordillera, Bolivia: implications for the uplift history of the Central Andes. American Journal of Botany 88: 1545–57.

Graham, A., and D. M. Jarzen. 1969. Studies in neotropical paleobotany. I. The Oligocene communities of Puerto Rico. Annals of the Missouri Botanical Garden 56: 308–57.

Graham, A., J. Nowicke, J. J. Skvarla, S. A. Graham, V. Patel, and S. Lee. 1985. Palynology and systematics of the Lythraceae. I. Introduction and genera *Adenaria* through *Ginoria*. American Journal of Botany 72: 1012–31.

———. 1987. Palynology and systematics of the Lythraceae. II. Genera *Haitia* through *Peplis*. American Journal of Botany 74: 829–50.

Graham, S. A. 2003. Biogeographic patterns of Antillean Lythraceae. Systematic Botany 28: 410–20.

———. 2013. Fossil records in the Lythraceae. Botanical Review 79: 48–145.

Graham, S. A., J. Hall, K. Systma, and S.-H. Shi. 2005. Phylogenetic analysis of the Lythraceae based on four gene regions and morphology. International Journal of Plant Sciences 166: 995–1017.

Gregory-Wodzicki, K. M., W. C. McIntosh, and K. Velásquez. 1998. Climate and tectonic implications of the late Miocene Jakokkota flora, Bolivian Altiplano. Journal of South American Earth Sciences 11: 533–60.

Grimsson, F., R. Zetter, and C.-C. Hofmann. 2011. *Lythrum* and *Peplis* from the Late Cretaceous and Cenozoic of North America and Eurasia: new evidence suggesting early diversification within the Lythraceae. American Journal of Botany 98: 1801–15.

Grow, N. B., S. Gursky-Doyen, and A. Krzton (eds.). 2014. High Altitude Primates. Springer, Berlin.

Gustafsson, A. L., C. F. Verola, and A. Antonelli. 2010. Reassessing the temporal evolution of orchids with new fossils and a Bayesian relaxed clock, with implications for the diversification of the rare South American genus *Hoffmannseggella* (Orchidaceae: Epidendroideae). BMC Evolutionary Biology. doi: 10.1186/1471-2148-10-177.

Guzmán, A., and P. R. Stevenson. 2011. A new hypothesis for the importance of seed dispersal in time. Revista de Biología Tropical 59 (4), online version.

Harley, M. M. 2006. A summary of fossil records for Arecaceae. Botanical Journal of the Linnean Society 151: 39–67.

Herrera, F., S. R. Manchester, C. Jaramillo, B. MacFadden, and S. A. da Silva-Caminha. 2010. Phytogeographic history and phylogeny of the Humiriaceae. International Journal of Plant Sciences 171: 392–408.

Herrera, F., S. R. Manchester, J. Vélez-Juarbe, and C. Jaramillo. 2014. Phytogeographic history of the Humiriaceae (part 2). International Journal of Plant Sciences 175: 828–40.

Hinsley, S. R. 2005 et seq. Malvaceae Info. http://www/malvaceae.info/fossil. (Accessed July 2016.)

Hoorn, C. 1994. An environmental reconstruction of the palaeo-Amazon river system (middle–late Miocene, NW Amazonia). Palaeogeography, Palaeoclimatology, Palaeoecology 112: 187–238.

Hoorn, C., et al. (+ 17 authors). 2010. Amazonia through time: Andean uplift, climate change, landscape evolution, and biodiversity. Science 330: 927–31.

Jabaily, R. S., and K. J. Sytsma. 2013. Historical biogeography and life-history evolution of Andean *Puya* (Bromeliaceae). Botanical Journal of the Linnean Society 171: 201–24.

Jaramillo, C., and M. Rueda. 2013. A morphological electronic database of Cretaceous-Tertiary fossil pollen and spores from northern South America. Version 2012–2013. Colombian Petroleum Institute and Smithsonian Tropical Research Institute, Panama City.

Jaramillo, C., et al. (+ 12 authors). 2014. Palynological record of the last 20 million years in Panama. In W. D. Stevens, O. M. Montiel, and P. H. Raven (eds.), Paleobotany and Biogeography, a Festschrift for Alan Graham in His 80th Year. Missouri Botanical Garden Press, St. Louis. Pp. 134–251.

Jia, L. B., S. R. Manchester, T. Su, Y. W. Xing, W. Y. Chen, Y. J. Huang, and Z. K. Zhou. 2015. First occurrence of *Cedrelospermum* (Ulmaceae) in Asia and its biogeographic implications. Journal of Plant Research 128: 747–61.

Jiménez-Mejías, P., E. Martinetto, A. Morohara, S. Popova, S. Y. Smith, and E. H. Roalson. 2016. A commented synopsis of the pre-Pleistocene fossil record of *Carex* (Cyperaceae). Botanical Review. doi: 10.1007/s12229-016-9169-7.

Jud, N. A., G. W. Rothwell, and R. A. Stockey. 2008. *Todea* from the Lower Cretaceous of western North America: implications for the phylogeny, systematics and evolution of modern Osmundaceae. American Journal of Botany 95: 330–39.

Katinas, L., J. V. Crisci, M. C. Tellería, V. Barreda, and L. Palazzesi. 2007. Early history of Asteraceae in Patagonia: evidence from fossil pollen grains. New Zealand Journal of Botany 45: 605–10.

Kay, K. M., P. A. Reeves, R. G. Olmstead, and D. W. Schemske. 2005. Rapid speciation and the evolution of hummingbird pollination in neotropical *Costus* subgenus *Costus* (Costaceae): evidence from nrDNA ITS and ETS sequences. American Journal of Botany 92: 1899–910.

Kay, K. M., and Y. Surget-Groba. 2013. Restricted gene flow within and between rapidly diverging neotropical plant species. Molecular Ecology 22: 4931–42.

Kennedy, J. P., et al. (+ 5 authors). 2016. Postglacial expansion pathways of red mangrove, *Rhizophora mangle*, in the Caribbean Basin and Florida. American Journal of Botany 103: 260–76.

Knight, C. L., and P. Wilf. 2013. Rare leaf fossils of Monimiaceae and Atherospermataceae (Laurales) from Eocene Patagonian rainforests and their biogeographic significance. Palaeotologia Electronica 16: article 26A, 21–39.

Krassilov, V. A. 1976. Tsagayan flora of Amur Province. Nauka, Moscow.

Kubitzki, K. (ed.). 1998. The Families and Genera of Vascular Plants. Springer, Berlin. [Flowering Plants-Costaceae (by K. Larsen), 128–32.]

Kubitzki, K., J. G. Rohwer, and V. Bittrich (eds.). 1993. The Families and Genera of Vascular Plants. Flowering Plants-Dicotyledons: Magnoliid, Hamamelid and Caryophyllid Families. Springer, Berlin.

Lebreton Anberrée, J., S. R. Manchester, J. Huang, S. Li, Y. Wang, and Z.-K Zhou. 2015. First fossil fruits and leaves of *Burretiodendron* s.l. (Malvaceae s.l.) in Southeast Asia: implications for taxonomy, biogeography, and paleoclimate. International Journal of Plant Sciences 176: 682–96.

Ledig, F. T., P. E. Smouse, and J. L. Hom. 2015. Postglacial migration and adaptation for dispersal in pitch pine (Pinaceae). American Journal of Botany 102: 2074–91.

Li, Y.-J., A. A. Oskolski, F. M. B. Jacques, and Z.-K. Zhou. 2015. New middle Miocene fossil wood of *Wataria* (Malvaceae) from southwest China. IAWA Journal 36: 345–57.

Liang, X.-Q., Y. Li, Z. Kvaček, V. Wilde, and C.-S. Li. 2013. Seeds of *Weigela* (Caprifoliaceae) from the early Miocene of Weichang, China, and the biogeographical history of the genus. Taxon 62: 1009–18.

Lötschert, W., and K. Mädler. 1975. Die plio-pleistozäne Flora aus dem Sisimico-Tal, El Salvador. Ein Beitrag zur Frage der Kontinuität tropischer Regenwälder im Quartär. Geologisches Jahrbuch 13: 97–191.

Magallón, S., and M. J. Sanderson. 2001. Absolute diversification rates in angiosperm clades. Evolution 55: 1762–80.

Manen, J.-F., M. C. Boulter, and Y. Naciri-Graven. 2002. The complex history of the genus *Ilex* L. (Aquifoliaceae): evidence from the comparison of plastid and nuclear DNA sequences and from fossil data. Plant Systematics and Evolution 235: 79–98.

Martínez, C., S. Madriñán, M. Zavada, and C. A. Jaramillo. 2013. Tracing the fossil pollen record of *Hedyosmum* (Chloranthaceae), an old lineage with recent neotropical diversification. Grana 52: 161–80.

McIver, E. E. 1994. An early *Chamaecyparis* (Cupressaceae) from the Late Cretaceous of Vancouver Island, British Columbia, Canada. Canadian Journal of Botany 72: 1787–96.

McIver, E. E., and K. R. Aulenbeck. 1994. Morphology and relationships of *Mesocyparis umbonata* sp. nov. fossil Cupressaceae from the Late Cretaceous of Alberta, Canada. Canadian Journal of Botany 72: 273–95.

McIver, E. E., and J. K. Bassinger. 1987. *Mesocyparis borealis* gen. et sp. nov.: fossil Cupressaceae from the early Tertiary of Saskatchewan, Canada. Canadian Journal of Botany 65: 2238–351.

Merckx, V. S. F. T., et al. (+ 6 authors). 2013. Phylogenetic relationships of the mycoheterotrophic genus *Voyria* and the implications for the biogeographic history of Gentianaceae. American Journal of Botany 100: 712–21.

Morris, J. A. 2007. A molecular phylogeny of the Lythraceae and inference of the evolution of heterostyly. Ph.D. diss., Kent State University, Kent, OH.

Olmstead, R. G. 2013. Phylogeny and biogeography in Solonaceae, Verbenaceae and Bignoniaceae: a comparison of continental and intercontinental diversification patterns. Botanical Journal of the Linnean Society 171: 80–102.

Perret, M., A. Chautems, A. Onofre de Araujo, and N. Salamin. 2013. Temporal and spatial origin of Gesneriaceae in the New World inferred from plastid DNA sequences. Botanical Journal of the Linnean Society 171: 61–79.

Poole, I., and H. Gottwald. 2001. Monimiaceae *sensu lato*, an element of Gondwana polar forests: evidence from the Late Cretaceous–early Tertiary wood flora of Antarctica. Australian Systematic Botany 14: 207–30.

Ramírez, S. R., B. Gravendeel, R. B. Singer, C. R. Marshall, and N. E. Pierce. 2007. Dating the origin of the Orchidaceae from a fossil orchid with its pollinator. Nature 448: 1042–45.

Renner, S. S., D. B. Foreman, and D. Murray. 2000. Timing transantarctic disjunctions in the Atherospermataceae (Laurales): evidence from coding and noncoding chloroplast sequences. Systematic Biology 49: 579–91.

Renner, S. S., J. S. Strijk, D. Strasberg, and C. Thébaud. 2010. Biogeography of the Monimiaceae (Laurales): a role for East Gondwana and long-distance dispersal, but not West Gondwana. Journal of Biogeography 37: 1227–38.

Richardson, J. E., L. W. Chatrou, J. B. Mols, R. H. J. Erkens, and M. D. Pirie. 2004. Historical biogeography of two cosmopolitan families of flowering plants: Annonaceae and Rhamnaceae. Philosophical Transactions of the Royal Society of London B (Biological Sciences) 359: 1495–508.

Richardson, J. E., R. T. Pennington, T. D. Pennington, and P. M. Hollingsworth. 2001. Rapid diversification of a species-rich genus of neotropical rain forest trees. Science 293: 2242–45.

Rodríguez-Reys, O., H. Falcon-Lang, P. Gasson, M. Collinson, and C. Jaramillo. 2014. Fossil woods (Malvaceae) from the lower Miocene (early to mid-Burdigalian) part of the Cucaracha Formation of Panama (Central America) and their biogeographic implications. Review of Palaeobotany and Palynology 209: 11–34.

Ruhfel, B. R., C, P. Bove, C. T. Philbrick, and C. C. Davis. 2016. Dispersal largely explains the Gondwanan distribution of the ancient tropical clusioid plant clade. American Journal of Botany 103: 1117–28.

Salariato, D. L., and F. O. Zuloaga. 2016. Climatic niche evolution in the Andean genus *Menonvillea* (Cremolobeae: Brassicaceae). Organisms Diversity and Evolution. Online version, 1–18. http://link.springer.com/article/10.1007/s13127-016-0291-5.

Salzman, S., H. E. Driscoll, T. Renner, T. André, S. Shen, and C. D. Sprecht. 2015. Spiraling into history: a molecular phylogeny and investigation of biogeographic origins and floral evolution for the genus *Costus*. HHS Public Access 40: 104–15. 10.1600/036364415x686404.

Sanín, M. J., et al. (+ 9 authors). 2016. The Neogene rise of the tropical Andes facilitated diversification of wax palms (*Ceroxylon*: Arecaceae) through geographical colonization and climatic niche separation. Botanical Journal of the Linnean Society 182: 303–17.

Soltis, D. E., C. D. Bell, S. Kim, and P. S. Soltis. 2008. Origin and early evolution of angiosperms. Annals of the New York Academy of Sciences 1133: 3–25.

Specht, C. D. 2006. Gondwanan vicariance or dispersal in the tropics? The biogeographic history of the tropical monocot family Costaceae (Zingiberales). In J. T. Columbis et al. (+ 5 eds.), Monocots: Comparative Biology and Evolution. Rancho Santa Ana Botanic Garden, Claremont, CA. Pp. 631–42.

Stevenson, P. R. 2000. Seed dispersal by wooly monkeys (*Lagothrix lagothricha*) at Tinigua National Park, Colombia: dispersal distance, germination rates, and dispersal quantity. American Journal of Primatology 50: 275–89.

Stull, G. W., F. Herrera, S. R. Manchester, C. Jaramillo, and B. H. Tiffney. 2012. Fruits of an "Old World" tribe (Phytocreneae; Icacinaceae) from the Paleogene of North and South America. Systematic Botany 37: 784–94.

Suetsugu, K., A. Kawakita, and M. Kato. 2015. Avian seed dispersal in a mycoheterotrophic orchid *Cyrtosia septentrionalis*. Nature Plants. doi: 10.1038/nplants.2015.52.

Surget-Groba, Y., and K. M. Kay. 2013. Restricted gene flow within and between rapidly diverging neotropical plant species. Molecular Ecology. doi: 10.1111/mec.12442.

Sytsma, K. J., et al. (+ 7 authors). 2004. Clades, clocks, and continents: historical and biogeographical analysis of Myrtaceae, Vochysiaceae, and relatives in the Southern Hemisphere. International Journal of Plant Sciences 165 (supplement): S85–S105. [II, appendix 2.2, 596–97.]

Terry, R. G., M. I. Pyne, J. A. Bartel, and R. P. Adams. 2016. A molecular biogeography of the New World cypresses (*Callitropsis, Hesperocyparis*; Cupressaceae). Plant Systematics and Evolution. doi: 10.1007/s00606-016-1308-4.

Thiede, D. A., and C. K. Augspurger. 1996. Intraspecific variation in seed dispersion of *Lepidium campestre* (Brassicaceae). American Journal of Botany 83: 856–66.

Trénel, P., et al. (+ 5 authors). 2007. Mid-Tertiary dispersal, not Gondwana vicariance explains distribution patterns in the wax palm subfamily (Ceroxyloideae: Arecaceae). Molecular Phylogenetics and Evolution 45: 272–88.

Tripp, E. A., and L. A. McDade. 2014. A rich fossil record yields calibrated phylogeny for Acanthaceae (Lamiales) and evidence for marked biases in timing and directionality of intercontinental disjunctions. Systematic Biology 63: 660–84.

Viljoen, J.-A., et al. (+ 7 authors). 2013. Radiation and repeated transoceanic dispersal of Schoeneae (Cyperaceae) through the Southern Hemisphere. American Journal of Botany 100: 2494–508.

Weir, J. T. 2006. Divergent timing and patterns of species accumulation in lowland and highland neotropical birds. Evolution 60: 842–55.

Wendel, J. F., C. L. Brubaker, and T. Seelanan. 2010. The origin and evolution of *Gossyp-*

ium. In J. M. Stewart, D. Costerhuis, J. J. Heitholt, and J. R. Mauney (eds.), Physiology of Cotton. Springer, Berlin. Pp. 1–18.

Westgate, J. W., and C. T. Gee. 1990. Paleoecology of a middle Eocene mangrove biota (vertebrate, plants, and invertebrates) from southwest Texas. Palaeogeography, Palaeoclimatology, Palaeoecology 78: 163–77.

Wickens, G. E. 2008. The Baobabs: Pachyculs of Africa, Madagascar and Australia. Springer, Berlin.

Wikström, N., V. Savolainen, and M. W. Chase. 2001. Evolution of the angiosperms: calibrating the family tree. Proceedings of the Royal Society of London B (Biological Sciences) 268: 2211–20.

Winkworth, R. C., J. Grau, A. W. Robertson, and P. J. Lockhart. 2002. The origins and evolution of the genus *Myosotis* L. (Boraginaceae). Molecular Phylogenetics and Evolution 24: 180–93.

Zona, S., and A. Henderson. 1989. A review of animal-mediated seed dispersal of palms. Selbyana 11: 6–21.

Additional References

Antonelli, A., C. F. Verola, C. Perisod, and A. L. S. Gustafsson. 2010. Climate cooling promoted the expansion and radiation of a threatened group of South American orchids (Epidendroideae: Laeliinae). Botanical Journal of the Linnean Society 100: 597–607.

Bardon, L., et al. (+ 10 authors). 2016. Unraveling the biogeographical history of Chrysobalanaceae from plastid genomes. American Journal of Botany 103: 1089–102.

Bartish, I. V., A. Antonelli, J. E. Richardson, and U. Swenson. 2010. Vicariance or long-distance dispersal: historical biogeography of the pantropical subfamily Chrysophylloideae (Sapotaceae). Journal of Biogeography 38: 177–90. [II, appendix 2.2, 600.]

Bremer, K., E. M. Friis, and B. Bremer. 2004. Molecular phylogenetic dating of asterid flowering plants shows early Cretaceous diversification. Systematic Biology 53: 496–505.

Burge, D. O., and S. R. Manchester. 2008. Fruit morphology, fossil history, and biogeography of *Paliurus* (Rhamnaceae). International Journal of Plant Sciences 169: 1066–85.

Collinson, M. E., M. C. Boulter, and P. R. Holmes. 1993. Magnoliophyta ("Angiospermae"). In M. J. Benton (ed.), The Fossil Record 2. Chapman and Hall, London. Pp. 809–41.

Conran, J. G., J. M. Bannister, and D. E. Lee. 2009. Earliest orchid macrofossils: early Miocene *Dendrolobium* and *Earina* (Orchidaceae: Epidendroideae) from New Zealand. American Journal of Botany 96: 466–74.

Costea, M., S. Stefanović, M. A. García, S. De La Cruz, M. L. Casazza, and A. J. Green. 2016. Waterfowl endozoochory: an overlooked long-distance dispersal mode for *Cuscuta* (dodder). American Journal of Botany 103: 957–62. [The study indicates at least 18 historical cases of long-distance dispersal within the mostly American Subgenus *Grammica* likely involving a transoceanic dispersal from South Africa to South America (957).]

Cracraft, J. 1985. Historical biogeography and patterns of differentiation within the South American avifauna: areas of endemism. Ornithological Monographs 36. Neotropical Ornithology: 49–84.

Doria, G., C. A. Jaramillo, and F. Herrera. 2008. Menispermaceae from the Cerrejón Formation, middle to late Paleocene, Colombia. American Journal of Botany 95: 954–73. [See also Herrera et al., 2004, re Menispermaceae.]

Drew, B. T., and K. J. Sytsma. 2013. The South American radiation of *Lepechinia* (Lamiaceae): phylogenetics, divergence times and evolution of dioecy. Botanical Journal of the Linnean Society 171: 171–90. [The authors note diversification of *Lepechinia* in South America beginning within the past 15 Ma (crown group originating in the mid-late Miocene) and accelerating at ca. 8–7 ma. This is attributed to cooling and drying, with an increase in fires, and uplift of the Northern Andes (Graham et al., 2001; Gregory-Wodzicki et al., 1998; Hoorn et al., 2010) creating the dry open habitats common to the genus. The fossil record for the family is sparse, but see II, app. 2.2, 593, Labiatae.]

Fritsch, P. W., S. R. Manchester, R. D. Stone, B. C. Cruz, and F. Almeda. 2015. Northern Hemisphere origins of the amphi-Pacific tropical plant family Symplocaceae. Journal of Biogeography 42: 891–901.

Graham, A. 1992. The current status of the legume fossil record in the Caribbean region. In P. S. Herendeen and D. L. Dilcher (eds.), Advances in Legume Systematics. Part 4. The Fossil Record. Royal Botanic Gardens, Kew. Pp. 161–67.

———. 1996. A contribution to the geological history of the Compositae. In D. J. N. Hind and H. J. Beenje (eds.), Compositae: Systematics. Proceedings of the International Compositae Conference, Kew, 1: 123–40.

———. 2008. Fossil record of the Rubiaceae. Annals of the Missouri Botanical Garden 96: 90–108. [See also II, app. 2.2, 599.]

Harris, AJ, M. Papes, Y.-D. Gao, and L. Watson. 2014. Estimating paleoenvironments using ecological niche models of nearest living relatives: a case study of Eocene *Aesculus* L. Journal of Systematics and Evolution 52: 16–34.

Herrera, F. A., C. A. Jaramillo, D. L. Dilcher, S. L. Wing, and C. Gómez-N. 2008. Fossil Araceae from a Paleocene neotropical rainforest in Colombia. American Journal of Botany 95: 1569–84. [Leaves; reports of fossil inaperturate striate pollen from the Barremian-early Aptian (~124–117 Ma) of Portugal are questioned; minimum age estimated for the family by nuclear analysis calibrated with fossils is 128–105 Ma, p. 1569.]

Herrera, F. A., S. R. Manchester, S. B. Hoot, K. M. Wefferling, M. A. Carvalho, and C. Jaramillo. 2004. Phytogeographic implications of fossil endocarps of Menispermaceae from the Paleocene of Colombia. American Journal of Botany 98: 2004–17. [See also Doria et al., 2008, re Menispermaceae.]

Ickert-Bond, S., and J. Wen. 2006. Phylogeny and biogeography of Altingiaceae: evidence from combined analysis of five non-coding chloroplast regions. Molecular Phylogenetics and Evolution 39: 512–28.

Jud, N. A., and C. W. Nelson. 2017. A liana from the lower Miocene of Panama and the fossil record of Connaraceae. American Journal of Botany 104: 685–93.

Lavin, M. 2006. Floristic and geographical stability of discontinuous seasonally dry tropical forests explains patterns of plant phylogeny and endemism. In R. T. Pennington, J. A. Ratter, and G. P. Lewis (eds.), Neotropical Savannas and Seasonally Dry Forests: Plant Biodiversity, Biogeography and Conservation. CRC Press, Boca Raton. Pp. 433–37.

Lavin, M., P. S. Herendeen, and M. F. Wojciechowski. 2005. Evolutionary rates analysis of Leguminosae implicates a rapid diversification of lineages during the Tertiary. Systematic Biology 54: 575–94.

Lewitus, E., and H. Morton. 2016. Natural constraints to species diversification. PLoS Biol. 14: e1002532. doi: 10.1371/journal.pbio.1002532.

Lu-Irving, P., and R. G. Olmstead. 2013. Investigating the evolution of Lantaneae (Verbenaceae) using multiple loci. Botanical Journal of the Linnean Society 171: 103–19. [II, app. 2.2, 601.]

Manchester, S. R. 1994. Inflorescence bracts of fossil and extant *Tilia* in North America, Europe, and Asia: patterns of morphologic divergence and biogeographic history. American Journal of Botany 81: 1176–85. [The presence of these fossils in western North America (late Eocene to Miocene), Europe (early Miocene to Pliocene), and Asia (middle and late Tertiary) suggests a North American origin with migration to Asia and Europe via the BLB and the NALB. The movement of *Tilia* along these routes at these times is consistent with the cool-temperate habitats in the mid- to late Tertiary.]

Manos, P. S. 2016. Systematics and biogeography of the American oaks. International Oaks 27: 23–36.

Márquez-Corro, J. I., M. Escudero, S. Martín-Bravo, T. Villaverde, and M. Luceño. 2017. Long-distance dispersal explains the bipolar disjunction in *Carex macloviana*. American Journal of Botany 104: 663–73.

Martín-Bravo, S., and T. F. Daniel. 2016. Molecular evidence supports ancient long-distance dispersal for the amphi-Atlantic disjunction in the giant yellow shrimp plant (*Barleria oenotheroides*). American Journal of Botany 103: 1103–16.

Morley, R. J., and C. W. Dick. 2003. Missing fossils, molecular clocks, and the origin of the Melastomataceae. American Journal of Botany 90: 1638–44. [See also Renner and Meyer, 2001, re Melastomeae.]

Müller, S., et al. (+ 7 authors). 2015. Intercontinental long-distance dispersal of Canellaceae from the New to the Old World revealed by a nuclear single copy gene and chloroplast loci. Molecular Phylogenetics and Evolution 84: 2-5-219.

Nauheimer, L., D. Metzler, and S. S. Renner. 2012. Global history of the ancient monocot family Araceae inferred with models accounting for past continental positions and previous ranges based on fossils. New Phytologist 195: 938–50.

Nicolas, A. N., and G. M. Plunkett. 2014. Diversification times and biogeographic patterns in Apiales. Botanical Review 80: 30–58.

Ogutcen, E., and J. C. Vamosi. 2016. A phylogenetic study of the tribe Antirrhineae: genome duplications and long-distance dispersals from the Old World to the New World. American Journal of Botany 103: 1071–81.

Pennington, R. T., and C. W. Dick. 2004. The role of immigrants in the assembly of the South American rainforest tree flora. Philosophical Transactions of the Royal Society London B (Biological Sciences) 359: 1611–22.

Pennington, R. T., M. Lavin, A. Oliveira-Filho. 2009. Woody plant diversity, evolution, and ecology in the tropics: perspectives from seasonally dry tropical forests. Annual Review of Ecology, Evolution, and Systematics 40: 437–57.

Pennington, R. T., M. Lavin, D. E. Prado, C. A. Pendry, S. K. Pell, and C. Butterworth. 2004. Historical climate change and speciation: neotropical seasonally dry forest plants show patterns of both Tertiary and Quaternary diversification. Philosophical Transactions of the Royal Society London B (Biological Sciences) 359: 515–37.

Pennington, R. T., M. Lavin, T. Särkinen, G. P. Lewis, B. B. Klitgaard, and C. E. Hughes. 2010. Contrasting plant diversification histories within the Andean biodiversity hotspot. Proceedings of the National Academy of Sciences USA 107: 13783–87.

Renner, S. S., G. W. Grimm, G. M. Scheeswiss, T. F. Stuessy, and R. E. Ricklefs. 2008. Rooting and dating maples (*Acer*) with an uncorrelated-rates molecular clock: implications for North American/Asian disjunctions. Systematic Botany 57: 795–808.

Renner, S. S., and K. Meyer. 2001. Melastomeae come full circle: biogeographic reconstruction and molecular clock dating. Evolution 55: 1315–24. [See also Morley and Dick, 2003, re Melastomataceae.]

Ruhfel, B. R., C. P. Bove, C. T. Philbrick, and C. C. Davis. 2016. Dispersal largely explains the Gondwanan distribution of the ancient tropical clusioid plant clade. American Journal of Botany 103: 1–12.

Salmaki, Y., S. Kattari, G. Heubl, and C. Bräuchler. 2016. Phylogeny of non-monophyletic *Teucrium* (Lamiaceae: Ajugoideae): implications for character evolution and taxonomy. Taxon 65: 805–22.

Särkinen, T., L. Bohs, R. G. Olmstead, and S. Knapp. 2013. A phylogenetic framework for evolutionary study of the nightshades (Solonaceae): a dated 1000-tip tree. BMC Evolutionary Biology 13: 214. http://www.biomedcentral.com/1471-2148/13/214.

Schneider, A. C., W. A. Freyman, C. M. Guilliams, Y. P. Springer, and B. G. Balwin. 2016. Pleistocene radiation of the serpentine-adapted genus *Hesperolinon* and other divergence times in Linaceae (Malpighiales). American Journal of Botany 103: 221–32. [Western flax is a western United States endemic related to *Linum* with most diversification taking place within the past 1–2 Ma. It would interesting to explore it through microsatellite analysis (as was done for *Rhizophora*; Kennedy et al., 2016); divergence times as recent as ca. 18,000 years could be detected for flax, allowing comparison with the crossing of Beringia at the time by migrating animals and the humans that followed them.]

Scholl, J. P., and J. J. Wiens. 2016. Diversification rates and species richness across the Tree of Life. Proceedings of the Royal Society London B (Biological Sciences) 283: 20161334. http://dx.doi.org/10.1098/rspb.2016.1334.

Sigel, E. M., M. D. Windham, C. H. Haufler, and K. M. Pryer. 2014. Phylogeny, divergence time estimates, and phylogeography of the diploid species of the *Polypodium vulgare* complex (Polypodiaceae). Systematic Botany 39: 1042–55.

Wikström, N., V. Savolainen, and M. W. Chase. 2001. Evolution of the angiosperms: calibrating the family tree. Proceedings of the Royal Society London B (Biological Sciences) 268: 2211–20.

Xi, Z., et al. (+ 6 authors). 2012. Phylogenetics and a posteriori data partitioning resolve the Cretaceous angiosperm radiation of Malpighiales. Proceedings of the National Academy of Sciences USA 109: 17519–24.

Xiang, Q.-Y., D. T. Thomas, W. Zhang, S. R. Manchester, and Z. Murrell. 2006. Species level phylogeny of the genus *Cornus* (Cornaceae) based on molecular and morphological evidence—implications for taxonomy and Tertiary intercontinental migration. Taxon 55: 9–30.

EIGHT

Summary and Conclusions

It is a long way from the Arctic to the Antarctic, and 100 Ma is a long time allowing for the probable and the improbable to occur. During this interval the Atlantic Ocean opened, separating the New World from Africa and Europe; mean annual temperatures varied by 3°–5°C in the tropics to 8°–15°C toward the poles; the Rocky Mountains, Sierra Nevada, and Andes Mountains rose; North America was joined to South America via the Isthmus of Panama; the Gulf Stream strengthened, bringing greater heat to the North Atlantic; palms grew within the Arctic Circle; the Amazon River reversed its course from the Pacific to the Atlantic; South America and Antarctica separated with formation of the Magellan Strait, the Drake Passage, and the Antarctic Circum-Current; glaciers formed, plants disappeared, oceans cooled, and the cold Humboldt Current flowing northward added to the dryness developing along the west coast of South America; sea levels rose and fell by 150 m, alternately inundating then exposing coastlines and continental interiors; humans crossed Beringia for the first time, moving from Asia into North America; plants evolved new ecological tolerances and pollination, dispersal, and defense mechanisms; CO_2 concentration varied from more than 1000 ppmv (or perhaps capped at 1000 ppmv; see McElwain et al., 2016) to less than 200 ppmv; and during all this time land bridges were modifying atmospheric and ocean circulation patterns and periodically joining the New World to and separating it from regions and biotas immediately adjacent and far beyond.

Attempting to reconstruct the broad outlines of this history is challenging to the point of privilege (see appendix 8 in online supplementary materials). The organizational framework has been to describe the details while considering broader conceptual issues along the way. Places are identified where future work is necessary and databases needed for storing and analyzing the information.

Events, Processes, and Responses

Bering Land Bridge

Use of the New World bridges is based on documentation provided by the inventory of fossil plants in the vicinity of the crossings and expectations from the global context of independent geological, geographical, and climatic histories. In the Paleogene, and in particular at the EECO, the MAT of surface waters in the North Pacific was an estimated 15 °C warmer than at present; ca. 10 ° – 11 °C warmer in the transitional times of the Oligocene; then decreasing episodically throughout the Neogene to present values (see table 8.1 in online supplementary materials). The MAP changed from wet to seasonally dry with the cooling temperatures. After the Cretaceous and onward there was moderate to more substantial physiographic relief, allowing for some zonation of habitats and vegetation. These trends provide a useful context for inferring use of the BLB to explain distribution of other modern plants where direct fossil evidence is meager. In the Paleogene, vegetation in west Beringia was warm-temperate and included plants like *Regnellidium, Araucaria, Byttneria*; in east Beringia *Anemia, Cyathea, Allophyllus*, and *Phytocrene*; and *Sabal* and *Litsea* in both regions. In the Oligocene through the middle Miocene there were temperate *Osmunda, Cedrus, Metasequoia, Pinus, Sequoia, Taxodium, Acer, Alnus, Betula, Cedrela, Fagus, Juglans, Nyssa, Pittosporum, Platanus, Quercus, Salix, Ulmus*, and *Zelkova*. Such an admixture is more compatible with a boreotropical flora rather than an Arcto-Tertiary Geoflora of near-exclusive temperate deciduous forest elements. By the late Miocene and Pliocene many of the warm-temperate plants were disappearing from Beringia (*Taxodium, Liriodendron, Nyssa*), while cold-temperate ones like *Abies, Picea, Larix, Tsuga, Alnus, Betula, Populus*, and *Salix* were becoming widespread. Tundra is first recognized as a community by ca. 3 Ma. From the end of the Cretaceous to the late Pliocene the Bering Land Bridge was expansive and functioning as an important means of exchange between the biotas of boreal lands. The 18–20 Milankovitch-regulated and CO_2-amplified glacial and interglacial intervals of the Quaternary, approximately 100,000 years and averaging 20,000 years long, respectively, permitted continued and episodic migrations. Intense cold developed on the vast plains of Siberia slightly earlier than in eastern Beringia, shifting the distribution of organisms, including humans, toward the east into the physiographically diverse and compartmentalized habitats of northwestern North America. The people followed a coastal route, given the new evidence available, and possibly near-contemporaneously an inland route, given the opportunities of an ice-free corridor.

The differences in landscape between Siberia and northwestern North America allow a point raised by Simpson (1940) over 75 years ago to be addressed. He cogently noted that "mammals [and plants] do not as a rule acquire new territory simply by traveling into it but by a less purposeful peripheral expansion in all possible directions. Similarly, they do not usually lose territory simply by traveling away from it, but by a complex sequence of attenuation and local extinction that can be called contraction." The apparent exception of a preferential west-to-east movement through Beringia can be explained by the development of colder climates and glaciers slightly earlier in the vast and comparatively flat source area of Siberia than in the less extensive and mountainous target area of northwestern North America. Simpson also considered a bridge (or isthmus) to be a filter to some extent. One factor involved in the filtering process, in addition to climate, topography, soil, and biological phenomena, is the spatial configuration of most bridges that imparts increased density and competition among organisms where a narrow bridge connects lands of continental extent.

North Atlantic Land Bridge

Continuous land extended from Europe to North America until ca. 59 Ma, when promulgation of the Mid-Atlantic Ridge reached the North Atlantic. By the early to middle Eocene the NALB was physically disrupted. Before this the "bridge" across the North Atlantic was wide—essentially the land of northern Laurasia between the Arctic Ocean and the Tethys Sea. For much of the Cretaceous and early Paleogene it was continuous, expansive, and less of a filter than the BLB. After the Eocene it became an extreme filter, while the BLB was a comparative causeway: "In the whole history of mammals there are exceedingly few cases (e.g., Lower Eocene between Europe and North America) where the evidence really warrants the inference of a wide-open corridor between two now distinct continental masses. The usual sort of connection is selective, not acting as a corridor or open door but as a sort of filter permitting some things to pass but holding back others" (Simpson, 1940). In other words, the filtering was different between the two bridges in timing and degree. They differed in other ways (see tables 8.1 and 8.2 in the online supplementary materials), reflecting a diverse history, and therein lies their collective capacity for accommodating a wide array of organisms and ecosystems into and out of the New World over a long interval of time.

As the EECO began to wane, cooling climate was added to the discontinuous land surface as a further barrier to migration across the NALB. In the Paleogene tropical plants like *Anemia, Azolla, Cinnamomon,* and *Nypa* were present, along with temperate trees and shrubs (*Ginkgo, Sequoia, Metasequoia, Acer, Carya, Fraxinus, Ilex, Magnolia*), and later in the Paleocene gymnosperms like *Abies, Picea,* and *Tsuga.* Even acknowledging the foibles of the fossil record, the flora (including observations on leaf physiognomy) suggests a boreotropical forest, as in Beringia, with less ecological definition and somewhat more intermingling of tropical and temperate elements allowing for some separation by habitat (lowland to upland, coastal to inland). Warm-temperate components are well represented and remained in the vicinity somewhat longer because of the presence of the warm Gulf Stream compared to Beringia and the colder north-flowing Japan Current.

Throughout the Late Cretaceous and Tertiary these were the plants and the conditions that prevailed across both the boreal lands. Collectively, they provide little evidence of an Arcto-Tertiary Geoflora of the kind envisioned by Chaney and others, or of an arid Madrean-Tethyan flora connecting or nearly connecting the dry areas of the Mediterranean and western North America as contemplated by Axelrod. Together, they suggest a less (but not randomly) ecologically differentiated flora through the region as described by Wolfe (see chapter 3 above, section on geofloras).

Antillean Land Bridge

A recent renovation in thought (since about the 1970s) regarding the vegetation of the equatorial zone, including the Caribbean, is that rather than being climatically stable, the tropical forest and associated communities experienced changes in both range and composition. These changes were due to muted but still evident fluctuations in climate, especially in temperature (see online tables 8.1, 8.2). At the EECO MATs were warmer by ca. 3°–5°C, the vegetation was tropical, and *Peliceria,* for example, and other plants expanded from northern coastal South America into the Antilles. In the late Tertiary and Quaternary the biota underwent changes in response to temperature fluctuations that were ca. 2°–4°C colder than at present. This variability was long suspected and eventually confirmed by results first from study of the middle Pliocene Paraje Solo flora near the northern limit of the tropical rainforest in southeastern Mexico and now recognized throughout the tropics generally. It has made the dispersal, speciation, and exceptionally high biodiversity of the region easier to explain than when it was assumed that environments were stable and unchanging.

Regarding the physical configuration, the ALB was a series of discontinuous island stepping stones throughout the Tertiary; that is, there is no definite geological evidence for a landspan through the Aves Ridge (see chapter 4, section "Stepping Stones or Lost Highway?"). Filtering was moderate, because even though the bridge was discontinuous, the islands were close together, land in the Greater and Lesser Antilles has been present since the latest Eocene, and some islands in the Greater Antilles were even connected at various times. The prevalent vegetation varied around warm-temperate with allowances for elevation and the moderate changes in climate.

Central American Land Bridge

In the Late Cretaceous and throughout most of the Tertiary, plants migrated between North America and Central America as far south as eastern Panama. *Alnus* and *Quercus* reached their southern limits in the late Miocene/Pliocene and remained in the moderate highlands, while ocean waters extended across far eastern Panama. First lowlands and later uplands appeared. Terrestrial animals requiring more or less continuous land for migration began the Great American Biotic Interchange at ca. 3.5 Ma, consistent with marine organisms beginning to show a reciprocal pattern of isolation between the Pacific Ocean and the Caribbean Sea. More extensive and better-drained uplands were present by ca. 2.8–2.5 Ma, and *Alnus* and *Quercus* crossed over at 1 Ma and 430 kyr, respectively.

There is a proposal that North America was connected to South America at 15–13 Ma. In light of the few organisms making the crossing, however, and the terrestrial and marine paleontology and ocean geochemistry records, a full, sustained, connection at ca. 3.5 Ma and uplands at ca. 2.5 Ma is still most convincing. In my opinion, the current geologic model of the CALB is supported by an array of independent evidence and proves satisfactory for reading the biogeographic history of New World equatorial regions. Filtering was moderate for lowland organisms and strong for terrestrial walkers needing continuous land, and for temperate plants moving along upland habitats, until the middle Pliocene. Afterwards a greater array of habitats and climate zones was available.

Magellan Land Bridge

In the northwestern Cono del Sur of South America, dry conditions developed along the coast and in the uplands through a combination of the cold Humboldt Current after ca. 40 Ma and especially the rise of the Andes to their maximum heights after ca. 10 Ma. Farther to the south, local open and dry habitats were available within the warm-temperate vegetation throughout the Cretaceous and early Cenozoic due to fires caused by volcanism and lightning. In the Paleogene of Patagonia and in Antarctica there were palms and other megathermal plants (*Lygodium, Cupania,* Lauraceae-*Nectandra* and *Ocotea, Sterculia,* and *Symplocos*), consistent with the global warmth at that time, trending later toward temperate vegetation mixed with the patches of drier open communities (*Ephedra, Acacia*). Throughout this interval *Araucaria, Nypa,* and various growth forms of *Nothofagus* (tall arborescent to stunted, depending on the climate) were present in Australasia, Antarctica, and southern South America, identifying the region as a biogeographic province with interchange occurring mostly from west to east and south to north—"a biotic gateway for dispersal across Gondwana." The vegetation of Antarctica was considerably reduced after the Oligocene, and by the Miocene/Pliocene it had nearly disappeared as a result of the cold and the spread of glaciers that covered much of the land. The discovery of *Nothofagus* in the Pliocene at the MPCO indicates the decline was episodic, but the trend was set, and soon afterward the MLB was closed to higher plants and animals (see table 8.3 in the online supplementary materials).

People first entered South America from the north at ca. 14.5 ka, possibly earlier, consistent with the view that they had crossed Beringia into North America soon after 18 ka following termination of LGM at 21 ka. They definitely used a coastal route and possibly sparsely glaciated interior pathways. It is likely that humans began modifying the modern analog vegetation used for reconstructing paleovegetation and ancient environments in Mexico after ca. 7–6 ka, in the Caribbean after ca. 6–5 ka, and in the Cono del Sur by ca. 5 ka.

Conceptual Issues and Future Needs

The name selected here for the units traversing the land bridges is *ecosystem* because it conveys the interaction between organisms, geography, geology, and climate better than biome or genome, which could imply undue emphasis on just the biological or the geological component of the system.

Neither the fossil record, nor morphological taxonomies, phylogenies, divergences, or migratory scenarios calculated on the basis of molecular evidence in the absence of sufficient fossils, nor statistical strategies alone are adequate in themselves for reconstructing patterns of historical biogeography. As stated at the outset, (1) an adequate fossil record does not exist for many organisms; (2) molecular clocks for establishing absolute time "are like Santa Claus—everyone wants to believe in them but nobody really does"; and (3) statistics can veer toward fine-tuning factors within the model rather than explaining the ground-truth situation based on firsthand, field-derived observations. Identification of fossils can be off by orders of magnitude (taxonomically and temporally) and molecularly calculated dates by millions of years. A capacious and cautious approach, combined with recognition of the inexact nature of the conclusions, is a great asset in tracing vegetation and land-bridge history where single approaches provide only a partial and provisional picture. In that regard:

My impression from the literature is that the most useful writings are ones where wisdom is at work, and among those practitioners with the patina of experience that often leads to wisdom there is a growing reluctance to accept scenarios of dispersal and diversification in the absence of a fossil record. The results can be important focal points for organizing data and suggesting future lines of research, but a growing sentiment is that they are rough and provisional. And further:

The modern flora, fauna, and fossil record are all partial and fragmentary, emphasizing the need for each, as well as geologic and climatic history, in attempts to unravel the origin and dispersal of lineages and the use of land bridges.

One benefit of a better definition of physiography, climate, age, and paleovegetation of land bridges is that it constrains scenarios—that is, it provides a limiting context for proposing histories and dispersal pathways for modern taxa where fossils of the taxon are lacking, meager, or unconfirmed. The mere fact that two regions were "connected in the past" is not sufficiently limiting to be useful.

Modern biotas are the analogs for reconstructing past ecosystems. As such, it is important to know the extent to which they have been altered through difficult-to-detect escaped and naturalized introductions; modified by restoration, reintroduction, rewilding, and other conservation activities (Callicott, 2011; Corlett, 2016; deLaplante et al., 2011; Higgs et al., 2014; Keulartz, 2016; Nogués-Bravo et al., 2016; Pimm and Raven, 2017; Sarkar, 2011); and/or simplified in commercial forests by replanting now required of most operations (such as single- or reduced-species pine plantations [for paper] of the US Gulf Coast; pine and *Eucalyptus* plantations [for charcoal] in forests of Bahia, Brazil—the well-named Recovery Debt, Moreno-Mateos et al., 2017). The long-term consequences are relevant to vegetation history and will become more so with the passage of time. In turn, that history is useful in planning conservation strategies through recognition that (a) ecosystems are dynamic and change in composition and distribution; and as such, (b) reserves set aside are essentially way-stations and must be designed and managed to accommodate the different communities that will be passing through them. "Reserves alone are not adequate for nature conservation" (Margules and Pressey, 2000), and as Paddy Woodworth reminds us (Missouri Botanical Garden Symposium, 2016), to assume otherwise "is a futile exercise in nostalgia." An example is the commendable and currently successful efforts now under way to protect rhinos in parts of Africa, but the same applies to endangered plant or animal species anywhere. The preserves are fenced in, and armed military guards prevent poaching. The question is: what efforts are being made to ensure preservation when climate change alters the ecosystems within the preserves, as it undoubtedly is doing now, and will be doing at an ever-increasing rate? The problem has been raised (Graham, 2015, 154) and the solution—possibly through continuous corridors, transplantation, and DNA/seed banks—is critical.

Modern distributions carry a past environmental imprint, and present-day occurrences, as a result of former reshufflings, from armadillos to hummingbirds to mangroves, rarely reflect the full ecological geographic potential of the organism. They include a legacy left by pathogens, extinction of dispersal and pollination vectors (ancient and recent), and time, as well as the existence and nature of land bridges.

Island biogeography—Some land bridges have remained separate islands throughout their history, while others have fused and separated to form lands of various sizes. This is relevant to the theory of island biogeography, where recent evidence suggests that in addition to size and isolation, other factors may be involved, such as changing climates, ocean currents, wind patterns, elevation, and fluctuating configurations. The question of whether the deeper past (pre–Late glacial/interglacials) leaves a residue of environmental memory with reference to volcanic/continental island biogeography is worth pursuing.

Large plant collections and mega-databases need improvement: "scores of museum specimens carry a name that isn't theirs"; "to what extent can we trust public databases?"; and the number of incorrect names and identifications in the older literature, and still being used, is legendary. Collaboration between morphological taxonomists, paleobotanists, molecular systematists, and statisticians is becoming standard, but the number of specialists in the two former fields is decreasing, and progress is slow considering the magnitude of the task. There should be a database of fossils examined at least preliminarily by specialists to establish a subset of specimens with accurate ages and identifications that can be used by nonspecialists in related fields. Even a casual appraisal of older museum specimens and reports in the literature would separate identifications that are indeterminable and/or blatantly wrong from those that are probably correct, and provide a cadre of fossils that have at least been examined since the time they were collected. It would reduce the former category from use and citation as "having been reported in the fossil record" paralleling "when land existed."

Another database listing dispersal vectors, pathways and distances these vectors travel, the propagules recovered, retention times, and survivability is needed for better explanation of distribution patterns. Much of this information is available but widely scattered, and coordination, even for a few representative model vectors and plants of wide occurrence and special biogeographic importance, would be useful.

Organisms have moved about during the past 100 million years in different ways. One is by direct migration across land bridges. These are shown by study of the connection's geologic history, climate, plants, and communities of plants known from the fossil record, or inferred as likely from floristic and phylogenetic consideration of the modern taxa. Over time numerous organisms have walked, hopped, hitchhiked, and migrated directly on land-bridge surfaces. However, to keep the importance of the connections in perspective, case studies show the many alternative means organisms have used for crossing barriers. The 45 examples presented in chapter 7, above, include passive transport through the ancient drift, collision, and fragmentation of continents, and the more recent separation of islands (vicariance); movement by wind, ocean currents, and animal vectors, extant and extinct, such as birds, fish, monkeys, bears, tortoises, tapirs, bats, and gomphotheres; chance sweepstakes dispersal; and different combinations simultaneously or at separate times and places. After arrival, by whatever means, the die for determining the ultimate distribution is cast by speciation, extinction, or, more rarely, evolutionary stasis. Of the case-study examples presented, 27 (60%) did use or are considered likely to have used land bridges in achieving their distribution, and 18 (40%)

probably dispersed by other means; at least there is no fossil evidence that they used land bridges.

In all cases, however (presence or absence of land bridges, utilization, circumvention), climate emerges as a powerful factor in regulating migration. For the 5 New World land bridges, topographic diversity and climate may have been more important than physical connections, given the relatively short distances involved and the vast time available. Climates were changing from tropical/subtropical and moist (Paleogene) to temperate (early Neogene) to seasonally cold-temperate and dry (late Neogene). The balance between these factors is the present arrangement of organisms over the surface of the Earth, and the present survey reveals it as a long, immensely complicated, and fascinating process.

References

Callicott, J. B. 2011. Postmodern ecological restoration: choosing appropriate temporal and spatial scales. In K. deLaplante, B. Brown, and K. A. Peacock (vol. eds.), Handbook of the Philosophy of Science. Vol. 11: Philosophy of Ecology. Elsevier, Amsterdam. Pp. 301–26.

Corlett, R. T. 2016. Restoration, reintroduction, and rewilding in a changing world. Trends in Ecology and Evolution 31: 453–62.

deLaplante, K., B. Brown, and K. A. Peacock (vol. eds.). 2011. Handbook of the Philosophy of Science. Vol. 11: Philosophy of Ecology. Elsevier, Amsterdam.

Graham, A. 2015. Past ecosystem dynamics in fashioning views on conserving extant New World vegetation. Annals of the Missouri Botanical Garden 100: 150–58.

Higgs, E., et al. (+ 8 authors). 2014. The changing role of history in restoration ecology. Frontiers in Ecological Environments 12. doi: 10.1890/110267.

Keulartz, J. 2016. Future directions for conservation. Environmental Values 25: 385–407.

Margules, C. R., and R. L. Pressey. 2000. Systematic conservation planning. Nature 405: 243–53.

McElwain, J. C., I. Montañez, J. D. White, J. P. Wilson, and C. Yiotis. 2016. Was atmospheric CO_2 capped at 1000 ppm over the past 300 million years? Palaeogeography, Palaeoclimateology, Palaeoecology 441: 653–58.

Moreno-Mateos, D., et al. (+ 9 authors). 2017. Anthropogenic ecosystem disturbance and the recovery debt. Nature Communications. doi: 10.1038/ncomms 14163.

Noguées-Bravo, D., D. Simberloff, C. Rahbek, and N. J. Sanders. 2016. Rewilding is the new Pandora's box in conservation. Current Biology 26: R87–R91.

Pimm, S. L., and P. H. Raven. 2017. The fate of the world's plants. Trends in Ecology and Evolution 32: 317–20.

Sarkar, S. 2011. Habitat reconstruction: moving beyond historical fidelity. In K. deLaplante, B. Brown, and K. A. Peacock (vol. eds.), Handbook of the Philosophy of Science. Vol. 11: Philosophy of Ecology. Elsevier, Amsterdam. Pp. 327–61.

Simpson, G. G. 1940. Mammals and land bridges. Journal of the Washington Academy of Sciences 30: 137–63. [Early Classics in Biogeography, Distribution, and Diversity Studies. http://people.wku.edu/charles.smith/biogeog/SIMP940B.htm.]

Additional References (Conservation)

Ahrends, A., et al. (+ 12 authors). 2011. Conservation and the botanists effect. Biological Conservation 144: 131–40.

Barber, C. V., and M. Parker-Forney. 2017. Rainforest CSI: how we're catching illegal loggers with DNA, machine vision and chemistry. World Resources Institute. http://www.wri.org.

Barnosky, A. D., et al. (+ 40 authors). 2017. Merging paleobiology with conservation biology to guide the future of terrestrial ecosystems. Science 355: eaah4787.

Bond, D. P. G., and S. E. Grasby (eds.). 2017. Mass extinction causality: records of anoxia, acidification, and global warming during Earth's greatest crises. Palaeogeography, Palaeoclimatology, Palaeoecology 478: 1–148.

Bonebrake, T. C. 2012. Conservation implications of adaptations to tropical climates from a historical perspective. Journal of Biogeography 40: 409–14.

Botanic Gardens Conservation International. 2017. How many tree species are there? BGCI. https://www.bgci.org.

Colles, A., L. H. Liow, and A. Prinzing. 2009. Are specialists at risk under environmental change? Neoecological, paleoecological and phylogenetic approaches. Ecology Letters 12: 849–63.

Cui, Y., et al. (+ 9 authors). 2011. Slow release of fossil carbon during the Palaeocene-Eocene Thermal Maximum. Nature Geoscience 4: 481–85. [Current carbon release to the atmosphere is 10 times faster than during the Paleocene-Eocene Thermal Maximum.]

Graham, A. 2015. Past ecosystem dynamics in fashioning views on conserving extant New World vegetation. Annals of the Missouri Botanical Garden 100: 150–58.

Hughes, A. C. 2017. Global roadless areas: hidden roads. Science 355: 1381.

Klare, M. T. 2017. Inaction on climate change equals human annihilation. TomDispatch.com.

Ladle, R. J., and R. J. Whittaker (eds.). 2011. Conservation Biogeography. Wiley-Blackwell, Hoboken. [See review by T. Newbold, 2011, Frontiers of Biogeography 3: 93–94.]

Norris, J. R., et al. (+ 5 authors). 2016. Evidence for climate change in the satellite cloud record. Nature 536: 72–75 (plus additional supplementary information).

Peel, G. T., et al. (+ 40 authors). 2017. Biodiversity redistribution under climate change: impacts on ecosystems and human well-being. Science 355: eaai9214.

Weller, R. J., C. Hoch, and C. Huang. 2017. Atlas for the end of the world. http://atlas-for-the-end-of-the-world.com/.

Wu, R., et al. (+ 6 authors). 2017. Global roadless areas: consider terrain. Science 355: 1381.

Additional References (Selected Classical Literature)

Carson, R. 1962. Silent Spring. Houghton Mifflin, Boston.

Darwin, C. 1859. The Origin of Species by Means of Natural Selection. John Murray, London. [With special reference to the role of islands and geographic isolation in biogeography and species diversification.]

Ehrlich, P. R. (and A. Ehrlich). 1968. The Population Bomb. Sierra Club, San Francisco; Ballantine Books, New York. [Wee Wikipedia for background notes on authorship and title.]

Humboldt, A. von. 1807. Essai sur la Géographie des Plantes. Levrault Schoell, Paris. [Reprint, 1977, Arno Press, New York.]

Leopold, A. L. 1949. A Sand County Almanac. Oxford University Press, Oxford.

Raven, P. H., and D. I. Axelrod. 1974. Angiosperm biogeography and past continental movements. Annals of the Missouri Botanical Garden 61: 539–673.

Wallace, A. R. 1892. Island Life. Macmillan, London.

Wegener, A. 1915. Entstehung der Kontinente. 4th ed. Friedr. Vieweg & Sohn, Berlin. [Origin of Continents and Oceans, translated by J. Biram, 2011, Dover, Mineola, NY.]

ACKNOWLEDGMENTS

Peter Raven, president emeritus of the Missouri Botanical Garden, has long
been an inspiration and a vast reservoir of innovative thought on biogeography. The information that flows through his office from around the
world and then is passed on is of inestimable value in keeping up with current activities and the global literature. Peter Wyse Jackson, president of the
Missouri Botanical Garden, James Miller, senior vice-president for research,
and Olga Martha Montiel, director of the Center for Conservation and Sus-
tainable Development (CCSD), continue to maintain an unexcelled environment for research and writing.

Stuart Pimm of Duke University, Laura Lagomarsino of University of
Missouri, and Roy Gereau, Robert Hart, Peter Jørgensen, George Schatz, and
other curators at the Missouri Botanical Garden provided helpful informa-
tion on plant diversity and diversification. Dolores Piperno, Smithsonian
National Museum of Natural History, Washington, DC, and Smithsonian
Tropical Research Institute (STRI), Panama, advised on Mesoamerican ar-
chaeology and ethnobotany; William Keegan, Florida Museum of Natural
History, Gainesville, on Antillean archaeology and indigenous people; Da-
vid Meltzer, Southern Methodist University, on early people in the New
World; and Carlos Jaramillo, STRI, on the vegetation history of Panama and
northern South America. Bruce Bennett, Yukon Conservation Data Centre,
Whitehorse, Yukon Territory; Matthew Carlson, Alaska National Heritage
Program, University of Alaska, Anchorage; and Bruce Ford, University of
Manitoba, Winnipeg, advised on the flora of Alaska and eastern Beringia.
James Zarucchi and Heidi Schmidt, Missouri Botanical Garden, advised on
the flora of North America. I am indebted to Rainer Bussman, Missouri
Botanical Garden, and to Jens-Christian Svenning, Aarhus University,
Denmark, for information on European vegetation. Also at the Missouri

Botanical Garden, Gerrit Davidse, Michael Grayum, and Warren D. Stevens answered questions about the vegetation of Mesoamerica, Costa Rica, and Nicaragua, respectively, and Olga Martha Montiel provided photographs of modern Nicaraguan vegetation. During a visit to Kamchatka, Siberia, and Lake Baikal in 2015, Victor Kuzevanov, director of the Irkutsk State University Botanic Garden, and Elena Kuzevanova, deputy director of the Lake Baikal Museum-Institute of the Russian Academy of Sciences, hosted field trips to western Beringia and Kamchatka. Bruce Baldwin (University of California, Berkeley) and Doug Soltis (University of Florida) advised about the possibility of detecting very recent lineage divergences from morphological and microsastellite data, and Andrew McDonald, St. Louis Science Center, reviewed the section on Antarctic dinosaurs. Burgund Bassuner prepared many of the original graphics; Mike Bloomberg downloaded others from diverse sources; Dawn Trog (CCSD) advised on a variety of computer matters; and Mary Stiffler, Vicki McMichael, Linda Oestry (MBG Library), Donna Rodgers (President Emeritus' Office), and Ulla-Karin Skälund (Naturhistoriska Riksmuseet, Stockholm) tracked down much of the far-flung literature. My daughter Alison Graham, a former long-term resident of Moscow, and Olga Fomina, from a small town in the Ural Mountains, later from St. Petersburg, and recently employed in the herbarium at Missouri Botanical Garden, translated portions of the Russian texts. Shirley Graham took many of the photographs and, along with two anonymous reviewers, provided numerous suggestions that improved the manuscript. Christie Henry, editorial director of science, social sciences, and reference; Miranda Martin, editorial associate; and Kelly Finefrock-Creed, at the University of Chicago Press; along with Lys Ann Weiss of Post Hoc Academic Publishing Services in Grafton, New Hampshire, capably guided the project through the complicated publication process. My sincere appreciation goes to everyone for their time and help.

INDEX

Alaska/Chukotka Block, 62
Alaska/Seward Peninsulas, 58 fig. 2.1; field
    work, 60; provinces, 46, 57–61
Aleutian Islands, 9, 19, 46, 57, 61 fig. 2.3;
    Aleutian Range, 63–64
Alexander Island, 239
Alleghenian Orogeny, 103
Amatignak Island, 21
Andes Mountains, 164–65, 215; attained
    current height, 216, 218, 298; Patagonian
    Andes, climate, 219
ANDRILL 2A (AND-2A) core, 243
Angiosperm Phylogeny Group website
    (APG), xviii, 242
Antarctica, 218, 226–34, 239–46; plate
    movements, blocks, 230 fig. 6.6, 239;
    Pliocene climate fluctuations, 232; prov-
    inces, 231; stability of the ice sheet, 233
Antarctic Circum-Current (ACC), xv, 226,
    231, 239
Anthropocene, 80–81, 94
Antillean Land Bridge (ALB), xvi, 7, 138
    fig. 4.3, 296–97; directional movement,
    159; distance between islands, 144–
    46; earthquake epicenters, 138 fig. 4.2;
    stepping stones?, 135–39, 143; time of
    separation of islands, 144–45; utilization,
    152–59, 157 fig. 4.13; vegetation, 146–
    49, 137 fig. 4.1
Appalachian Mountains, 37, 46, 91, 92, 103
aquatic apes (human origins), 225
Arctic Ocean/Sea (White Sea), Arctic Is-
    lands, 19, 29, 32, 45, 46, 57, 62, 104

assumptions/approaches (molecular, mor-
    phological, paleobotanical, statistical),
    2, 4–7, 299; faunal, 9–10; fragmented
    record, 10, 299; need for improvement/
    databases, xviii, 70, 184, 294, 300–301
asteroid/bolide impact, 47, 176, 242. *See
    also* meteorite event (Tunguska)
Attu Island (Near Island Group), 21
austrotropical flora, 243
Aves Ridge, 135–36
Azuero Peninsula, 179

Bahia Blanca, climate, 216
Basin and Range Province, 216; affinities
    with monte, Argentina, 216
Beagle Channel, 217
Bently Subglacial Trench, 226
Beringia, Bering Sea, Bering Strait, eastern/
    western Beringia, island, 6, 19, 20
    figs. P1.1–P1.2, 29, 46, 57; Preserve, 21
Beringian Standstill Model, 80
Bering Land Bridge (BLB), xv, xvi, 8, 19, 62,
    69 fig. 2.5, 294–95; directional move-
    ment (west to east), 79, 295; extent, 76
    fig. 2.10
Bluefields, 178
boreotropical, 113, 121, 243, 296

Cahokia Mounds, 176
camels, 78
Canadian Basin, 42, 46, 62
Canto Fault Zone, 144
Cape Horn, 218

carbon dioxide (CO_2)/atmospheric concentration, 7, 94, 231, 234, 235; Antarctica, 232, 233, 239, 243, 244, 293, 294

Central American Land Bridge (CALB)/ seaway, xvi, xix, 7, 90, 165 fig. 5.1, 178, 297; completion, 179; directional migrations, 195; indigenous people, 189–90; utilization, 190–97, 192 fig. 5.9; vegetation, 166–67, 184–89, 185–87 figs. 5.3–5.7

Central European portion (NALB), 101–2; vegetation recovery, 102–3

Central Siberian Plateau, 32

Cerro de Chirripó Grande, 176

Chocó, 165

chrons, xix

Chukchi Peninsula (Chukotka), 33, 46

climate intervals, 7; changes/concerns, 94–95

coal fires, 5–6

Cocos subplate, 179

Commander Islands, 21

Committee on Antarctic Research, 232

conceptual issues, 298–302

conferences (land bridges, vicinity), early studies, 22–25

Cono del Sur, xix, 213, 215, 219, 235, 240

conservation. See vegetation

Cordillera Darwin, 219

Cordillera de los Cuchumantanes, 176, 179

Cordillera de Talamanca, 178, 179

cratons/terranes, 46, 61, 62, 63, 144, 145

crown/stem group(s), xviii, 242

Darién Province, Panama, 165, 178, 179

databases. See assumptions/approaches

Davis Mountains, 176

Denali mountain/fault, 63

differential (directional) movement/migrations, 195. See also Antillean Land Bridge (ALB); Bering Land Bridge (BLB); Central American Land Bridge (CALB); Magellan Land Bridge (MLB); North Atlantic Land Bridge (NALB)

distribution types/patterns, residuals of ancient continental positions (vicariance), 46

Drake Passage, xvi, 213, 216, 218, 227, 229, 231, 245, 293

Early Eocene Climatic Optimum (EECO), 47, 91, 112, 189, 197, 231, 243, 264, 265, 280, 294, 296

eastern Asia/eastern North America vegetation affinities/floristic relationship, 109, 117–19

eastern US/eastern Mexico vegetation affinities/floristic relationship, 177

ecosystem (biome, geome), xviii, 298

edaphic. See soils

Ellsworth Mountains, 226

Eurasian (Siberian) climates, 34–35

explorers/explorations, rulers: Balboa, xx; V. J. Bering, 21, 29, 30 fig. 1.1, 32; V. P. Bering, 30, 31 fig. 1.2; Bligh, 146; Darwin, 217; Drake, 217, 226; Harrison, 21; Heer, 31, 105 fig. 3.9; Hultén, 22, 22 P1.3; Kotzebue, 19; Magellan, xx, 216; Peter the Great, 29, 32; Shackleton, xx, 31, 32, 227, 228 fig. 6.5; Vikings, xx, 122–23

faunas, 51, 180, 196, 236, 240, 260 fig. 7.1. See also assumptions/approaches

fossil floras/plants

 ALB: Guys Hill (Eocene, Jamaica), San Sebastian (Oligocene, Puerto Rico), 2, 152–56, 153–56 figs. 4.6–4.12; Saramaguacan (Eocene, Cuba), 144 fig. 4.4

 BLB: Alaska, 71–73 figs. 2.6–2.9; Kamchatka, 44–45 figs. 1.12–1.13

 CALB: Cerrejón (Colombia), 164; Paraje Solo (Mexico), 164, 297

 MLB: Laguna del Hunco, 238; Rio Pichileufú, 238; Santa Cruz Formation, 236; Sarmiento Formation, 236, 237–38 figs. 6.7–6.8, 241–42 figs. 6.9–6.11, 263 fig. 7.2, 274 fig. 7.3, 277 fig. 7.4, 280 fig. 7.5

 NALB (Brandon, Brandywine, London Clay, Potomac Group, Wealden), 89, 113–15 figs. 3.13–3.16

GAARlandia, 135–36

geofloras, 119–21

geologic interval(s), xix

geologic vs. biogeographic history, 136, 176, 180–81

Great American Biotic Interchange, 180, 182, 297

Gulf Stream, xv, 90, 99–100, 101, 103, 146, 147, 181, 293, 296

Hess Escarpment, 179

human impact/presence/early remains, artifacts, xvi, 41, 298; Antilles, 149; Beringia, 76–81; NALB, 102–3, 122–23

Humboldt Current, 100, 184, 216, 217, 219, 232, 293, 298

Iceland, xvi, 89, 90, 94–99, 96 figs. 3.4–3.5, 98–99 figs. 3.6–3.8, 103, 104, 106, 122, 123; Iceland(ic) Sagas, 278

indigenous people: Alaska/vicinity, 68; Antilles, 150–52; CALB, 191 fig. 5.8; MLB, 223, 224 fig. 6.4; NALB, 122; Siberia, 41

International Date Line, 21

Irkutsk, 21

island biogeography, 234–35, 300

isostacy. See sea level(s)

isotopes, 36, 63, 81, 233, 236

Isthmian Link, 179

Kamchatka, Kamchatka Peninsula, 6, 19, 29, 32, 42–52, 43 figs. 1.10–1.11, 57

Kayak Island, 31

Kazakhstania Block, 37

Kronotsky Nature Preserve, 48–49

Kuril-Kamchatka Trench, 20 fig. P1.1, 44, 46

Lake Baikal, 33, 34 fig. 1.4

land bridges: alternatives, 8, 168, 301; differences, 6; limitations determining use, xvi, 10, 70, 168, 300; location, xvii fig. 0.1; transportation, field work, 50–51 figs. 1.17–1.19

languages, effect of barriers, 8, 9, 150, 151, 223, 224 fig. 6.4

Largo Nicaragua, 178, 179

longitude, determining, dead reckoning, 21

Madrean-Tethyan hypothesis, 8, 119–21

Magellan Land Bridge (MLB), xvi, xix, 7, 214–15 figs. 6.1–6.2; directional

movement, 240, 242, 245–46; Eocene climate, 238; utilization, 235–39; vegetation, 219–23, 220 fig. 6.3, 298. See also Antarctica

Magellan Strait, xvi, 216

magnetic anomalies, xix

mammoth(s), 10, 23, 47, 75

Marine Isotope Stages (MIS), 94

meteorite event (Tunguska), 35 fig. 1.5, 103. See also asteroid/bolide impact

Mid-Atlantic Ridge, 42, 62, 89, 91, 97, 98, 104, 106, 111, 142, 143, 295

Middle Miocene Climatic Optimum (MMCO), 8, 35, 47, 69, 90, 100, 112, 117, 119, 166, 177, 194, 196, 216, 219, 231, 235–37, 244, 246, 270, 271

Middle Pliocene Climatic Optimum (MPCO), 35–36, 47, 69, 112, 174, 239, 244, 246, 298

migration/transport by animals, 5, 168–74, 169 fig. 5.2, 216

Milankovitch variations, 233, 234, 239, 243, 294

monte, 216

mountain orientation/migrations, 101–2

Mt. St. Elias, 30

Nazca subplate, 179

North Atlantic Land Bridge (NALB), xv, xvi, xix, 8, 62, 295–96; biodiversity/vegetation density, 116–17; Central Europe portion, 101–2; directional migration/movement, 103; disruption, 105–7, 106 fig. 3.10; fossils, 113–13 figs. 3.13–3.16; modernization, 116; physiographic provinces, 90 fig. 3.1, 91–103, 93 figs. 3.2–3.3; utilization, 110–17, 111 fig. 3.12

Nuclear Central America, 179

Ob River, 33

ocean currents, dispersals, xv, xix, 4, 8, 146, 147 fig. 4.5, 168, 174, 181, 234, 260 fig. 7.1, 261, 265, 275, 300, 301

Omolon Massif (block), 62

Oymyakon, 34

Paleocene-Eocene Thermal Maximum (PETM), 47, 69, 103, 104, 167, 234, 243, 244, 264

pampas, 215, 216
Patagonia/Pathagoni/Patagonians, 214, 216; climates, 217; Patagonian Platform, 218
permafrost, 5, 37, 39, 41, 58, 59 fig. 2.2, 60
Pinar Fault zone, 144
plate tectonics: MLB, 229–31, 230 fig. 6.6; NALB, 95–97. *See also* Wegener, Alfred
Puerto Montt, climate, 216
Puerto Rican Trench, 144, 156
Putoran Mountains, 33

Rochas Verdes Basin, 231
Rocky Mountains, 1, 46, 59, 117, 293
Ross Island/Sea, 232, 239, 240, 243
Rotational Opening Model, 62
Russian-American Company, 32
ruttiers, 217

Sabana de Bogotá, 165
Sacramento Mountains, 176
Santa Cruz Suture, 179
Santa Elena fault zone, 179
Scotia Ridge/Sea, 218
sea level(s)/curves, xix, 19, 30, 35, 43, 62–64, 68 fig. 2.5, 69, 77, 90–95, 103–5, 110 fig. 3.12, 137, 140, 145, 157 fig. 4.13, 165, 175–78, 192 fig. 5.9, 219, 223, 226, 233–35, 243, 246, 293
Sea of Okhotsk, 6
Seymour Island, 239
Siberia, xviii, xix, 1, 5, 6, 10, 19, 29, 30, 32–35, 38, 40–42, 47–50, 52, 57, 58, 61, 62, 65, 66, 69, 72, 77, 79, 80, 89, 92, 110, 122, 169, 226, 266, 281, 294, 295, 306
Sierra Madre del Sur, 176
Sierra Madre Occidental, 176, 177
Sierra Madre Oriental, 176, 177
soils, paleosoils, 5, 8, 36, 38, 47, 110, 140, 238; Antarctica, 231; Arctic (Russia), 36 fig. 1.6. *See also* permafrost
stability, tropical environments/vegetation, 188, 296
steppe, 3, 6, 8, 32, 33 fig. 1.3, 38, 40, 47, 48, 72, 110, 215

Taitao Platform, 218
Tethys Sea, 46
Tierra del Fuego, 216
"T in O" maps, 217
Torres del Paine, climate, 216
Totten Glacier, 233
Transantarctic Mountains, 226, 231
Transvolcanic Belt, 176, 177
TROPICOS, xviii, 184

Uralian Orogeny, 37
Ural Mountains, 8, 29, 32, 33, 35, 37, 38, 57, 61, 306
utilization/list of taxa: BLB, 68, 69 fig. 2.5, 72–75; directional movement (west-east) BLB, 78; restrictions, xvi

vegetation: Alaska/east Beringia, 65–67, 66 fig. 2.4; Antilles, 146–52; classification, 3–4, 37–38; conservation/destruction/modification, xx, 50 fig. 1.16, 300; Kamchatka, 48–53, 48–49 figs. 1.14–1.15; NALB, 90 fig. 3.1, 107–10; Russian (Siberia), 38–41; use as modern analogs for fossils, communities, xix–xx
Verkhoyansk, 34
vicariance, 136, 137, 267, 268, 279, 281, 301. *See also* land bridges: alternatives
Vinson Massif/block, 226
Volcán Barú (Volcán Chiriquí), 178
Volcanoes of Kamchatka (UNESCO sites), 50
Volcán Tajumulco, 176
Vostok core/Station, 226, 233

Wegener, Alfred, 96–97, 96 fig. 3.5; Wegener Fault, 103
West Antarctic Rift System, 231, 239
West Siberian Plain, 32, 34
Wrangell Mountains, 63

Yakutsk, 21, 34, 226; Yakutat block, microplate, 63

Zircon crystals (closure, CALB), 183